Titanium Dioxide: Diverse Applications

Titanium Dioxide: Diverse Applications

Edited by
Bruce Walter

STATES
ACADEMIC PRESS
www.statesacademicpress.com

Published by States Academic Press,
109 South 5th Street,
Brooklyn, NY 11249, USA

ISBN: 978-1-63989-525-0

Cataloging-in-Publication Data

Titanium dioxide : diverse applications / edited by Bruce Walter.
p. cm.
Includes bibliographical references and index.
ISBN 978-1-63989-525-0
1. Titanium dioxide. 2. Titanium compounds. 3. Oxides.
I. Walter, Bruce.
QD181.T6 T58 2022
546.512--dc23

For information on all States Academic Press publications
visit our website at www.statesacademicpress.com

STATES
ACADEMIC PRESS

Contents

Preface VII

Chapter 1 **1D Titania Nanoarchitecture as Bioactive and Photoactive Coatings for Modern Implants** 1
Aleksandra Radtke

Chapter 2 **Anticorrosive, Antimicrobial, and Bioactive Titanium Dioxide Coating for Surface-modified Purpose on Biomedical Material** 31
Hsi-Kai Tsou and Ping-Yen Hsieh

Chapter 3 **Photocatalytic Treatment Techniques using Titanium Dioxide Nanoparticles for Antibiotic Removal from Water** 51
António Armando Lima Sampaio Duarte and Maria Teresa Pessoa Amorim

Chapter 4 **Recent Overview on the Abatement of Pesticide Residues in Water by Photocatalytic Treatment using TiO_2** 73
Nuria Vela, Gabriel Pérez-Lucas, José Fenoll and Simón Navarro

Chapter 5 **Heat Transfer Applications of TiO_2 Nanofluids** 104
Hafiz Muhammad Ali, Muhammad Usman Sajid and Adeel Arshad

Chapter 6 **Titanium Dioxide as Food Additive** 130
Marie-Hélène Ropers, Hélène Terrisse, Muriel Mercier-Bonin and Bernard Humbert

Chapter 7 **Titanium Dioxide in Sunscreen** 149
Megha Trivedi and Jenny Murase

Chapter 8 **Application of Nanotechnology in Agriculture: Assessment of TiO_2 Nanoparticle Effects on Barley** 160
Alessandro Mattiello and Luca Marchiol

Chapter 9 **TiO_2: A Critical Interfacial Material for Incorporating Photosynthetic Protein Complexes and Plasmonic Nanoparticles into Biophotovoltaics** 177
Yiqun Yang and Jun Li

Chapter 10 **Composite Calcium Phosphate/Titania Scaffolds in Bone Tissue Engineering** 201
Massimiliano Dapporto, Anna Tampieri and Simone Sprio

Permissions

List of Contributors

Index

Preface

It is often said that books are a boon to mankind. They document every progress and pass on the knowledge from one generation to the other. They play a crucial role in our lives. Thus I was both excited and nervous while editing this book. I was pleased by the thought of being able to make a mark but I was also nervous to do it right because the future of students depends upon it. Hence, I took a few months to research further into the discipline, revise my knowledge and also explore some more aspects. Post this process, I begun with the editing of this book.

Titanium dioxide is a naturally occurring oxide of titanium, which is generally sourced from minerals like ilmenite, rutile, and anatase. It has a wide range of applications. Titanium dioxide is used in paint, sunscreen, and food coloring. Some of its other important applications are in areas like varnishes, paper, printing ink, rubber and plastics. It is also used in the production of technical pure titanium, glass and glass ceramics, electrical ceramics, metal patinas, catalysts, electric conductors, and chemical intermediates. It can be produced using the chloride process or the sulphate process. There has been rapid progress in the field of titanium dioxide and its applications are finding their way across multiple industries. Some of the diverse topics covered here address the varied applications of this compound. This book is a vital tool for all researching and studying this field.

I thank my publisher with all my heart for considering me worthy of this unparalleled opportunity and for showing unwavering faith in my skills. I would also like to thank the editorial team who worked closely with me at every step and contributed immensely towards the successful completion of this book. Last but not the least, I wish to thank my friends and colleagues for their support.

Editor

1

1D Titania Nanoarchitecture as Bioactive and Photoactive Coatings for Modern Implants

Aleksandra Radtke

Abstract

The research efforts in understanding the influence of TiO_2 1D nanoarchitecture structure and morphology on its biological and photocatalytic activity absorbed a lot of attention during last few years. Nowadays, the application of TiO_2 coatings in biomedical technologies (e.g., in modern implantology) requires the material of strictly defined structure and morphology, possessing both high biocompatibility, as well as antimicrobial properties. The presented review is a compilation of interdisciplinary knowledge about the application of 1D TiO_2 nanostructural coatings (nanotubes, nanofibres, nanowires) in biomedical technologies. The methods and parameters of their synthesis, and the physicochemical techniques used in the characterization of their structure and morphology, are discussed. Moreover, their ability to be applied as innovative coatings for modern implants is presented.

Keywords: 1D nanoarchitecture, TiO_2, nanotubes, nanofibers, nanowires, implant

1. Introduction

The use of modern materials in biomedical technologies requires detailed knowledge on the impact of the structure and the physicochemical properties of produced systems on their bioactivity. There are a lot of biomaterials, which can be applied in the human body, such as metals, ceramics, synthetic and natural polymers [1–8]. Among them titanium and its alloys have become extremely popular, especially in the implantology [9–15]. Commercially pure titanium (cpTi) is one of the dominant materials used for dental implants [16–20]. Ti6Al4V alloy plays an important role for orthopaedic applications [21–25]. Nickel-titanium alloy (Nitinol, shape memory alloy) has been used in the treatment of cardiovascular implants [26–30]. The combination of the high corrosion resistance, the tensile strength, the flexibility, and the biocompatibility, is the reason of widespread and successful application of titianium and its alloys in modern implantology [31–34]. The mentioned outstanding corrosion resistance of Ti and its alloys is a consequence of the passivation oxide layer spontaneous formation. This oxide layer is in fact responsible for the biocompatibility of Ti/Ti alloy. The thickness and the composition of natural oxide coatings, which appear in the presence of air or oxidizing media, and which are based on mainly TiO_2, Ti_2O_3 or TiO, depend on the environmental conditions [13–15, 35–37]. However, the stoichiometric defects and the low stability of this film can lead, in the case of implants, to their delamination and loosening. The key parameter in the success of bone implants (orthopaedics and dentistry) and the clinical goal is the establishment of a strong and long-lasting connection between the implant surface and peri-implant bone, in other word, achieving the optimal osseo-integration [38–43]. Even if Ti and its alloys are biocompatible materials, as they are biostable and biologically inert, the human body recognizes them as foreign ones and tries to isolate them using thin nonmineral, soft tissue layer. Instead, the mechanical interlocking of the titanium surface asperities and the bones pores leads to the formation of the bond between the implant and the bone, that is to the successful osseointegration [44–47]. The mechanical, chemical, and physical methods have been reported to improve the bioactivity of titanium and the bone conductivity (**Table 1**) [12, 48–81].

Modification method	**Modified surface layer**	**The aim of modification**	**Ref.**
Mechanical methods			[12]
Machining	Rough or smooth surface formed by subtraction process	Production of the specific surface topography; surface cleaning and roughening, improvement of adhesion in bonding	[49]
Grinding			[50–52]
Polishing			[53]
Blasting			[50, 51, 54, 55]
Chemical methods			[48]
Acidic treatment	<10 nm of surface oxide layer	Removal of oxide scales and contamination	[56, 57]
Alkaline treatment	~1 μm of titanate gel	Improvement of biocompatibility, bioactivity and bone conductivity	[57, 58]
H_2O_2 treatment	~5 nm of dense inner	Improvement of biocompatibility, bioactivity and bone conductivity	[59]
Sol-gel	~10 μm of thin film, such as calcium phosphate, TiO_2 and silica	Improvement of biocompatibility, bioactivity and bone conductivity	[60–62]
Anodic oxidation	~10 nm to 40 μm of TiO_2 nanotubular or porous layer, adsorption and incorporation of electrolyte anions	Production of the specific surface topography; improvement of biocompatibility, bioactivity and bone conductivity, improvement of corrosion resistance	[63–66]
CVD (chemical vapor deposition)	~1 μm of TiN, TiC, TiCN, diamond and diamond-like carbon thin films	Improvement of wear resistance, corrosion resistance, and blood compatibility	[67–69]

Modification method	Modified surface layer	The aim of modification	Ref.
Biochemical methods	Modification through silanized titania, photochemistry, self-assembled monolayers	Inducing the specific cell and tissue response by means of surface immobilized peptides, proteins, or growth factor	[70, 71]
Physical methods			[48]
Thermal spray Flame spray Plasma spray	~30 μm–200 μm of coatings such as titanium, hydroxyapatites, calcium silicates, Al_2O_3, ZrO_2, TiO_2	Improvement of wear resistance, corrosion resistance, and biological properties	[72–76]
Physical vapor deposition Evaporation Ion plating Sputtering	~1 μm of TiN, TiC, TiCN, diamond, and diamond-like carbon thin film; hydroxyapatite coating by sputtering	Improvement of wear resistance, corrosion resistance, and blood compatibility	[77, 78]
Ion implantation and deposition	~10 nm of surface modified layer	Modification of surface composition, improvement of wear resistance, corrosion resistance, and biocompatibility	[79, 80]
Glow discharge plasma treatment	~1 to ~100 nm of surface modified layer	Cleaning, sterilization, and oxidation of the surface, surface nitridation, removal of the native oxide layer	[81, 82]

Table 1. Overview of surface modification methods for Ti and Ti alloys.

A review of the Ti/Ti alloys surface chemical modification processes, which led to the formation of 1-D titanium dioxide nanostructures and improve the bioactivity of modified titanium materials, is the aim of this chapter. The relationship between the structure and the morphology of titania nanotubes (TNT), titania nanofibres (TNF), and titania nanowires (TNWs) and their bioactivity is discussed in the next subsections. Moreover, it cannot be forgotten that from chemical point of view, described 1D titania nanostructures are in the whole sense of the word, TiO_2, which is very known semiconducting photocatalyst. So, the aspects associated with the photoactivity of TNT, TNF, and TNW coatings are also discussed. This feature of titania can be treated as additional attribute to use, for example, in the process of implant surface antibacterial disinfection, with the use of UV light. But, the reader should bear in mind the fact that the photoactivity of TiO_2 is strongly dependent on its structure and morphology.

2. Structural characterization of 1D titania coatings

Titania is well known to exist mainly in three crystalline modifications: tetragonal rutile, tetragonal anatase, and orthorhombic brookite [82]. The structure of the titania lattice is determined by the way in which TiO_6 octahedra are linked [83]. These polymorphic forms of TiO_2 characterize different structural stability, different photo- and bioactivity, as well as different electrical and optical properties [84–86]. This fact causes that the determination of the structure of materials basing on TiO_2 is especially important for their further, for example, biomedical, applications.

In order to determine the titania structure, X-ray diffraction studies are often carried out. Characteristic set of 2Θ [°] signals: 25.33 (101), 37.80 (004), 48.08 (200), 55.12 (211) indicates the presence of anatase form, whereas the signals: 27.50 (110), 36.17 (101), 41.50 (111), 54.46 (211) prove that rutile form is our studied sample [87–90].

Raman spectroscopy is another very useful method to recognize and characterize the titania structures. Basing on the group analysis, there are six active Raman modes (A_{1g}, $2B_{1g}$, $3E_g$) for anatase D_{4h} ($I4_1/amd$) and four active Raman modes (A_{1g}, B_{1g}, B_{2g}, and E_g) for rutile D_{4h} ($P4_2/mnm$) [91, 92]. According to previous reports, the set of bands, which appear in Raman spectra at 197, 339, 519 and 639 cm^{-1}, indicates the formation of TiO_2 anatase form. For TiO_2 rutile form, the base of identification is the detection of bands at 447 and 612 cm^{-1} [91–94]. It should be pointed out that very often the fabrication of materials based on TiO_2 on nanoscale causes the formation of amorphic systems or amorphic systems containing small amount of crystalline phases [95, 96]. In such cases, Raman mapping of whole sample surface can help in the detailed determination of the structure.

The structure of thin titania films can be determined by the use of transmission electron microscopy studies, on the base of selective area electron diffraction (SAED) and the determination of *d*-spacing from HRTEM images of nano-TiO_2-sample [97, 98].

The use of mentioned in this subsection instrumental methods led to state about the structure of 1D-titania nanostructures. Even if they are not perfectly crystalline just like in case of amorphous samples possessing some crystalline islands in the structure.

3. Titania nanotubes (TNTs)

The choice of titania nanotubes as the first of the described 1D-titania nanostructures is not accidental, as the methodology of their production is well known. The electrochemical oxidation process of Ti and Ti alloys, in which TiO_2 nanotubes are produced, is one of the most popular methods to produce controlled and strictly defined structures on the surface of implants [99–106]. Sol-gel techniques, hydrothermal and solvothermal methods with or without templates, and atomic layer deposition (ALD) into the template are among other methods used in the production of TiO_2 nanotubes [107–111]. However, tailoring the process conditions and the possibility to obtain strictly defined morphology of nanotubes, caused that electrochemical anodization of titanium substrate, is particularly actively reviewed and practically used [67, 112–115]. The anodic oxidation process includes electrode reactions and metal and oxygen ions diffusion. The combination of these actions leads to the formation of oxide layer on the surface of Ti/Ti alloy, which acts as an anode (**Figure 1**).

This process is usually carried out by applying a constant voltage between 1 and 30 V in aqueous electrolyte or 5–150 V in nonaqueous electrolytes, containing in both cases approximately

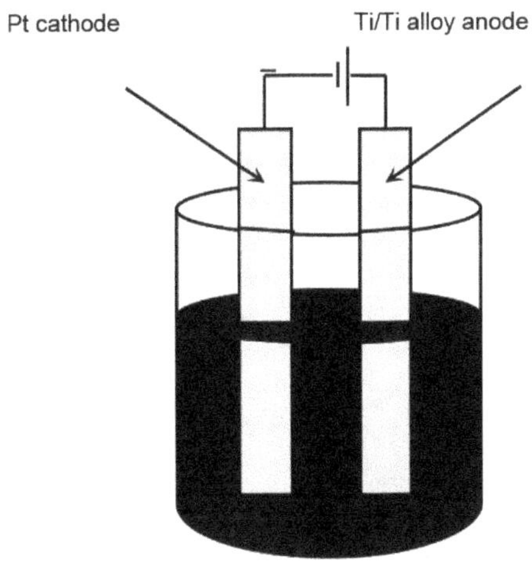

Figure 1. Scheme of the anodization setup.

0.1–1wt% fluoride ions. The presence of F^- ions in the electrolyte is absolutely essential, as they are, in fact, responsible for the creation of nanotubes morphology. Without them, a compact oxide layer would be formed on the titanium surface. Fluorides form water-soluble $[TiF_6]^{2-}$ species, both in the process of the complexation, which occurs with Ti^{4+} ions ejected at the oxide-electrolyte interface as well as by chemical attack of the formed TiO_2. The fluoride concentration is crucial for this process, and up to it, three very different electrochemical characteristics can be obtained. In case of fluoride content ≤0.5 wt%, stable compact oxide layer is formed. If the concentration is high, approximately 1 wt%, no oxide formation is observed, as all the Ti^{4+} ions formed in the oxidation of Ti, immediately react with the abundant fluorides, forming soluble complex anions $[TiF_6]^{2-}$. And for intermediate fluoride concentrations, between 0.5 wt% and 1 wt%, porous oxide or nanotubes formation can be observed as a consequence of a competition between oxide formation and Ti^{4+} solvatization [116, 117].

By tailoring the anodization parameters, such as applied voltage, anodization time, and concentrations of chemicals, TiO_2 nanotubes of different diameters (from 15 nm up to 300 nm) and lengths are possible to obtain (**Figure 2**a). And in such stable conditions, regular nanoscope pores/tubes, open on the top and close at the bottom, are formed. Our studies showed that the use of very small potential (below 4 V) led to the formation of regularly nanoporous coatings. The use of the potential from the region 4–12 V led to the creation of nanotubular coatings, in which nanotubes still posses common walls. Only the use of potential higher than 12 V gave the coatings composed of independent and separated nanotubes (**Figure 2**b–d) [118].

Some modifications in the tube geometry can be achieved by changing the anodization voltage during the tube growth process. Applying voltage steps, pulsing between two appropriate voltages can be used to generate tube stacks, bamboo nanotubes, or nanolaces [119–121]. The use of organic electrolytes, such as ethylene glycol, DMSO, glycerol, or ionic liquids, leads to the formation of nanotubes of different morphology and composition. The presence of lower water content in the electrolyte, which controls tube splitting, determines the synthesis of coatings with very long tubes and smooth walls, as well as with large diameters (up to 700 nm) [122–125].

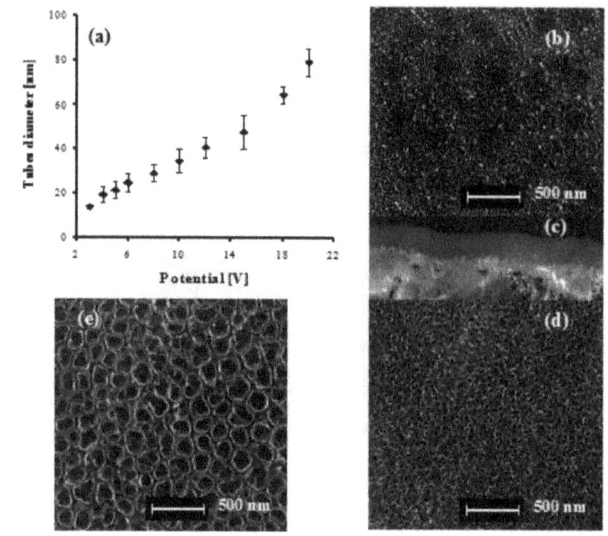

Figure 2. (a) The dependence between the applied voltage and diameter of nanotubes, SEM images of TNT3 (b), TNT10 (c), and TNT18 (d).

Regardless of whether an aqueous or nonaqueous electrolyte was used in the anodic oxidation process, TiO_2 nanotubes product is typically amorphous. It is necessary to anneal them in order to obtain crystalline product. Phase transformation to anatase takes place at 300–400°C, and from anatase to rutile at temperatures 500–700°C. Elevated temperatures (700–800°C) lead to the sintering and the collapsing of TNT [126]. The conversion temperature is strictly dependent on the several factors, such as nanotubes diameter, morphology, and impurities. It was proved that for small diameters (below 30 nm) rutile rather than anatase appears during annealing of amorphous nanotubes. Moreover, our studies showed that at very low potential (5 V), it was possible to obtain product in the rutile form, without annealing the sample. Both GAXRD and HRTEM proved this polymorphic form [118].

Taking into account the fact that titania nanotube arrays are one of the most promising candidates for coating of Ti and Ti alloys surface in implants fabrication (even for 3-D implants, as it can be seen in **Figure 3**), it is worth to consider the TNT structure and morphology impact on the cellular response.

There is a clear effect of the TNT diameter on the cell adhesion, proliferation and differentiation. Diameters of 15–20 nm are optimal and in case of such TNT presence on the surface of titanium implants, increase in adhesion and proliferation of several types of living cells, such as fibroblasts, osteoblasts, osteoclasts, mesenchymal stem cells, hematopoietic stem cells, and endothelial cells, was confirmed [127–134]. Higher tube diameters (>100 nm) have no positive influence on the increasing of adhesion or proliferation and some authors have shown that TNT of such diameters led to apoptosis, i.e., programmed cell death [135]. The size effect can be explained by the fact that integrin clustering in the cell membrane leads to a focal adhesion complex with the size of about 10 nm in diameter, perfectly fitted to nanotubes with about 15 nm diameters [134]. Gongadze et al. suggest that nanorough titanium surfaces with edges and spikes could promote the adhesion of living cells, especially osteoblasts [129].

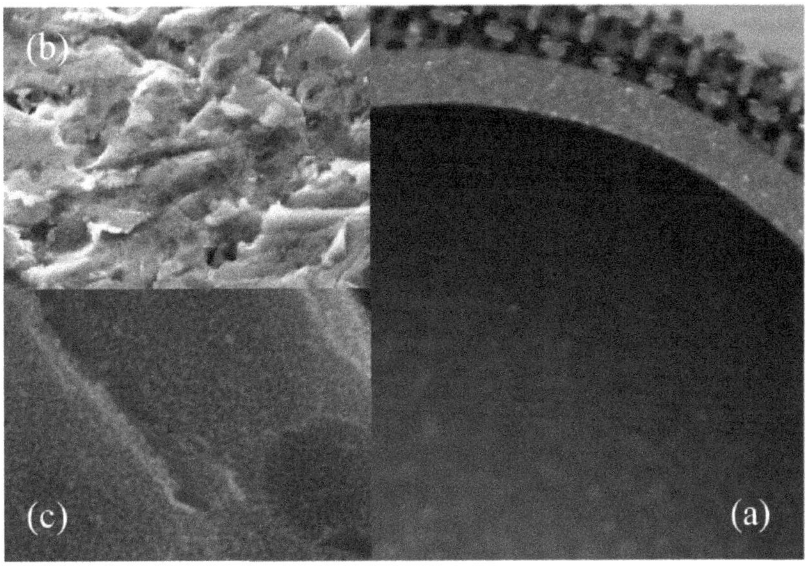

Figure 3. The surface of 3-D implant (a), uncoated (b) and coated by nanotubes (c).

A small diameter nanotube surface has more sharp convex edges per unit area than a large one. This fact can explain stronger cellular binding affinity on the surface of small diameter nanotubes, than on the surface of TNT with larger diameters. All studied samples mentioned above were annealed before carrying the adhesion and proliferation studies, so they were in crystalline form, mostly in anatase form. In our work, we studied the biological answer of as-obtained nanotubes, mostly amorphous, containing crystalline islands. We have noticed that even without further annealing, titania nanotubes showed better osseointegration than pure titanium. The adhesion and the proliferation of fibroblasts were different for the nanotubes of different diameters. The best biological answer was visible for nanotubes obtained at 5 V, whose diameters were ~20 nm [136]. This result is in accordance with earlier reports [127–134], however, our investigations revealed that the annealing of amorphous samples was not so indispensable [136].

Successful osseointegration is an important clinical goal but the important thing is the reduction of the bacterial biofilm formation on the surface of implant. Regardless of the type of biomaterial used, the initial inflammation response is always present and it may turn into an acute inflammation or even chronic inflammation. So, it is not surprising that the possibilities of a bacteria-repellent surface modification are investigated. Literature data and own studies showed that controlled diameter nanotubes displayed significantly changed responses to *Staphylococcus aureus* and *Staphylococcus epidermis* [137–140]. The size-effect exists for bacteria but also the structure of TNT influences the direction of the changes. According to Puckett et al., the use of larger diameter nanotubes decreased the number of live bacteria as compared to lower diameter ones and pure titanium [141]. But it is worth to know that analyzed nanotubes coatings were crystalline, in the form of anatase, as they were posttreated after anodization process by annealing. According to results of studies on amorphic nanotubes, which were not postannealed, the best antibacterial properties against *S. aureus* were seen for the nanotubes with small diameter but possessing the rutile form [118].

TiO_2 is one of the most photocatalytically active material used to decompose the organic pollutants and also bacteria [142–145]. The reason for this high activity is the band-edge positions in relation to typical environments, for example H_2O. The energy band gap is adequate to the ultraviolet light energy and UV light promotes electrons from the valence band to the conduction band. As a result, the holes are formed in the valence band. Separated holes and electrons reach the semiconductor-environment interface, and react with appropriate redox species. Because of water presence, several highly reactive species are generated by charge exchange at the valence band ($H_2O + h^+ \rightarrow OH^\bullet$) and at the conduction band ($O + e^- \rightarrow O_2^-$). These radicals and peroxo ions are able to virtually oxidize all organic materials to CO_2 and H_2O. Furthermore, at the valence band, direct h^+ transfer to adsorbed species to initiate the decomposition may also be considered [146–150].

It should be pointed out that in all of the photocatalytic applications, a higher overall reaction rate is achieved using high-surface-area geometries. Ordered nanotube arrangements offer various advantages over nanoparticulate assemblies, as their defined geometry provides strictly determined retention times in nanoscopic photoreactors. Moreover, the 1-D geometry may allow a fast carrier transport and thus less unwanted recombination losses [151, 152].

The earlier researches showed that TiO_2 nanotube coatings can indeed have the higher photocatalytic reactivity than a comparable nanoparticulate layer. The optimized reaction geometry for charge transfer, UV absorption characteristics over the tube, and solution diffusion effects are the main factors, which may be responsible for this effect. They allow an improved photoabsorption, longer electron lifetime, and diffusion length in TiO_2 nanotubes in comparison with nanoparticles [152, 153]. Photoactivity of TNT depends on the dimensions and wall thickness of the nanotubes, their crystallinity, and their packing density, because the separation and transport of charge, as well as the grain boundary effect, would greatly hinge on such factors [154]. It was shown that silver or gold particle decoration led to a significantly higher photocatalytic activity [155]. The same increasing of photoactivity was also obtained by applying an external anodic voltage [156]. These facts suggest that in case of investigated objects, a valence-band mechanism dominates, and the observed accelerating effects have a common origin in increased band bending, either by applied voltage or by the junction formation [157].

Our studies showed that amorphic titania nanotubes possessing some crystalline impurities indicated very high photoactivity in the reaction of methylene blue and acetone degradation in the presence of UV light [118]. The clear influence of the tubes diameter, and at the same time, of specific surface area, on the value of observed rate constants, was also visible. Additional information which we have obtained from our research was the fact that it was not possible to make a clear comparison of TNT photoactivity in the degradation of different organic pollutant patterns (water-soluble methylene blue and volatile aceton), as the mechanisms of their degradation were completely different and they depended on variable parameters (the size of pattern molecules which affects the reactant molecule adsorption, pH of solution). However, the surface of the implant modified by anodic oxidation can possibly be disinfected/sterilized only with the use of UV light, as it reveals higher activity than unmodified Ti surface possessing only the natural passivation oxide film. It is an important property because the same coating plays a dual role for the implants—increases osseointegration process and creates optimal conditions for carrying out the process of implant surface sterilization/disinfection with the use of UV light.

4. Titania nanofibrous coatings (TNFs)

The electrospinning is a technique mostly used in the fabrication of the TNF coatings [158–160]. It is a very simple and convenient method for the preparation of polymer fibers and ceramic fibers, which are extremely long, uniform in diameter, raging from tens nanometer to several micrometers, and diversified in compositions [161]. The electrospinning process involves a high voltage source connected to a needle and a metallic collector where the fibers are deposited. The needle, which is attached to injection pump, represents the positive electrode. The collector is connected to the negative electrode, thus creating a potential difference. Electric field created in this way stretches the drop that forms on the needle tip, which is then deformed into a conical shape (Taylor cone). When the applied electric field exceeds the surface tension of the drop, the solution is ejected in the form of an electrically charged jet, reaching the negative electrode, which is the collector. During this process, the solvent is evaporated resulting in the deposition of nanofibers over the collector. The diameter of the fibers can be adjusted by varying the rheological properties of the solution and turning the processing parameters [162, 163]. The scheme of the electrospinning setup is given in **Figure 4**. Electrospun TiO_2 nanoarchitecture is formed by electrospinning titania precursor (e.g., titanium(IV) alkoxides) along with adequate polymer and subsequent polymer burning in high temperature sintering process. The synthesis involves the following four steps: (1) the preparation of the titania precursor sol, (2) mixing of the sol with the polymer template to obtain the solution for electrospinning, (3) electrospinning of the solution with the use of the apparatus showed in **Figure 4**, and (4) the calcination of as-prepared TNF to obtain crystalline titania nanofibers [164]. The morphology and the diameter of the electrospun titania depend on the following parameters related to: (a) the solution, (b) the process, and (c) the ambient (**Table 2**), while the structure of formed nanofibers is strictly associated with the postcalcination process [165–173].

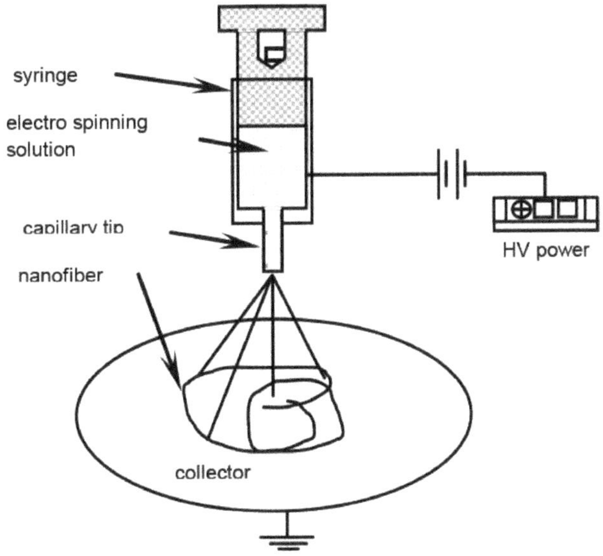

Figure 4. Scheme of electrospinning setup.

Parameter	Effect on TNF morphology	Ref.
Solution parameters		
Polymer concentration	Increase in TNF diameter with increase of polymer concentration	[165–167]
Titania precursor concentration	Increase in TNF diameter with increase of titania precursor concentration	[165, 166, 168, 169]
Viscosity	Low viscosity leads to bead formation, higher viscosity leads to disappearance of bead but nanofiber with larger diameter is produced	[165,170]
Solvent volatility	High volatility leads to formation of nanofiber in concave morphology	[171]
Dielectric constant	Low dielectric constant results in the formation of larger size nanofibers	[171]
Processing parameters		
Applied voltage, electric field	TNF diameter decreases with increasing applied voltage; when the applied voltage is above 1.6 kV/cm, the diameter increases, because of the jet instability	[165, 166]
Feeding rate	Increase in feeding rate results in the increase of nanofiber diameter	[165]
Collector geometry	The geometry affects the directionality of the formed nanofibers	[170]
Ambient parameters		
Calcination temperature	Higher temperature results in the reduction of nanofiber size, because of the loss of polymer and the crystallization process of titania	[165, 168, 171–173]

Table 2. Summarization of the electrospinning parameters affected the nanofibers morphology.

Another technique to produce titania nanofibrous coatings is the laser ablation, proposed by Tavangar et al. [174]. During the laser irradiation of titanium substrate, the illuminated region is heated up and vaporized, producing the plasma plume. The plume expands outwards and its temperature and pressure decreases. The next process is the condensation of plasma plume leading to the formation of liquid droplets in saturated vapor, which is responsible for the nucleation. Continuous irradiation pulses maintain the plasma plume formation, which in turn generates a continuous flow of vapor plume increasing the density of formed nucleus. Hugh amount of nuclei favors the growth of nanoparticles, which come in contact and aggregate to form interwoven nanofibrous structure [175].

Also, the anodization was used in order to obtain titania nanofibrous coatings, but only in some very special conditions. Lim and Choi reported that such fibers were obtained on the top of 20 nm in diameter nanotube array, which were more than 10 μm in length [176]. In another report, Chang et al. presented novel method to synthesize nanofibrous coatings—rotating of titanium anode with as-formed nanotube arrays, with the speed of 30 rpm in the same solution, in which the anodization of TNT has taken place (ethylene glycol solution containing 0.3 wt% NH_4F and 2 wt% of H_2O) for next 3 h [177].

Very interesting morphology, from the medical application point of view, has been shown for titania nanofibrous coatings obtained in the process of Ti/Ti alloy chemical oxidation with the use of hydrogen peroxide with or without simple inorganic salts: NaCl, Na_2SO_4, $CaCl_2$, in elevated temperature (80°C) [118].

Wang et al. studied the influence of titania electrospun nanofiber dimensions and microstructure pattern on the adhesion and proliferation of human osteoblasts MG63. Bioactivity of two types of nanofiber dimensions (184 ± 39 nm and 343 ± 98 nm) and two different ways of TNF alignment (flat and patterned TiO_2 nanofibers) were checked and the obtained results indicated that cell morphology was not sensitive to the differences in nanofiber diameter and in microscale structure [178]. These results are in contradiction to some other researchers' studies, which showed preferential cell attachment along patterned TNF. The same authors proved that the combination of microroughness and the nanotopography can be used to modify the differentiation of osteoblasts and generate an osteogenic environment [179–181].

Considering the fact that the rate of osseointegration is strictly related to the efficiency of bone-like apatite formation on the implants, Tavanger et al. used the nanofibers obtained in femtosecond laser ablation process to evaluate the apatite-inducing ability of nanofibrous titania [174]. SEM studies showed that all TNF coatings obtained by Tavanger et al. were covered by dense and homogeneous apatite precipitation layer, after soaking them in simulated body fluid (SBF) for 3 days. EDX results proved that Ca/P ratio was around 1.63, which was attributed to hydroxyapatite, possessing a composition similar to the bone. Moreover, the wettability tests of TNF were performed and very low contact angle (<9.2°), and almost complete spreading of H_2O droplets was observed on all the titania nanofibrous surface samples. Conclusion of the studies was the thesis that TiO_2 nanofibrous structure with the rapid apatite-inducing capability is expected to improve bone formation during *in vivo* implantation [174].

Chang et al. made a bioactivity comparison of different nanomorphology titania coatings: TiO_2 flat, TNT, and TNF, using human osteoblasts MG63. SEM studies revealed that the cell attached to flat TiO_2 possessed a round morphology, whereas these ones attached to nanotubes and nanofibers, showed polygonal shape and extending filopodia. Higher ratio of cell attachment and clearly visible lamellipodia were seen on nanofibrous coatings than on nanotubular ones [177]. This suggests that titania nanofibers surface, because of its rough and porous morphology, provides optimal environment for the cell adhesion, proliferation, and differentiation. The results of bioactivity studies of nanofibrous coatings obtained during Ti/Ti alloy chemical oxidation with the use of hydrogen peroxide with or without simple inorganic salts: NaCl, Na_2SO_4, $CaCl_2$, in elevated temperature (80°C), confirm this [118]. Results of fibroblasts adhesion and proliferation studies carried out with the use of MTT test are shown in **Figure 5**.

Taking into account the different nanotopography and the structure of TNF coatings presented in Ref. [118] (analysis of GAXRD and DRIFT data revealed that TNF obtained in the presence of H_2O_2 and HCl was amorphous with anatase islands; TNF/H_2O_2, TNF/ H_2O_2 + Na_2SO_4, and TNF/ H_2O_2 + NaCl – amorphous with anatase and rutile islands, TNF/ H_2O_2 + $CaCl_2$ – amorphous titania), it can be stated that both above-mentioned attributes influence the bioactivity toward osseointegration processes taking place on their surface.

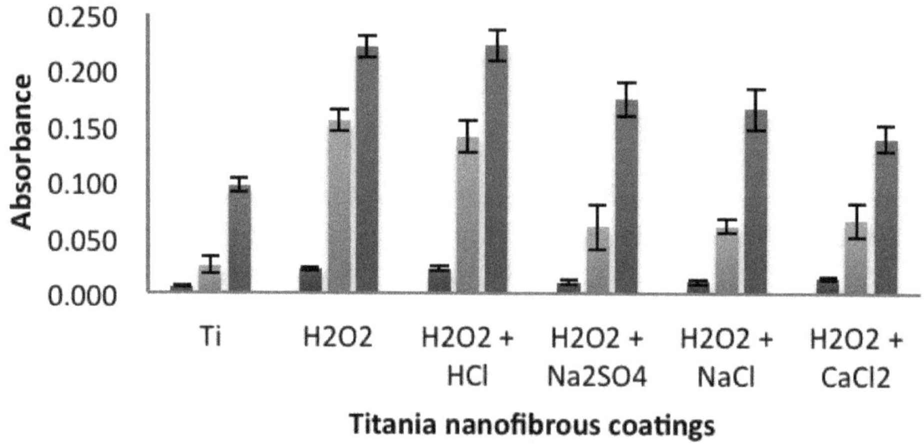

Figure 5. Results of MTT assay carried out on TNF coatings, obtained from different oxidation mixtures, with the use of murine fibroblasts L929 (adhesion after 24 h, proliferation after 72 h and differentiation after 5 days).

Photocatalytic activity of titania nanofibers, obtained by electrospinning method and annealed in 550°C, was studied with the use of dye degradation tests (basic blue 26, basic green 4, basic violet 4) [182]. The concentration of dye solution was measured in relation to UV irradiation time, using UV-Vis spectrophotometer. Additionally, the same tests were done for TiO_2 nanoparticles obtained by sol-gel method and for composite coatings consisting of titania nanofibers and TiO_2 nanoparticles. Doh et al. showed that after 3 h of UV illumination, 25.3% of basic blue 26 was degraded by titania nanofibers. This degradation efficiency was almost the same as for TiO_2 nanoparticles obtained by sol-gel method, for which the value was 23.7%. In comparison, the photoactivity of composite material (TiO_2 nanofibers and TiO_2 nanoparticles) was much higher, as basic blue 26 was degraded by 78.7%. Rate constant calculated for titania nanofibers, titania nanoparticles, and composite material, on the base of simplified equation of the Langmuir-Hinshelwood kinetic model was equal adequately 15.7×10^{-4} min^{-1}, 14.3×10^{-4} min^{-1}, and 85.4×10^{-4} min^{-1}. The values of kinetic rates obtained for composite material but during the degradation processes of dyes: basic green 4 and basic green violet were 81.2×10^{-4} min^{-1} and 67.4×10^{-4} min^{-1}, respectively. Authors concluded that composite materials consisted of TiO_2 nanofibers and TiO_2 nanoparticles had high photocatalytic activity due to their high active surface area and due to complex pore structure. They stated that such materials could be suitable for the application to the degradation of organic dye pollutant.

The photocatalytic activity of TiO_2 nanofibrous coatings obtained during Ti/Ti alloy chemical oxidation with the use of hydrogen peroxide with or without simple inorganic compounds: HCl, NaCl, Na_2SO_4 in 80°C, was analysed also on the base of degradation of two organic pollutant patters: acetone (A) and methylene blue (MB) [118]. Based on the same simplified equation of the Langmuir-Hinshelwood kinetic model, it was possible to calculate the kinetic rates. The values of calculated kinetic rates are showed in **Table 3**.

The values obtained during these studies are very close to those, obtained by Doh et al., however, titania nanofibers obtained in the process of chemical oxidation were not post annealed and they were not enriched by the titania nanoparticles [118, 182]. What should be pointed out clearly is the fact that the observed rate constants do not inform about the real nature of the appropriate processes. However, they provide basic information about change in the

	10^4 k_{obs}^{MB} (min^{-1})	10^4 k_{obs}^{AF} (min^{-1})
H_2O_2/Na_2SO_4	81.8 ± 1.0	24.0 ± 0.1
H_2O_2	67.0 ± 3.2	13.5 ± 0.1
H_2O_2/HCl	25.3 ± 0.7	21.3 ± 01
H_2O_2/NaCl	93.6 ± 3.2	14.0 ± 0.1

Table 3. The observed rate constant values for the degradation of MB (k_{obs}^{MB}) and acetone (k_{obs}^{A}) on titania nanofibers under UV light.

degradation rate, depending on the structure, morphology, and active surface of the TiO_2 coatings. The different degradation rate values result from the completely different steps, which occur during the degradation processes: the multistep degradation process of methylene blue (MB) and formation and degradation of some dimeric forms like mesityl oxide, which are observed during acetone degradation.

The results of carried out photocatalytic activity studies showed the potential usefulness of TNF coatings as photoactive ones. Even if they show lower kinetic rate values than titania nanotubes for which these values were almost in all cases five to ten times higher, comparing to titania nanofibers [118].

5. Titania nanowires (TNWs)

From the geometrical point of view, nanowires offer exceptional properties, such as flexibility and fatigue resistance, and the possibility to integrating them in large area with controlled pattern. Sometimes, nanofibers are also used to describe nanowires morphology, especially when nanowires (NW) are very long and not single crystalline. However, in order to achieve 1D NW morphology, it is crucial to obtain one rapid growth direction during the evolution of nanocrystals. This requirement is quite often fulfilled for some crystals, due to the strong anisotropic property of their crystal structures. For example, wurtzite crystals naturally have rapid growth along the [0001] direction and because of this fact NW is one of the preferred morphologies during self-assembly growth. For some other crystals, titania, for example, such anisotrophic behavior is less evident, and they demand additional support to create 1-D nanotopography [159]. Among auxiliary activities, the surface functionalization, introducing dislocations, applying catalysts, and increasing the building block concentration should be mentioned. More detailed strategies for crystal morphology control are given in review articles [182–186]. Plenty of TNW synthesis routes have been examined. Bottom up approaches include a large variety of solution- and vapor-based growth strategies [187–189]. Top down procedures, such as direct oxidation and electrochemical etching techniques, have also been explored for nanowires synthesis [190–192]. The widespread utilization of nanostructured TiO_2, including TNW, is often hindered by the opposite demands for precise control of well-ordered surface features and low-cost rapid production. Taking this into consideration, the synthetic routes, which engaged inexpensive techniques to produce nanowire coatings, are presented in this review.

Most of wet chemical methods require many steps, which increase production costs. Moreover, titania nanostructures, including nanowires, obtained in anodization process or by the electrospinning method, are amorphous, and they need to be annealed in order to endow them in high crystalline form. A method, which can be applied to the formation of titania nanowires, is the thermal oxidation of Ti/Ti alloy surface under a limited supply of oxygen, in argon atmosphere [193–195]. The exposure or pure titanium to argon, which contains ppm of oxygen, at a flow rate of 200 cm^3/min for 8 h at 600°C resulted in the growth of titania nanowires of the length 50–400 nm. Such obtained nanowires posses the structure of rutile. What should be pointed out is the visible impact of the argon flow rate on the morphology of nanostructural TiO_2. Increasing of the flow rate from 200 cm^3/min to 1000 cm^3/min caused the disappearance of nanowires. This effect was even more intensified by the increase of the temperature. Higher flow rate and higher temperature promoted growth of not nanowires but platelets and faceted oxide crystals. It can be concluded that the window of high aspect ratio nanowire growth is very narrow for pure titanium. The situation looks much better in case of titanium alloy—Ti6Al4V. The same procedure of thermal oxidation in the presence of argon both in low and high flow rates gives the same results—titania nanowires. However, the effect of increasing temperature is similar to adequate one for pure titanium. The optimal temperature of the titanium alloy oxidation in order to obtain nanowires is 700°C. At 800°C, a mixture of nanowires and well-faceted crystals is possible to obtain and at 900°C, only platelets and crystals are the results of process. The higher temperature impact on the formation of well-faceted equiaxed crystals indicates that one-dimensional growth at the low temperature is the result of oxidation reaction anisotrophy with the growth preferential on certain crystal faces. At higher temperatures, this anisotrophy decreases and the growth on other surface is promoted, allowing the formation of facetted crystals [194, 195].

My experience with the formation of nanowires, during the titanium direct oxidation, is similar to above mentioned. The use of lower argon flow rate (30 cm^3/min) led to obtain well morphologically oriented nanowires at 475–550°C (**Figure 6a**). Above this temperature, maintaining the same gas flow, facetted crystals were obtained (**Figure 6b**). The addition of H_2O_2 vapors to the carrier gas caused that the formation of nanowires was much slower and less efficient (**Figure 6c**). Prolongation of the process time improved the quality of the nanowires, but only a little, but failed to get a system similar to those received for pure argon (**Figure 6d**).

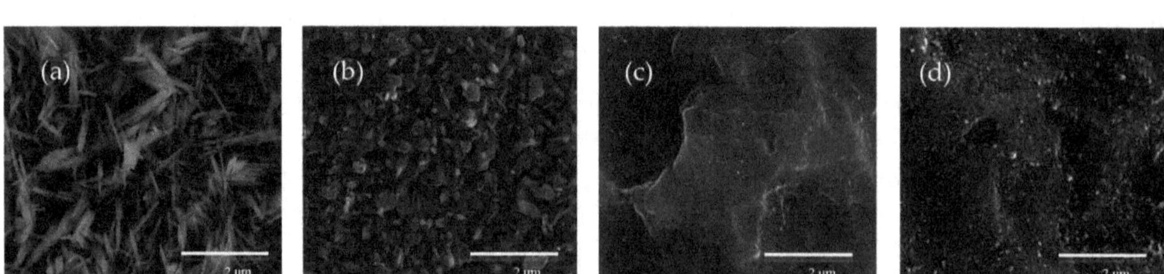

Figure 6. SEM images of titania nanowires obtained in the process of thermal oxidation with the use of pure argon, at 450°C (a), at 600°C (b), and with the use of vapors Ar/H_2O_2, at 500°C, but in 2 h (c) and in 6 h (d).

Daothong et al. found that in the presence of adequate organic vapour, titanium can be directly oxidized and TNWs were formed. They presented a size-controlled growth of titania nanowires in the presence of ethanol vapor at high temperature (650–850°C) and low pressure (~10 Torr). The nanowire length was proved to be directly proportional to the oxidation time, and their diameter was strictly connected with applied temperature [196].

The aspect ratio and the shape of titania nanowires were described as resembling the needle-like shape of crystalline hydroxyapatite and collagen fibers found in the bone. Such environment was reported as the place, which ensures the proper cell organization, their vitality, and functionality. Studies on osteoblasts adhesion and growth assessment carried out by Tan et al. showed that osteoblast was able to adhere and spread on the nanowire coating [197]. Osteoblasts adhered to the surface, exhibited an oval shape on the first day and polygonal shape with some protruding lamellipodia on days 3–7. On day 14, osteoblasts formed intercellular network and indicated the cell-to-cell communication, by stretching out their filopodia toward each other. Osteoblasts growth was determined with the use of AlmarBlue assay, and it was proved that osteoblasts were able to proliferate on nanowires coatings. The cell number of osteoblasts on TNW coatings was increased significantly over the level of pure titanium sample. Degree of osteoblasts differentiation was investigated with the use of examination of intracellular ALP (key marker) activity. The test showed that until 7 days, the ALP level was significantly higher than for adequate pure titania coatings. Extracellular matrix mineralization of osteoblasts was evaluated on the base of Alizarin Reds. Also, this assay showed that greater number of discrete mineralized nodules in greater abundance was seen on the nanowire surface.

Our studies on the adhesion and proliferation of fibroblasts show that these processes proceed more efficiently on nanowires, but there are no significant difference between nanowire coatings and pure Ti/Ti alloy (**Figure 7**).

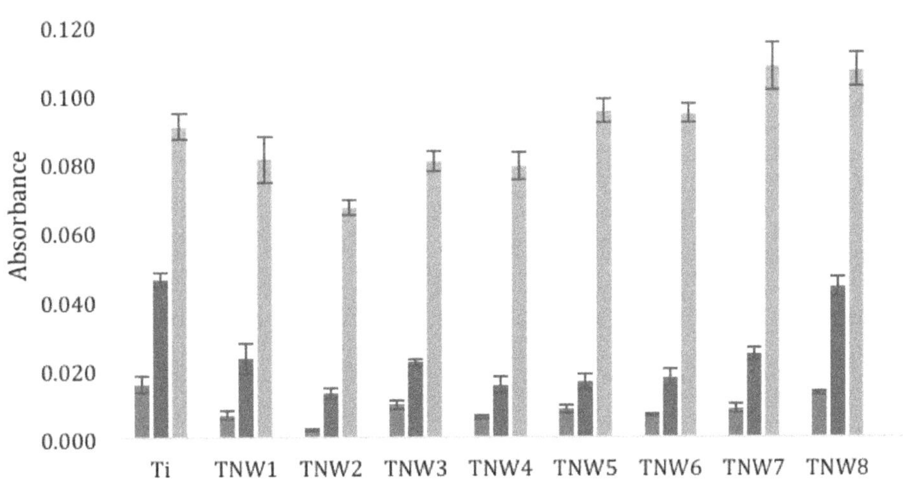

Figure 7. Adhesion (after 24 h) and proliferation (after 72 h and 5 days) of fibroblasts on the surface of titania nanowires obtained at different conditions of temperature and Ar rate flow: TNW1–TNW4 (475°C), TNW5–TNW8 (500°C), TNW1, TNW3, TNW5, TNW7 (30 cm^3/min), TNW2, TNW4, TNW6, TNW8 (100 cm^3/min).

Assuming the biological activity of all reviewed titania nanostructures, it can be stated that at the moment, titania nanotubes are the strongest used in biomedical applications, as the procedure of their fabrication is the most predictable and easy. However, the necessity of their further annealing may lead to the increase of the interest to systems, e.g., TNW, during the production of which, a specific crystalline structure is formed.

The photoactivity of titania nanowires should also be taken into the consideration as such activity gives the possibility to use these coatings in the process of UV-activated sterilization. Most of earlier reports showed the usefulness of titania nanowires rather in the processes of photoelectrochemical water splitting than in the degradation of organic pollutants. In case of sooner, the photoactivity of Au-decorated TiO_2 nanowires electrodes for photoelectrochemical water oxidation, was enhanced in the entire UV-VIS region by the manipulation of the shape of decorated Au nanostructures: nanoparticles and nanorods [198]. The titania nanowires photoactivity in the degradation of 4-chlorophenol was studied by Stengl et al. [199] They calculated the degradation rate constant assuming the reaction kinetic of the first order. The obtained values were in the range 0.0045–0.0083 min^{-1}. So, they were comparable with those obtained for titania nanofibers. The authors noticed that the photocatalytic activity of the annealed samples gradually increased from the temperature of 350°C (0.0045 min^{-1}) to 750°C (0.0129 min^{-1}), and for the samples annealed to temperatures 900°C and 1000°C, respectively, the photoactivity decreased (0.0104 and 0.0083 min^{-1}). They assumed that the initial photocatalytic activity growth for samples annealed in the range 350–750°C corresponds with enlargement of the anatase crystalline phase in consequence of annealing. The decrease of photocatalytic activity of the sample heated above 750°C, they associated this with the transformation of anatase to rutile phase and also with the lowering of surface area. In case of samples obtained in our lab during the thermal oxidation of titanium, in which the rutile phase was present, the calculated rate constants for the methylene blue photodegradation process were in the range 0.0001–0.0002 min^{-1}, so much lower than in case of Ref. [199]. The rutile structure and the low surface area (see **Figure 6c** and **d**) were burdened of such low activity reason.

Author details

Aleksandra Radtke

Address all correspondence to: aradtke@umk.pl

Faculty of Chemistry, Nicolaus Copernicus University in Toruń, Torun, Poland

References

[1] Williams D. On the nature of biomaterials. Biomaterials. 2009;**30**:5897-5909

[2] Mihov D, Katerska B. Some biocompatible materials used in medical practice. Trakia Journal of Sciences. 2010;**8**:119-125

[3] Ratner BD, Hoffman AS, Schoen FJ, Lemons JE, editors. Biomaterials Science—An Introduction to Materials in Medicine. New York; Academic Press, Elsevier; 2013. pp. 37-132

[4] Bhat S, Kumar A. Biomaterials and bioengineering tomorrow's healthcare. Biomatter. 2013;**3**(3):e24717

[5] Bae H, Chu H, Edalat F, Cha JM, Sant S, Kashyap A, Ahari AF, Kwon CH, Nichol JW, Manoucheri S. Development of functional biomaterials with micro-and nanoscale technologies for tissue engineering and drug delivery applications, Journal of Tissue Engineering and Regenerative Medicine. 2014;**8**:1-14

[6] Davis JR. Overview of biomaterials and their use in medical devices. In: Davis JR, editor. Handbook of Materials for Medical Devices. Illustrated edition, Ohio: ASM International; 2003. pp. 1-11

[7] Ramakrishna S, Mayer J, Wintermantel E, Leong KW. Biomedical applications of polymer-composite materials: A review. Composites Science and Technology. 2001;**61**:1189-1224

[8] Williams D. An introduction to medical and dental materials. In: Williams D, editor. Concise Encyclopedia of Medical & Dental Materials. Pergamon Press, Oxford and The MIT Press Cambridge, 1990

[9] Williams D. Titanium in medicine: Material science, surface science. In: Brunette DM, Tengvall P, Textor M, Thompson P, editors. Engineering, Biological Responses and Medical Applications. Berlin and Heidelberg: Springer-Verlag; 2001. pp. 13-24

[10] Geetha M, Singh A, Asokamani R, Gogia A. Ti based biomaterials, the ultimate choice for orthopaedic implants—A review. Progress in Materials Science. 2009;**54**:397-425

[11] Ratner B. Titanium in medicine: Material science, surface science. In: Brunette DM, Tengvall P, Textor M, Thompson P, editors. Engineering, Biological Responses and Medical Applications. Berlin and Heidelberg: Springer-Verlag; 2001. pp. 1-12

[12] Liu X, Chu P, Ding C. Surface modification of titanium, titanium alloys, and related materials for biomedical applications. Materials Science and Engineering R. 2004;**47**:49-121

[13] Adya M, Alam M, Ravindranath T, Mubeen A, Saluja B. Corrosion in titanium dental implants: Literature review. The Journal of Indian Prosthodontic Society. 2005;**5**:126-131

[14] Donachie MJ. Titanium: A Technical Guide. ASM International, Material Park, OH, USA, 2000

[15] Brown S, Lemons J. Medical Applications of Titanium and Its Alloys: The Material and Biological Issues. ASTM special series publication, West Conshohocken, PA, USA. 1272; 1996

[16] Oldani C, Dominguez A. Titanium as a biomaterial for implants. In: Fokter S, editor. Recent Advances in Arthroplasty. InTech, Rijeka, Croatia; 2012. ISBN: 978-953-307-990-5. Available from: http://www.intechopen.com/books/recent-advances-in-arthroplasty/titanium-as-a-biomaterial-for-implants

[17] Van Noort R. Titanium: The implant material for today. Journal of Material Science. 1987;**22**:3801-3811

[18] Bardos DI. Titanium and titanium alloys. In: M'illiams D, editor. Concise Encyclopedia of Medical and Dental Materials. Oxford: Pergamon Press; 1990. pp. 360-365

[19] Sykaras N, Lacopino AM, Marker VA, Triplett RG, Woody RD. Implant materials, designs, and surface topographies: Their effect on osseointegration. A literature review. The International Journal of Oral & Maxillofacial Implants. 2000;**15**:675-690

[20] Adell R, Eriksson B, Lekholm U, Brånemark PI, Jemt T. A long-term follow-up study of osseointegrated implants in the treatment of totally edentulous jaws. International Journal of Oral & Maxillofacial Implants. 1990;**5**:347-359

[21] Niinomi M. Mechanical properties of biomedical titanium alloy. Materials Science and Engineering: A. 1998;**243**:231-236

[22] Branemark R, Branemark PI, Rydevik B, Myers RR. Osseointegration in skeletal reconstruction and rehabilitation: A review. Journal of Rehabilitation Research and Development. 2001;**38**:175-181

[23] Wang W, Poh CK. Titanium alloys in orthopaedics. In: Sieniawski J, editor. Titanium Alloys—Advances in Properties Control. InTech, Rijeka, Croatia; 2013. DOI: 10.5772/55353. Available from: http://www.intechopen.com/books/titanium-alloys-advances-in-properties-control/titanium-alloys-in-orthopaedics

[24] Shah FA, Trobos M, Thomsen P, Palmquist A, Commercially pure titanium (cp-Ti) *versus* titanium alloy (Ti6Al4V) materials as bone anchored implants—Is one truly better than the other? Materials Science and Engineering: C. 2016;**62**:960-966

[25] Wolfle JV, Fiedler J, Durselen L, Reichert J, Scharnweber D, Forster A, Schwenzer B, Reichel H, Ignatius A, Brenner RE. Improved anchorage of Ti6Al4V orthopaedic bone implants through oligonucleotide mediated immobilization of BMP-2 in osteoporotic rats. PLoS One. 2014;**9**:e86151

[26] Ryhänen J, Niemi E, Serlo W, Niemelä E, Sandvik P, Pernu H, Salo T. Biocompatibility of nickel-titanium shape memory metal and its corrosion behavior in human cell cultures. Journal of Biomedical Materials Research. 1997;**35**:451-457

[27] Meisner LL, Markov AB, Proskurovsky DI, Rotshtein VP, Ozur GE, Meisner SN, Yakovlev EV, Poletika TM, Girsova SL, Semin VO. Effect of inclusions on cratering behavior in TiNi shape memory alloys irradiated with a low-energy, high-current electron beam. Surface and Coatings Technology. 2016;**302**:495-506

[28] Mohamed DS, Lotfy KM. Biocompatibility of nickel titanium versus nitrided nickel titanium in bone and bone marrow of male New Zealand rabbits. The Egyptian Journal of Histology. 2015;**38**:629-636

[29] Nasakina EO, Sevost'yanov MA, Gol'dberg MA. Long-term corrosion tests of nanostructural nitinol of (55.91 wt% Ni, 44.03 wt% Ti) composition under static conditions: Ion release, Inorganic Materials: Applied Research. 2015;**6**:59-66

[30] Shabalovskaya S. On the nature of the biocompatibility and medical applications of NiTi shape memory and superelastic alloys. Bio-Medical Materials and Engineering. 1996;**6**:267-289

[31] Sieniawski J, Ziaja W, Kubiak K, Motyka M. Microstructure and mechanical properties of high strength two-phase titanium alloys. In: Sieniawski J, editor. Titanium Alloys—Advances in Properties Control. InTech, Rijeka, Croatia; 2013. DOI: 10.5772/56197. Available from: http://www.intechopen.com/books/titanium-alloys-advances-in-properties-control/microstructure-and-mechanical-properties-of-high-strength-two-phase-titanium-alloys

[32] Hosseini S, Limooei MB. Investigation of fatigue behavior and notch sensitivity of Ti-6Al-4V. Applied Mechanics and Materials. 2001;**7**:80-81

[33] Hosseini S, Arabi H, Tamizifar M, Zeyaei A. Effect of tensile strength on behavior and notch sensitivity of Ti-6Al-4V. Iranian Journal of Materials Science and Engineering. Vol. 3. Winter & Spring; 2006. pp. 12-16

[34] Steinemann SG. Corrosion of Titanium and Titanium Alloys for Surgical Implants, Titanium 84 Science and Technology. Vol.2, Deutsche Gesellschaft Fur Metallkunde EV, Munich, Germany: 1985. pp. 1373-1379

[35] Hiromoto S, Hanawa T, Asami K. Composition of surface oxide film of titanium with culturing murine fibroblasts L929. Biomaterials. 2004;**25**:979-986

[36] Hanawa T, Asami K, Asaoka K. Repassivation of titanium and surface oxide film regenerated in simulated bioliquid. Journal of Biomedical Materials Research. 1998;**40**:530-538

[37] Nasakina EO, Sevost'yanov MA, Gol'dberg MA, Demin KY, Baikin AS, Goncharenko BA, Cherkasov VA, Kolmakov AG, Zabolotnyi VT. Long-term corrosion tests of nanostructural nitinol of (55.91 wt% Ni, 44.03 wt% Ti) composition under static conditions: Composition and structure before and after corrosion. Inorganic Materials: Applied Research. 2015;**6**:59-66

[38] Le Gué hennec L, Soueidan A, Layrolle P, Amouriq Y. Surface treatments of titanium dental implants for rapid osseointegration. Dental Materials. 2007;**23**:844-854

[39] Carlsson L, Röstlund T, Albrektsson B, Albrektsson T, Brånemark P-I. Osseointegration of titanium implants. Acta Orthopaedica Scandinavica. 1986;**57**:285-289

[40] Elias CN, Meirelles L. Improving osseointegration of dental implants. Expert Review of Medical Devices. 2010;**7**:241-256

[41] Hao J, Li Y, Li B, Wang X, Li H, Liu S, Liang C, Wang H. Biological and mechanical effects of micro-nanostructured titanium surface on an osteoblastic cell line in vitro and osteointegration in vivo. Applied Biochemistry and Biotechnology. 2017 Mar 20. DOI: 10.1007/s12010-017-2444-1

[42] Branemark P, Hansson B, Adell R, Breine U, Lindstrom J, Hallen O, Ohman A. Osseointegrated implants in the treatment of edentulous jaw. Scandinavian Journal of Plastic and Reconstructive Surgery Supplement. 1977;**16**:1-132

[43] Coathup MJ, Blunn GW, Mirhosseini N, Erskine K, Liu Z, Garrod DR, Li L. Controlled laser texturing of titanium results in reliable osteointegration. Journal of Orthopaedic Research. 2016 Jun 16. DOI: 10.1002/jor.23340

[44] Mavrogenis AF, Dimitriou R, Parvizi J, Babis GC. Biology of implant osseointegration. Journal of Musculoskeletal and Neuronal Interactions. 2009;**9**:61-71

[45] Schatzker J. Osseointegration of metal. Canadian Journal of Surgery. 1995;**38**:49-54

[46] Wróbel E, Witkowska-Zimny M, Przybylski J. Biological mechanisms of implant osseo-integration. Ortopedia Traumatologia Rehabilitacja. 2010;**12**:401-409

[47] Junker R, Dimakis A, Thoneick M, Jansen JA. Effects of implant surface coatings and composition on bone integration: A systematic review Clinical Oral Implants Research. 2009;**4**:185-206

[48] Molitor P, Barron V, Young T. Surface treatment of titanium for adhesive and adhesives. International Journal of Adhesion and Adhesives. 2001;**21**(2):129-136

[49] Citeau A, Guicheux J, Vinatier C, Layrolle P, Nguyen TP, Pilet P, Daculsi G. In vitro biological effects of titanium rough surface obtained by calcium phosphate grid blasting. Biomaterials. 2005;**26**:157-165

[50] Liang CY, Yang XJ, Wei Q, Cui ZD. Comparison of calcium phosphate coatings formed on femtosecond laser-induced and sand-blasted titanium. Applied Surface Science. 2008;**255**:515-518

[51] Gbureck U, Masten A, Probst J, Thull R. Tribochemical structuring and coating of implant metal surfaces with titanium oxide and hydroxyapatite layers. Materials Science and Engineering C. 2003;**23**:461-465

[52] Hryniewicz T, Rokicki R, Rokosz K. Corrosion and surface characterization of titanium biomaterial after magnetoelectropolishing. Surface & Coatings Technology. 2009;**203**:1508-1515

[53] Strnad J, Strnad Z, Sestak J. Physico-chemical properties and healing capacity of potentially bioactive titanium surface. Journal of Thermal Analysis and Calorimetry. 2007;**88**(3):775-779

[54] Mohammadi Z, Ziaei-Moayyed AA, Sheikh-Mehdi Mesgar A. Grit blasting of Ti-6Al-4V alloy: Optimization and its effect on adhesion strength of plasma-sprayed hydroxyapatite coatings. Journal of Materials Processing Technology. 2007;**194**:15-23

[55] Lu X, Zhao Z, Leng Y. Biomimetic calcium phosphate coatings on nitric-acid-treated titanium surfaces. Materials Science and Engineering C. 2007;**27**:700-708

[56] Yousefpour M, Afshar A, Chen J, Xingdong Z. Bioactive layer formation on alkaline-acid treated titanium in simulated body fluid. Materials and Design. 2007;**28**:2154-2159

[57] Pattanayak DK, Kawai T, Matsushita T, Takadama H, Nakamura T, Kokubo T. Effect of HCl concentrations on apatie-forming ability of NaOH-HCl- and heat-treated titanium metal. Journal of Materials Science: Materials in Medicine. 2009;**20**:2401-2411. http://www.springerlink.com/content/m7l108m885m83056

[58] Assis SL, Costa I. The effect of hydrogen peroxide on the electrochemical behaviour of Ti-13Nb-13Zr alloy in hanks' solution. Materials Research. 2006;**9**(4):425-429

[59] Han JY, Zu ZT, Zhou L. Hydroxyapatite/titania composite bioactivity coating processed by sol-gel method. Applied Surface Science. 2008;**255**:455-458

[60] Nguyen HQ, Deporter DA, Pilliar RM, Valiquette N, Yakubovich R. The effect of sol-gel formed calcium phosphate coatings on bone ingrowth and osteoconductivity of porous-surfaced Ti alloy implants. Biomaterials. 2004;**25**(5):865-876

[61] Wierzchon T, Czarnowska E, Krupa D. Inżynieria powierzchni w wytwarzaniu biomateriałów tytanowych. Warszawa: Oficyna Wyd. Politechniki Warszawskiej; 2004

[62] Wilks RG, Santos E, Kurmaev EZ, Yablonskikh MV, Moewes A, Kuromoto NK, Soares GA. Characterization of oxide layers fordem on electrochemically treated Ti by Rusing soft X-ray absorption measurements. Journal of Electron Spectroscopy and Related Phenomena. 2009;**169**:46-50

[63] Diamanti MV, Pedeferri MP. Effect of anodic oxidation parameters on the titanium oxides formation. Corrosion Science. 2007;**49**:939-948

[64] Cui X, Kim H-M, Kawashita M, Wang L, Xiong T, Kokubo T, Nakamura T. Preparation of bioactive titania films on titanium metal via anodic oxidation. Dental Materials. 2009;**25**:80-86

[65] Bauer S, Park J, Mark K, Schmuki P. Improved attachment of mesenchymal stem cells on super-hydrophobic TiO_2 nanotubes. Acta Biomaterialia. 2008;**4**:576-1582

[66] Goto T. Surface coating technology for biomaterials—morphology and nono-structure control. International Congress Series. 2005;**1284**:248-256

[67] Sevilla P, Aparicio C, Planell JA, Gil FJ. Comparison of the mechanical properties between tantalum and nickel-titanium foams implant materials for bone ingrowth applications. Journal of Alloys and Compounds. 2007;**439**:67-73

[68] Trommer RM, Santos LA, Bergmann CP. Alternative technique for hydroxyapatite coatings. Surface & Coatings Technology. 2007;**201**:9587-9593

[69] Kim DS, Han SJ, Kwak S-Y. Synthesis and photocatalytic activity of mesoporous TiO_2 with the surface area, crystallite size, and pore size. Journal of Colloid and Interface Science. 2007;**316**:85-91

[70] Baram N, Starosvetsky D, Starosvetsky J, Epshtein M, Armon R, Ein-Eli Y. Enhanced inactivation of *E. coli* bacteria using immobilized porous TiO_2 photoelectrocatalysis. Electrochimica Acta. 2009;**54**:3381-3386

[71] Lewis G, McVay B. Effect of thermal spray process for deposition hydroxyapatite coating on a titanium alloy on its fatigue performance. Proceedings of the 17th Southern Biomedical Engineering Conference 1998, p.119

[72] Gledhill HC, Turner IG, Doyle C. In vitro dissolution behavior of two morphologically different thermally sprayed hydroxyapatite coatings. Biomaterials. 2001;**22**:695-700

[73] Li H, Khor KA. Characteristics of the nanostructures in thermal sprayed hydroxyapatite coatings and their influence on coating properties. Surface & Coatings Technology. 2006;**201**:2147-2154

[74] Lima RS, Khor KA, Li H, Cheang P, Marple BR. HVOF spraying on nanostructured hydroxyapatite for biomedical applications. Materials Science and Engineering A. 2005;**396**:181-187

[75] Goana M, Lima RS, Marple BR. Influence of particle temperature and velocity on the microstructure and mechanical behavior of high velocity oxy-fuel (HVOF)—sprayed nanostructured titania coatings. Journal of Materials Processing Technology. 2008;**198**:426-435

[76] Hoseini M, Jedenmalm A, Boldizar A. Tribological investigation of coatings for artificial joints. Wear. 2008;**264**:958-966

[77] Chiu S-M, Chen Z-S, Yang K-Y, Hsu Y-L, Gan D. Photocatalytic activity of moped TiO2 coatings prepared by sputtering deposition. Journal of Materials Processing Technology. 2007;**192-193:**60-67

[78] Krupa D, Baszkiewicz J, Rajchel B, Barcz A, Sobczak JW, Biliński A, Borowski T. Effect of calcium-ion implantation on the corrosion resistance and bioactivity of the Ti6Al4V Allom. Vacuum. 2007;**81**:1310-131

[79] Xie Y, Liu X, Huang A, Ding CH, Chu PK. Improvement of surface bioactivity on titanium by water and hydrogen plasma immersion ion implantation. Biomaterials. 2005;**26**:6129-6135

[80] Jo YJ, Lee CM, Jang HS, Lee NS, Suk J-H, Lee WH. Mechanical properties of fully porous and porous-surfaced Ti-6Al-4V implants fabricated by electro-discharge-sintering. Journal of Materials Processing Technology. 2007;**194**:121-125

[81] An YB, Lee WH. Synthesis of porous titanium implants by environmental- electro-discharge-sintering process. Materials Chemistry and Physics. 2006;**95**:242-247

[82] Hanaor DAH, Sorrell CHC. Review of the anatase to rutile phase transformation. Journal of Materials Science. 2011;**46**:855-874

[83] Gateshki M, Yin S, Ren Y, Petkov V. Titania polymorphs by soft chemistry: Is there a common structural pattern? Chemistry of Materials. 2007;**19**:2512-2518

[84] Diebold U. Structure and properties of TiO_2 surfaces: A brief review. Applied Physics A. 2003;**76**:681-687

[85] Wang B, Qi H, Wang H, Cui Y, Zhao J, Guo J, Cui Y, Liu Y, Yi K, Shao J. Morphology, structure and optical properties in TiO2 nanostructured films annealed at various temperatures. Optical Materials Express. 2015;**5**:1410-1418

[86] Park J-Y, Lee CH, Jung K-W, Jung D. Structure related photocatalytic properties of TiO_2. Bulletin of the Korean Chemical Society. 2009;**30**:402-404

[87] Alemany LJ, Bañares MA, Pardo E, Martín-Jiménez F, Blasco JM. Morphological and structural characterization of a titanium dioxide system. Materials Characterization. 2000;**44**:271-275

[88] Thamaphat K, Limsuwan P Ngotawornchai B. Phase Characterization of TiO2 Powder by XRD and TEM. Kasetsart Journal (Natural Science) 2008;**42**:357-361

[89] Sakurai K, Mizusawa M. X-ray diffraction imaging of anatase and rutile. Analytical Chemistry. 2010;**82**:3519-3522

[90] Thangavelu K, Annamalai R, Arulnandhi D. Synthesis and characterization of nanosized TiO_2 powder derived from a sol-gel process in acidic conditions. International Journal of Engineering Sciences & Emerging Technologies. 2013;**4**:90-95

[91] Rao CNR, Turner A, Honig JM. The effect of impurities on the anatase-rutile transformation. The Journal of Physical Chemistry. 1959;**11**:173-174

[92] Oshaka T, Izumi F, Fujiki Y. Raman spectrum of anatase TiO_2. Journal of Raman Spectroscopy. 1978;**7**:321-324

[93] Deo G, Turek AM, Wachs IE, Machej T, Haber J, Das N, Edkert H, Hirt AM. Physical and chemical characterization of surface vanadium-oxide supported on titania. Influence of the titania phase (anatase, rutile, brookite A and B). Applied Catalysis A. 1992;**A91**:27-42

[94] Alemany LJ, Bañares MA, Larrubia MA, Jiménez MC, Delgado F, Blasco JM. Vanadia-titania systems—Morphological and structural properties. Materials Research Bulletin. 1996;**31**:513-520

[95] Zeng G, Li K-K, Yang H-G, Zhang Y-H. Micro-Raman mapping on an anatase TiO_2 single crystal with a large percentage of reactive (0 0 1) facets. Vibrational Spectroscopy. 2013;**68**:279-284

[96] Foucher F, Guimbretière G, Bost N, Westall F. Petrographical and Mineralogical applications of Raman mapping. In: Maaz K, editor. Raman Spectroscopy and Applications. InTech, Rijeka, Croatia; 2017. ISBN 978-953-51-2908-0, Print ISBN 978-953-51-2907-3

[97] Jianliang LJ, BoWang B, Sproul WD, Ou Y, Dahan I. Anatase and rutile TiO_2 films deposited by arc-free deep oscillation magnetron sputtering. Journal of Physics D: Applied Physics. 2013;**46**:084008 (9 pp)

[98] Dai S, Wu Y, Sakai T, Du Z, Sakai H, Abe M. Preparation of highly crystalline TiO_2 nanostructures by Acid-assisted hydrothermal treatment of Hexagonal-structured nanocrystalline titania/cetyltrimethyammonium bromide nanoskeleton. Nanoscale Researh Letter. 2010;**11:**1829-1835

[99] Roy P, Berger S, Schmuki P. TiO_2 Nanotubes: Synthesis and applications. Angewandte Chemie International Edition. 2011;**50**:2904-2939

[100] Regonini D, Bowen CR, Jaroenworaluck A, Stevens R. A review of growth mechanism, structure and crystallinity of anodized TiO_2 nanotubes. Materials Science and Engineering: R: Reports. 2013;**74**:377-406

[101] Wang Y, Wu Y, Qin Y, Xu G, Hu X, Cui J, Zheng H, Hong Y, Zhang X. Rapid anodic oxidation of highly ordered TiO_2 nanotube arrays. Journal of Alloys and Compounds. 2011;**509**:157-160

[102] Zhao J, Wang X, Chen R, Li L. Fabrication of titanium oxide nanotube arrays by anodic oxidation. Solid State Communications. 2005;**134**:705-710

[103] Sulka GD, Kapusta-Kołodziej J, Brzózka A, Jaskuła M. Anodic growth of TiO_2 nanopore arrays at various temperatures. Electrochimica Acta. 2013;**104**:526-535

[104] Diamanti MV, Spreafico FC, Pedeferri MP. Production of anodic TiO_2 nanofilms and their characterization. Physics Procedia. 2013;**40**:30-37

[105] Komiya S, Sakamoto K, Ohtsu N. Structural changes of anodic layer on titanium in sulfate solution as a function of anodization duration in constant current mode. Surface Science. 2014;**296**:163-168

[106] Kuromoto NK, Simão RA, Soares GA. Titanium oxide films produced on commercially pure titanium by anodic oxidation with different voltages. Materials Characterization. 2007;**58**:114-121

[107] Kasuga T, Hiramatsu M, Hoson A, Sekino T, Niihara K. Formation of titanium oxide nanotube. Langmuir. 1998;**14**:3160-3163

[108] Sander MS, Cote MJ., Gu W, Kile BM, Tripp CP. Template-Assisted fabrication of dense, aligned arrays of titania nanotubes with Well-Controlled dimensions on substrates. Advanced Materials. 2004;**16**:2052-2057

[109] Adachi M, Murata Y, Harada M, Yoshikawa S. Formation of titania nanotubes with high photo-catalytic activity. Chemical Letters. 2000;**29**:942-943

[110] Lakshmi BB, Dorhout PK, Martin CR. Sol-gel template synthesis of semi-conductor nanostructures. Chemistry of Materials. 1997;**9**:857-862

[111] Kasuga T, Hiramatsu M, Hoson A, Sekino T, Niihara K. Titania nanotubes prepared by chemical processing. Advanced Materials. 1999;**11**:1307-1311

[112] Kulkarni M, Mazare A, Schmuki P, Iglic A. Influence of anodization parameters on morphology of TiO_2 nanostructured surfaces. Advanced Materials Letters. 2016;**7**:23-28

[113] Lim Y-Ch, Zulkarnain Zainal Z, Wee-Tee Tan W-T, Hussein MZ. Anodization parameters influencing the growth of titania nanotubes and their photoelectrochemical response. International Journal of Photoenergy.2012;**2012**. 9. Article ID 638017. DOI: 10.1155/2012/638017

[114] Haring A, Morris A, Hu M. Controlling morphological parameters of anodized titania nanotubes for optimized solar energy applications. Materials. 2012;5:1890-1909

[115] Amin Yavari S, Chai YC, Böttger AJ, Wauthle R Schrooten J, Weinans H, Zadpoor AA. Effects of anodizing parameters and heat treatment on nanotopographical features, bioactivity, and cell culture response of additively manufactured porous titanium. Materials Science and Engineering C: Materials for Biological Applications. 2015;**51**:132-138

[116] Kaczmarek A, Klekiel T, Krasicka-Cydzik E. Fluoride concentration effect on the anodic growth of self-aligned oxide nanotube array on Ti6Al7Nb alloy. Surface and Interface Analysis. 2010;**42**:510-514

[117] Habazaki H, Fushimi K, Shimizu K, Skeldon P, Thompson GE. Fast migration of fluoride ions in growing anodic titanium oxide. Electrochemistry Communications. 2007;**9**:1222-1227

[118] Radtke A, Piszczek P, Topolski A, Lewandowska Ż, Talik E, Andersen IH, Nielsen LP, Heikkilä M, Leskelä M. The structure and the photocatalytic activity of titania based nanotube and nanofiber coatings. Applied Surface Science. 2016;**368**:165-172

[119] Lee K, Mazare A, Schmuki P. One-Dimensional titanium dioxide nanomaterials: Nanotubes. Chemical Reviews. 2014;**19**:9385-9454

[120] Albu SP, Kim D, Schmuki P. Growth of aligned TiO_2 bamboo-type nanotubes and highly ordered nanolace. Angewandte Chemie International Edition. 2008;**47**:1916-1919

[121] Kim D, Ghicov A, Albu SP, Schmuki P. TiO_2 bamboo-type nanotubes: Improved conversion efficiency in dye–sentisized solar cells. Journal of the American Chemical Society. 2008;**130**:16454-16455

[122] Ruan CH, Paulose M, Varghese OK, Mor GK, Grimes CA. Fabrication of highly ordered TiO_2 nanotube arrays using an organic electrolyte. Journal of Physical Chemistry B. 2005;**109**:15754-15759

[123] Yoriya S, Mor GK, Sharma S, Grimes CA. Synthesis of ordered arrays of discrete, partially crystalline titania nanotubes by Ti anodization using diethylene glycol electrolytes. Journal of Materials Chemistry. 2008;**18**:3332-3336

[124] Yoriya S, Grimes CA. Self-assembled TiO_2 nanotube arrays by anodization of titanium in diethylene glycol: Approach to extended pore widening. Langmuir. 2010;**26**(1):417-420

[125] Hassan FMB, Nanjo H, Tetsuka H, Kanakubo M, Aizawa T, Nishioka M, Ebina T. Formation of self ordered TiO_2 nanotubes by electrochemical anodization of titanium in 2-Propanol/NH4F. ECS Transactions. 2009;**16**:35-47

[126] Xiong JY, Wang Y, An Y, Wen M, Ding YF, Li YC, Hodgson P. Morphology evolution during annealing and electrical conductivity of titania nanotube films. Advanced Materials Research. 2012;**399-401**:548-551

[127] Bayram C, Demirbilek M, Calişkan N, Demirbilek ME, Denkbaş EB. Osteoblast activity on anodized titania nanotubes: Effect of simulated body fluid soaking time. Journal of Biomedical Nanotechnology. 2012;**8**:482-490

[128] Minagar S, Wang J, Berndt CC, Ivanova EP, Wen C. Cell response of anodized nanotubes on titanium and titanium alloys. Journal of Biomedical Materials Research Part A. 2013;**101**:2726-2739

[129] Gongadze E, Kabaso D, Bauer S, Park J, Schmuki P, Iglič A. Adhesion of osteoblasts to a vertically aligned TiO_2 nanotube surface. Mini Reviews in Medicinal Chemistry. 2013;**13**:194-200

[130] Tan AW, Pingguan-Murphy B, Ahmad R, Akbar SA. Review of titania nanotubes: Fabrication and cellular response. Ceramics International. 2012;**38**:4421-4435

[131] Smith BS, Yoriya S, Johnson T, Popat KC. Dermal fibroblast and epidermal keratinocyte functionality on titania nanotube arrays. Acta Biomaterialia. 2011;**7**:2686-2696

[132] Bariana M, Dwivedi P, Ranjitkar S, Kaidonis JA, Losic D, Anderson PJ. Biological response of human suture mesenchymal cells to titania nanotube-based implants for advanced craniosynostosis therapy. Colloids and Surfaces B: Biointerfaces. 2017;**1**:59-67

[133] Burns K, Yao C, Webster TJ. Increased chondrocyte adhesion on nanotubular anodized titanium. Journal of Biomedical Materials Research Part A. 2009;**88A**:561-568

[134] Park J, Bauer S, Schlegel KA, Neukam FW, von der Mark K, Schmuki P. TiO_2 nanotube surfaces: 15 nm—an optimal length scale of surface topography for cell adhesion and differentiation. Small. 2009;**5**:666-671

[135] Park J, Bauer S, von der Mark K, Schmuki P. Nanosize and vitality: TiO_2 nanotube diameter directs cell fate. Nano Letters. 2007;**7**(6):1686-1691

[136] Lewandowska Ż, Piszczek P, Radtke A, Jędrzejewski T, Kozak W, Sadowska B. The evaluation of the impact of titania nanotube covers morphology and crystal phase on their biological properties. Journal of Materials Science: Materials in Medicine. 2015;**26**:163

[137] Ercan B, Taylor E, Alpaslan E, Webster T. Diameter of titanium nanotubes influences anti-bacterial efficacy. Journal of Nanotechnology. 2011;**22**:295102

[138] Ercan B, Kummer KM, Tarquinio KM, Webster TJ. Decreased *Staphylococcus aureus* biofilm growth on anodized nanotubular titanium and the effect of electrical stimulation. Acta Biomaterialia. 2011;**7**:3003-3012

[139] Popat KC, Eltgroth M, Latempa TJ, Grimes CA, Desai TA. Decreased *Staphylococcus epidermis* adhesion and increased osteoblast functionality on antibiotic-loaded titania nanotubes. Biomaterials. 2007;**28**:4880-4888

[140] Cipriano AF, Miller C, Liu H. Anodic growth and biomedical applications of TiO_2 nanotubes. Journal of Biomedical Nanotechnology. 2014;**10**:2977-3003

[141] Puckett DS, Tayler E, Raimondo T, Webster TJ. The relationship between the nanostructure of titanium surfaces and bacterial attachment. Biomaterials. 2010;**31**:706-713

[142] Akira Fujishima A, Rao TN, Tryk DA. Titanium dioxide photocatalysis. Journal of Photochemistry and Photobiology C: Photochemistry Reviews. 2000;**1**:1-21

[143] Diebold U. The surface science of titanium dioxide. Surface Science Reports. 2003;**48**:53-229

[144] Carp O, Huisman CL, Reller A. Photoinduced reactivity of titanium dioxide. Progress in Solid State Chemistry. 2004;**32**:33-177

[145] Fujishima A, Zhang X, Tryk DA. TiO_2 photocatalysis and related surface phenomena. Surface Science Reports. 2008;**63**:515-582

[146] Schneider J, Matsuoka M, Takeuchi M, Zhang J, Horiuchi Y, Anpo M, Bahnemann DW. Understanding TiO_2 photocatalysis: Mechanisms and materials. Chemical Reviews. 2014;**114**:9919-9986

[147] Paz Y. Application of TiO_2 photocatalysis for air treatment: Patents' overview. Review Article. Applied Catalysis B: Environmental. 2010;**99**(3-4):448-460

[148] Pelaez M, Nolan NT, Pillai SC, Seery MK, Falaras P, Kontos AG, Dunlop P, Hamilton J, Byrne JA, O'Shea K, Entezari MH., Dionysiou D. A review on the visible light active titanium dioxide photocatalysts for environmental applications. Applied Catalysis B: Environmental. 2012;**125**:331-349

[149] Park H, Park Y, Kim W, Choi W. Surface modification of TiO_2 photocatalyst for environmental applications. Journal of Photochemistry and Photobiology C: Photochemistry Reviews. 2013;**15**:1-20

[150] Herrmann J-M. Fundamentals and misconceptions in photocatalysis. Journal of Photochemistry and Photobiology A: Chemistry. 2010;**216**:85-93

[151] Feng T, Feng GS, Yan L, Pan JH. One-Dimensional nanostructured TiO_2 for photocatalytic degradation of organic pollutants in wastewater. International Journal of Photoenergy. 2014;**14**. Article ID 563879

[152] Albu SP, Ghicov A, Macak JM, Hahn R, Schmuki P. Self-organized, free-standing TiO_2 nanotube membrane for flow-through photocatalytic applications Nano Letters. 2007;**7**:1286-1289

[153] Macak JM, Zlamal M, Krysa J, Schmuki P. Self-organized TiO_2 nanotube layers as highly efficient photocatalysts. Small. 2007;**3**:300-304

[154] Lee K, Mazare A, Schmuki P. One-dimensional titanium dioxide nanomaterials: Nanotubes. Chemical Reviews. 2014;**114**:9385-9454

[155] Paramasivam I, Macak JM, Ghicov A, Schmuki P. Enhanced photochromism of Ag loaded self-organized TiO_2 nanotube layers. Chemical Physics Letters. 2007;**445**:233-237

[156] Zlamal M, Macak JM, Schmuki P, Krysa J. Electrochemically assisted photocatalysis on self-organized TiO_2 nanotubes. Electrochemistry Communications. 2007;**9**:2822-2826

[157] Paramasivam I, Jha H, Liu N, Schmuki P, A Review of Photocatalysis using Self-organized TiO2 Nanotubes and Other Ordered Oxide Nanostructures, Small 2012;**8**: 3073-3103

[158] Teo WE, Ramakrishna S. A review on electrospinning design and nanofibre assemblies. Nanotechnology. 2006;**17**:89-106

[159] Thoppey NM, Bochinski JR, Clarke LI, Gorga RE. Edge electrospinning for high throughput production of quality nanofibers, Nanotechnology. 2011;**22**. Article ID 345301

[160] Panda PK, Ramakrishna S. Electrospinning of alumina nanofibers using different precursors. Journal of Materials Science. 2007;**42**:2189-2193

[161] Ramaseshan R, Sundarrajan S, Jose R, Ramakrishna S. Nanostructured ceramics by electrospinning. Journal of Applied Physics. 2007;**102**. Article ID 111101

[162] Watthanaarun J, Pavarajarn V, Supaphol P. Titanium(IV) oxide nanofibers by combined sol-gel and electrospinning techniques: Preliminary report on effects of preparation conditions and secondary metal dopant. Science and Technology of Advanced Materials. 2005;**6**:240-245

[163] Chronakis IS. Novel nanocomposites and nanoceramics based on polymer nanofibers using electrospinning process—A review. Journal of Materials Processing Technology. 2005;**167**:283-293

[164] Caratão B, Carneiro E, Sá P, Almeida B, Carvalho S. Properties of Electrospun TiO_2 nanofibers. Journal of Nanotechnology. 2014;**5**. Article ID 472132

[165] Li D, Xia Y. Fabrication of titania nanofibers by electrospinning. Nano Letters. 2003;**3**: 555-560

[166] Lee J-S, Lee Y-I, Song H, Jang D-H, Choa Y-H. Synthesis and characterization of TiO_2 nanowires with controlled porosity and microstructure using electrospinning method. Current Applied Physics. 2011;**11**:210-214

[167] Wang X, Zhu J, Yin L, Liu S, Zhang X, Ao Y, Chen H, Evaluation of the Morphology and Osteogenic Potential of Titania-Based Electrospun Nanofibers, Journal of Nanomaterials, 2012; **2012**:7. Article ID 959578, DOI:10.1155/2012/959578

[168] Ding B, Kim CK, Kim HY, Se MK, Park SJ. Titanium dioxide nanofibers prepared by using electrospinning method. Fibers and Polymers. 2004;**5**:105-109

[169] Chen J-Y, Chen H-C, Lin J-N, Kuo C. Effects of polymer media on electrospun mesoporous titania nanofibers. Materials Chemistry and Physics. 2008;**107**:480-487

[170] Lee S-J, Cho N-I, Lee DY. Effect of collector grounding on directionality of electrospun titania fibers. Journal of the European Ceramic Society. 2007;**27**:3651-3654

[171] Chandrasekar R, Zhang LF, Howe JY, Hedin NE, Zhang Y, Fong H. Fabrication and characterization of electrospun titania nanofibers. Journal of Materials Science. 2009;**44**:1198-1205

[172] Khalil KA, Kim SW, Kim KW, Dharmaraj N, Kim HY. A novel bio-nanocomposites composed of hydroxyapatite reinforced with TiO_2 electrospun nanofiber consolidated using high-frequency induction heating. International Journal of Applied Ceramic Technology. 2011;**8**:523-531

[173] Park S-J, Chase G, Jeong K-U, Kim H. Mechanical properties of titania nanofiber mats fabricated by electrospinning of sol-gel precursor. Journal of Sol-Gel Science and Technology. 2010;**54**:188-194

[174] Tavangar A, Tan B, Venkatakrishnan K. Synthesis of bio-functionalized three dimensional titania nanofibrous structures using femtosecond laser ablation. Acta Biomaterialia. 2011;**7**:2726-2732

[175] Tavangar A, Tan B, Venkatakrishnan K. Study of the formation of 3-D titania nanofibrous structure by MHz femtosecond laser in ambient air. Journal of Applied Physics. 2013;**113**:023102-023110

[176] Lim JH, Choi J. Titanium oxide nanowires originating from anodically grown nanotubes: The bamboo-splitting model. Small. 2007;**3**:1504-1507

[177] Chang CH, Lee HC, Chen CC, Wu Y-H, Hsu Y-M, Chang Y-P, Yang T-I, Fang H-W.A novel rotating electrochemically anodizing process to fabricate titanium oxide surface nanostructures enhancing the bioactivity of osteoblastic cells. Journal of Biomedical Materials Research Part A. 2012;**100A**:1687-1695

[178] Wang X, Gittens RA, Song R, Tannenbaum R, Olivares-Navarrete R, Schwartz Z, Chen H, Boyan BD. Effects of structural properties of electrospun TiO_2 nanofiber meshes on their osteogenic potential. Acta Biomaterialia. 2012;**8**:878-885

[179] Ma JY, He XZ, Jabbari E. Osteogenic differentiation of marrow stromal cells on random and aligned electrospun poly(l-lactide) nanofibers. Annals of Biomedical Engineering. 2011;**39**:14-25

[180] Menel AJ, Kubow KE, Klotzsch E, Garcia-Fuentes M, Smith ML, Vogel V. Optimization strategies for electrospun silk fibroin tissue engineering scaffolds. Biomaterials. 2009;**30**:3058-3067

[181] Murugan R, Ramakrishna S. Design strategies of tissue engineering scaffolds with controlled fiber orientation. Tissue Engineering. 2007;**13**:1845-1866

[182] Doh SJ, Kim CH, Lee SG, Lee SJ, Kim H. Development of photocatalytic TiO_2 nanofibers by electrospinning and its application to degradation of dye pollutants. Journal of Hazardous Materials. 2008;**134**:118-127

[183] Jun Y-W, Choi J-S, Cheon J. Shape control of semiconductor and metal oxide nanocrystals through nonhydrolytic colloidal routes. Angewandte Chemie International Edition. 2006;**45**:3414-3439

[184] Jun Y-W, Lee J-H, Choi J-S, Cheon J. Symmetry-controlled colloidal nanocrystals: Nonhydrolytic chemical synthesis and shape determining parameters. Journal of Physical Chemistry B. 2005;**109**:14795-14806

[185] Niederberger M. Nonaqueous sol-gel routes to metal oxide nanoparticles. Accounts of Chemical Research. 2007;**40**:793-800

[186] Cozzoli PD, Pellegrino T, Manna L. Synthesis, properties and perspectives of hybrid nanocrystal structures. Chemical Society Reviews. 2006;**35**:1195-1205

[187] Lee JC, Park KS, Kim TG, Choi HJ, Sung YM. Controlled growth of high-quality TiO_2 nanowires on sapphire and silica. Nanotechnology. 2006;**17**:4317-4321

[188] Ma R, Fakuda K, Sasaki T, Osada M, Bando Y. Structural features of titanate nanotubes/nanobelts revealed by Raman, X-ray absorption fine structure and electron diffraction characterizations. Journal of Physical Chemistry B. 2005;**109**:6210-6214

[189] Amin SS, Nicholls AW, Xu TT. A facile approach to synthesize single crystalline rutile TiO_2 one-dimensional nanostructures. Nanotechnology. 2007;**18**:445609

[190] Mukherjee K, Teng TH, Jose R, Ramakrishna S. Electron transport in electrospun TiO_2 nanofiber dye-sensitized solar cells. Applied Physics Letters. 2009;**95**:012101

[191] Fujihara K, Kumar A, Jose R, Ramakrishna S. Spray deposition of electrospun TiO_2 nanorods for dye-sensitized solar. Nanotechnology. 2007;**18**:365709

[192] Kisumi T, Tsujiko A, Murakoshi K, Nakato Y. Crystal-face and illumination intensity dependences of the quantum efficiency of photoelectrochemical etching, in relation to those of water photooxidation, at n-TiO_2 (rutile) semiconductor electrodes. Journal of Electroanalytical Chemistry. 2003;**545**:99-107

[193] Lee H, Dregia S, Akbar S, Alhoshan M. Growth of 1-D TiO_2 nanowires on Ti and Ti alloys by oxidation. Journal of Nanomaterials. 2010;**7**. Article ID 503186

[194] Dinan B, Akbar S. One dimensional oxide nanostructures by gas-phase reaction. Functional Nanomaterials Letters. 2009;**2**:87-94

[195] Huo K, Zhang X, Fu J. Synthesis and filed emission properties of rutile TiO_2 nanowires arrays grown directly on a Ti metal self-source substrate. Journal of Nanoscience and Nanotechnology. 2009;**9**:3341-3346

[196] Daothong S, Songmee N, Thongtem S, Singjai P. Size-controlled growth of TiO_2 nanowires by oxidation of titanium substrates in the presence of ethanol vapor. Scripta Materialia. 2007;**57**:567-570

[197] Tan AW, Ismail R, Chua KH, Ahmad R, Akbar SA, Pingguan-Murphy B. Osteogenic potential of in-situ TiO_2 nanowire surfaces formed by thermal oxidation of titanium alloy surface. Applied Surface Science. 2014;**320**:161-170

[198] Pu Y-Ch, Wang G, Chang K-D, Ling Y, Lin Y-K, Fitzmorris BC, Liu Ch-M, Lu X, Tong Y, Zhang JZ, Yung-Jung Hsu Y-J, Li Y. Au Nanostructure-decorated TiO_2 nanowires exhibiting photoactivity across entire UV-visible region for photoelectrochemical water splitting. Nano Letters. 2013;**13**:3817-3823

[199] Stengl V, Bakardjieva S, Murafa N, Vecernıkova E, Subrt J, Balek V. Preparation and characterization of titania based nanowires. Journal of Nanoparticle Research. 2007;**9**:455-470

2

Anticorrosive, Antimicrobial, and Bioactive Titanium Dioxide Coating for Surface-modified Purpose on Biomedical Material

Hsi-Kai Tsou and Ping-Yen Hsieh

Abstract

A multifunctional titanium dioxide (TiO_2) coating was used to provide anticorrosive, antimicrobial, and bioactive properties for the surface modification of biomedical materials because TiO_2 has a stable bonding structure, photocatalytic characteristics, and negatively charged surfaces in nature. For successful deposition, an arc ion plating technique was adopted to deposit the TiO_2 coating. The antimicrobial activity values of anatase-TiO_2-coated stainless steel specimens against *Staphylococcus aureus* and *Escherichia coli* were 3.0 and 2.5, respectively, which are far beyond the value designated in JIS Z2801:2000 industrial standard. TiO_2 coatings on stainless steel also generate an increased (i.e., less negative) corrosion potential and decreased corrosion current in a sodium chloride solution, showing a reduced tendency and rate of substrate dissolution as well as a reduced coating of species into the electrolyte. In addition, TiO_2 coatings, especially with rutile phase, satisfied the requirements for activating the biological property of a polymeric polyetheretherketone surface. Therefore, TiO_2 is a promising surface modification for the biomedical materials used in surgical instruments and implants.

Keywords: anticorrosive, antimicrobial, bioactive, titanium dioxide, biomedical material

1. Introduction

Biomedical material is any substance that has been engineered to interact with biological systems for a medical purpose, which may be therapeutic (i.e., to treat, augment, repair, or replace malfunctioning tissue in the body) or diagnostic. Among the various types of biomedical materials, metallic materials are the most widely used because of their high load-supporting capacity, desirable qualities of wear and friction, and acceptable biocompatibility. Stainless steel, titanium, and their alloys are considered especially promising materials for surgical instruments and implants of many types and sizes. Polymeric materials have also garnered considerable interest in research and development as soft- and hard-tissue replacements, on the basis of the ease of manufacturing and modifying such materials, and their appropriate physical, chemical, and mechanical properties.

When biomedical materials come in contact with physiological tissue and body fluids, various interactions, such as corrosive reaction, inflammation, and host response, are triggered. For this reason, knowing and understanding the surface properties of biomedical materials are crucial. Unfortunately, metallic materials are easily influenced by corrosion damage due to electrochemical reactions; additionally, the bioinertness and hydrophobic surface properties render polymeric materials unfavorable for cell adhesion. Long-term clinical experiments have also indicated that the primary causes of implant failure include not only unstable implant fixation to bone tissue, but also bacterial infection.

To overcome the aforementioned problems, a surface modification technique that uses a multifunctional titanium dioxide (TiO_2) coating is introduced to provide anticorrosive, antimicrobial, and bioactive properties for the underlying biomaterial. These versatile natural features of TiO_2 are attributed to its stable bonding structure, photocatalytic characteristics, and negatively charged surfaces. In this paper, a brief overview of TiO_2 coating modification in the field of biomedical material is provided. The two main topics discussed in the next section are as follows:

- Antimicrobial and anticorrosive titanium dioxide coating on stainless steel to reduce hospital-acquired infection.
- Bioactive titanium dioxide coating on polyetheretherketone for spinal implant application.

2. Antimicrobial and anticorrosive titanium dioxide coating on stainless steel to reduce hospital-acquired infection

2.1. Background

The increasing incidence and host risk of device-related infections that result in morbidity and even mortality have been noted for some time, particularly regarding the spread of antibiotic-resistant bacteria, such as methicillin-resistant *Staphylococcus aureus* and bursting *Clostridium difficile.* These hospital-acquired infections are a worldwide problem [1]. The outbreaks of SARS and avian influenza have also drawn attention to novel preventative measures, including the development and application of antimicrobial materials, to enhance the conventional disinfection concept. This movement compelled us to develop an antimicrobial technique for medical implements in clinical use.

Antimicrobial or antibacterial refers to the inhibition of bacterial growth and reproduction [2]. Antimicrobial functions can be performed by essential materials themselves or through the use of coating materials. One example of an essential antimicrobial alloy material is stainless steel that has been doped with copper. This material forms when ε-copper precipitates in a steel matrix; specifically, copper ions can be dissolved into a surface-passivated chromium oxide film. Such creates an antimicrobial effect on the stainless steel surface, resulting in the inhibition of bacterial growth [3]. The similar antimicrobial metal alloys, such as copper-containing ferritic stainless steel [4], martensitic stainless steel [5], and austenitic stainless steel [6], were also developed. On the other hand, for the antimicrobial purpose on coating materials, the idea of coatings containing with copper, silver, zinc, and other antimicrobial active metals was considered [7]. Unfortunately, such substance may induce the corrosion reaction because of the undesired Galvanic effect between two metals, which may be unsustainable during service. In this regard, TiO_2 with anatase (A-TiO_2) phase may be the promising candidates for antimicrobial purposes.

The antimicrobial effects of TiO_2 are attributed to its photocatalytic characteristics, as discovered by Fujishima and Honda [8]. The photocatalytic process of TiO_2 involves the generation of electron-hole pairs when the material is exposed to light that emits energy exceeding the band gap energy of TiO_2. The aggressive superoxide ions (O^{2-}) are generated by the electron attack, and the holes accelerate hydroxyl radical ($^{\bullet}OH$) formation on the material surface [9, 10]. These active radicals subsequently inhibit the growth of germs and bacteria that are known to be antimicrobially active through the direct oxidation of intracellular coenzyme, reducing the respiratory activity and thereby causing cell death [11].

In the present study, arc ion plating (AIP) was used to deposit a TiO_2 coating on common medical-grade AISI 304 stainless steel. The antimicrobial efficacy of the TiO_2-coated stainless steel specimens was then evaluated according to the JIS standard. The corrosion resistance of the TiO_2 coating was also examined to determine whether such films can be stable in a physiological environment. The results suggest that this modification may be effective as an antimicrobial surface coating for medical implements to reduce the risk of hospital-acquired infections.

2.2. Preparation of antimicrobial and anticorrosive TiO_2 films

TiO_2 deposition was conducted using a typical AIP technique and involved three steps: argon ion bombardment, bottom titanium layer deposition, and TiO_2 coating deposition. The ion bombardment was performed to clean and mildly preheat the substrate, followed by the bottom titanium layer deposition, which enhanced the adhesion between the substrate and TiO_2 coating.

The wide acceptance indicates that an A-TiO_2 phase structure is the key factor for maximizing the antimicrobial efficiency of TiO_2. This corresponds to a specific condition with 100% oxygen pressure at 0.5 Pa by using the AIP technique with a cathode target voltage of 20 V and a cathode target current of 90 A. Under this optimized deposition condition, the proportion of A-TiO_2 in the TiO_2 coating has been reported to be 76.8% [12–14].

2.3. Antimicrobial characteristics of TiO_2-coated stainless steel

The JIS Z2801:2000 [15] was employed as a standard to test the antimicrobial efficacy of TiO_2-coated stainless steel specimens. The bacterial strains used in this test were Gram-positive *Staphylococcus aureus* (*S. aureus*, ATCC 6538P) and Gram-positive *Escherichia coli* (*E. coli*, ATCC 8739) with an initial concentration of 4.0×10^5 bacteria/mL. In the antimicrobial test, the specimens were divided into three groups: group A and group B consisted of uncoated stainless steel specimens, and group C consisted of TiO_2-coated stainless steel specimens. The specimens in group A immediately underwent serial dilution and plate culture after inoculation, while the specimens of groups B and C were incubated with exposure to fluorescent lighting for 24 h. The fluorescent lamp used was a regular daily-living light source that emitted mainly visible light and had a weak emission of 365 nm. Antimicrobial activity (R) of the specimens in all three groups was then calculated.

As revealed in **Figure 1** [13], the petri dishes corresponding to groups A and B (the uncoated stainless steel specimens) presented significant numbers of *S. aureus* and *E. coli* bacterial colonies, respectively; by contrast, the TiO_2-coated stainless steel specimens in group C did not show a significant amount of bacterial colonies. This qualitatively describes the antimicrobial ability of the TiO_2 coating. Although only one out of the three petri dishes corresponding to each group is pictured in **Figure 1**, those not shown revealed a similar situation; this confirms the statistical accuracy of the antimicrobial test.

For both *S. aureus* and *E. coli*, the numbers of viable bacteria for groups A, B, and C are compared in **Figure 2** [13]. The group A specimens showed 2.85×10^5 and 1.06×10^5 viable bacteria cells, respectively, for *S. aureus* and *E. coli*, whereas the group B specimens showed 1.04×10^4 and 1.36×10^4 viable bacteria cells, respectively, for *S. aureus* and *E. coli*. By contrast, the group C specimens showed no bacterial colonies (10 bacteria cells) for *S. aureus* and 4.30×10^1 viable bacteria cells for *E. coli*. Based on these results, the TiO_2-coated stainless steel specimens presented R values of 3.0 and 2.5, respectively, for *S. aureus* and *E. coli*. Such values are far beyond the index of 2 stipulated for the JIS test standard.

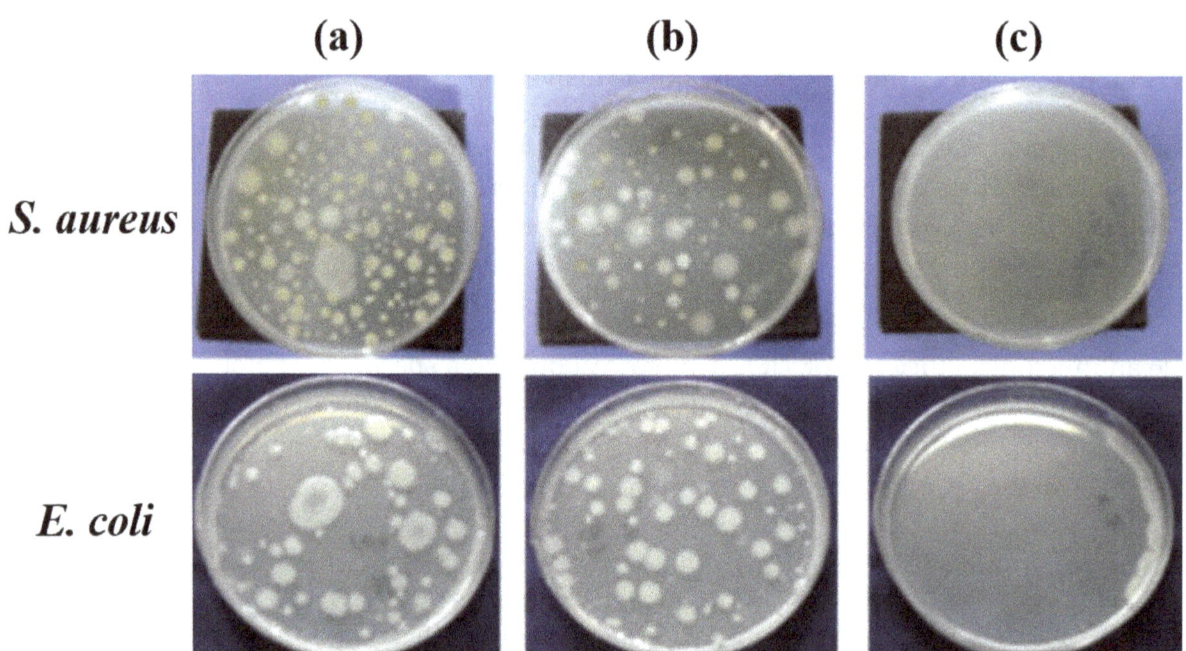

Figure 1. *S. aureus* and *E. coli* colonies formed on petri dishes after 24 h on the (a) group A stainless steel specimens, (b) group B stainless steel specimens, and (c) TiO_2-coated stainless steel specimens [13].

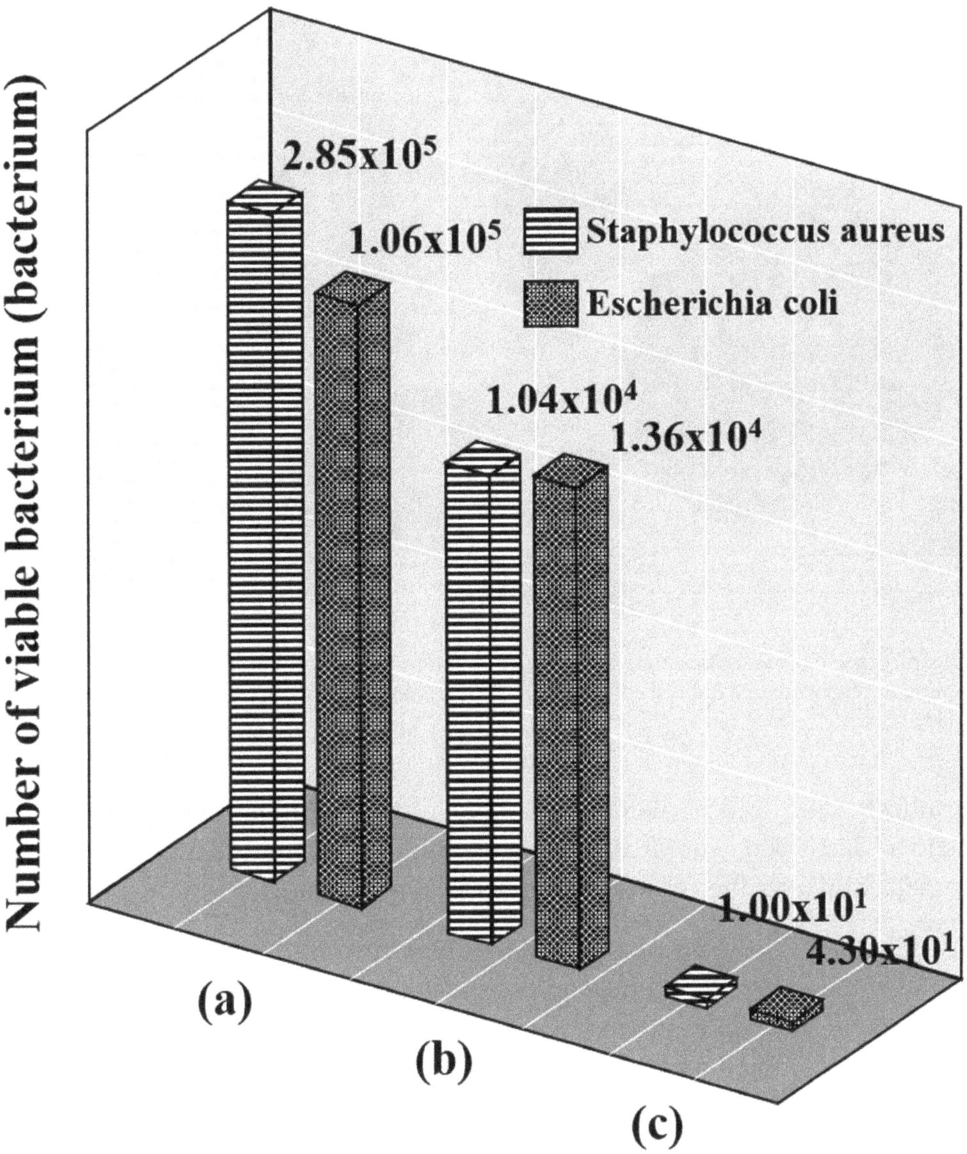

Figure 2. Viable bacteria numbers of *S. aureus* and *E. coli* for (a) group A stainless steel specimens, (b) group B stainless steel specimens, and (c) TiO_2-coated stainless steel specimens [13].

To further investigate the antimicrobial mechanism of a TiO_2 coating, the bacterial microstructure was observed using transmission electron microscopy (TEM; JEOL JEM-1230). This closer examination revealed that most of the *S. aureus* cells were retained their integrity as the cells were inoculated on bare stainless steel with the exposure to fluorescent light for 24 h; moreover, the complete cell structure, including the cell wall, cytoplasmic membrane, cytoplasma, and nucleoid, was observed. The cells were undergoing mitosis, as presented in **Figure 3(a)** [14], was also found. These results indicate that the inoculated *S. aureus* cells on bare stainless steel were not deactivated by the fluorescent light. However, for the *S. aureus* cells on the TiO_2-coated stainless steel specimens, detachment of the cell wall from the cell membrane was frequently observed in the microscopic field (**Figure 3(b)** [14]). As has been noted elsewhere [16–18], the cell walls in these specimens are attacked by superoxide ions and

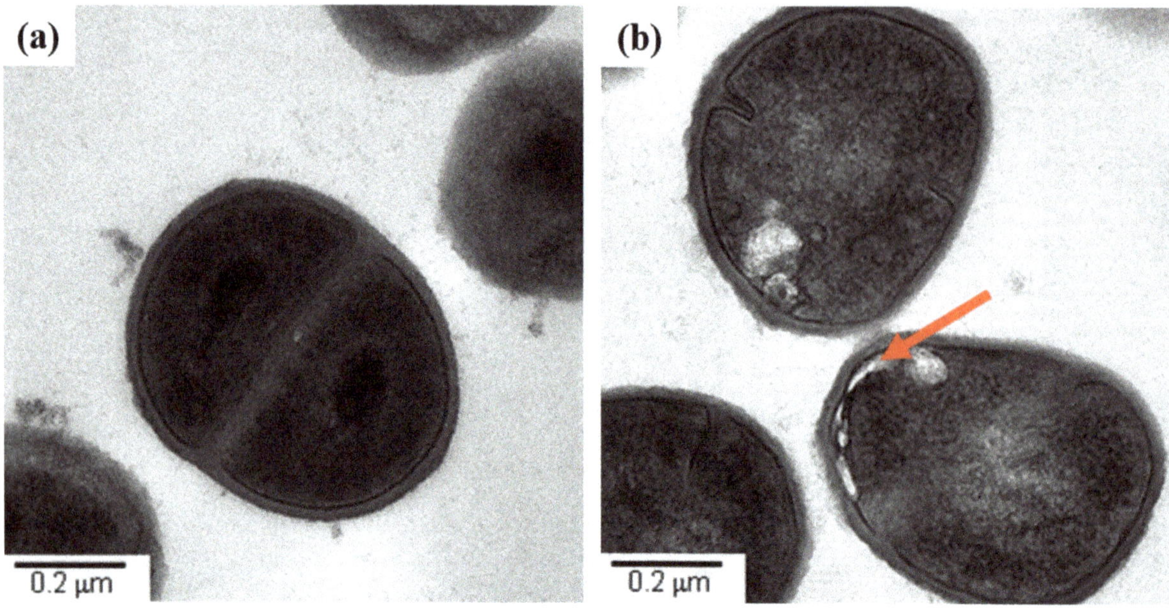

Figure 3. Cell structures of *S. aureus* inoculated on (a) bare stainless steel and (b) TiO_2-coated stainless steel specimens, following continuous exposure to a fluorescent lamp for 24 h. (The arrow indicates detachment of cell wall from the cell membrane.) [14].

hydroxyl radicals, and lipid peroxidation caused polyunsaturated phospholipids in the cell membrane to be destroyed; similarly, the degeneration of the membranes in the present study caused the detachment of the cell walls from the cell membranes.

A high percentage of the *E. coli* cells inoculated on bare stainless steel and exposed to fluorescent light for 24 h also retained their integrity, as depicted in **Figure 4(a)** [14]. By contrast, a large amount of *E. coli* cell fragments were observed following inoculation on TiO_2-coated stainless steel specimens and exposure to fluorescent light for 24 h, as presented in **Figure 4(b)** [14]. This occurred because *E. coli* cell walls are too thin to protect against attack by superoxide ions and hydroxyl radicals, resulting in massive death. A closer examination of the *E. coli* cells reveals that the nucleoid structures in the cytoplasma tend to give way to features of condensation, as indicated by the arrow in **Figure 4(b)** The degeneration of *E. coli* in response to photocatalysis found in the present study is similar to the degeneration that was observed in response to the antimicrobial effects of silver ions [16].

2.4. Anticorrosive characteristics of TiO_2-coated stainless steel

A potentiodynamic polarization test was carried out in a potentiostat (EG&G 263 A) according to the ASTM G44–99 standard [19] to evaluate the corrosion resistance of a TiO_2 coating in a 3.5 wt.% sodium chloride electrolyte. A saturated silver/silver chloride electrode was used as the reference, with a platinum counter electrode; a TiO_2-coated stainless steel specimen was inserted as the working electrode.

Figure 5 illustrates the potentiodynamic polarization curves of bare stainless steel and TiO_2-coated stainless steel specimens [20]. The corrosive potential (E_{corr}) and corrosive current (I_{corr}) were −0.54 V and 6.0×10^{-8} A/cm^2, respectively, for the bare stainless steel

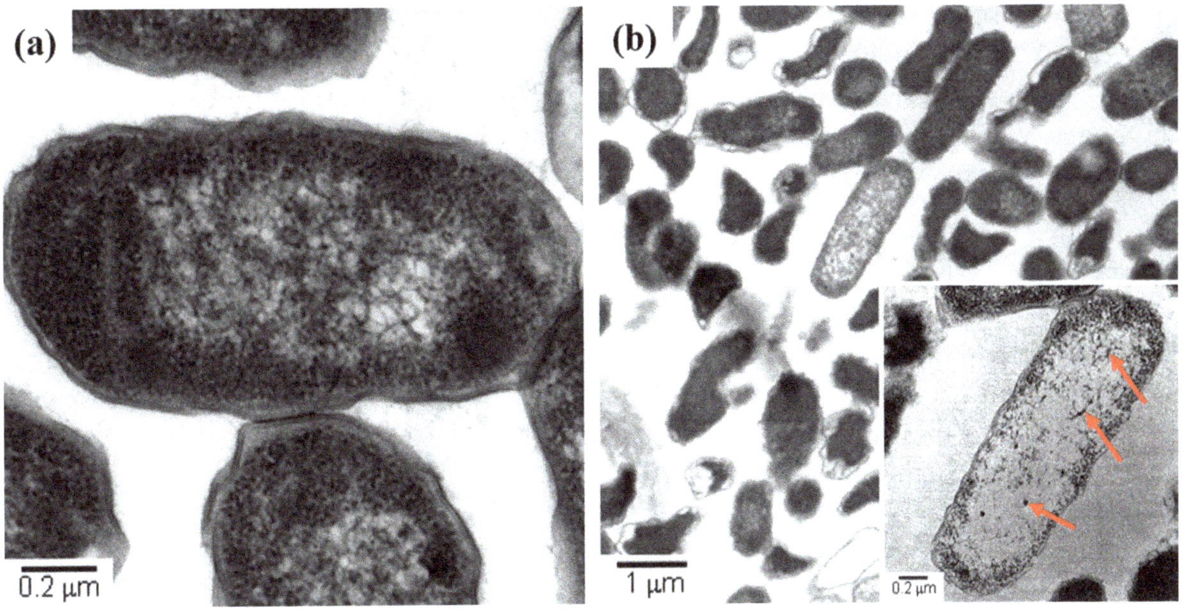

Figure 4. Cell structures of *E. coli* inoculated on (a) bare stainless steel and (b) TiO_2-coated stainless steel specimens, following continuous exposure to a fluorescent lamp for 24 h. (The arrows indicate the condensation features of the nucleoid) [14].

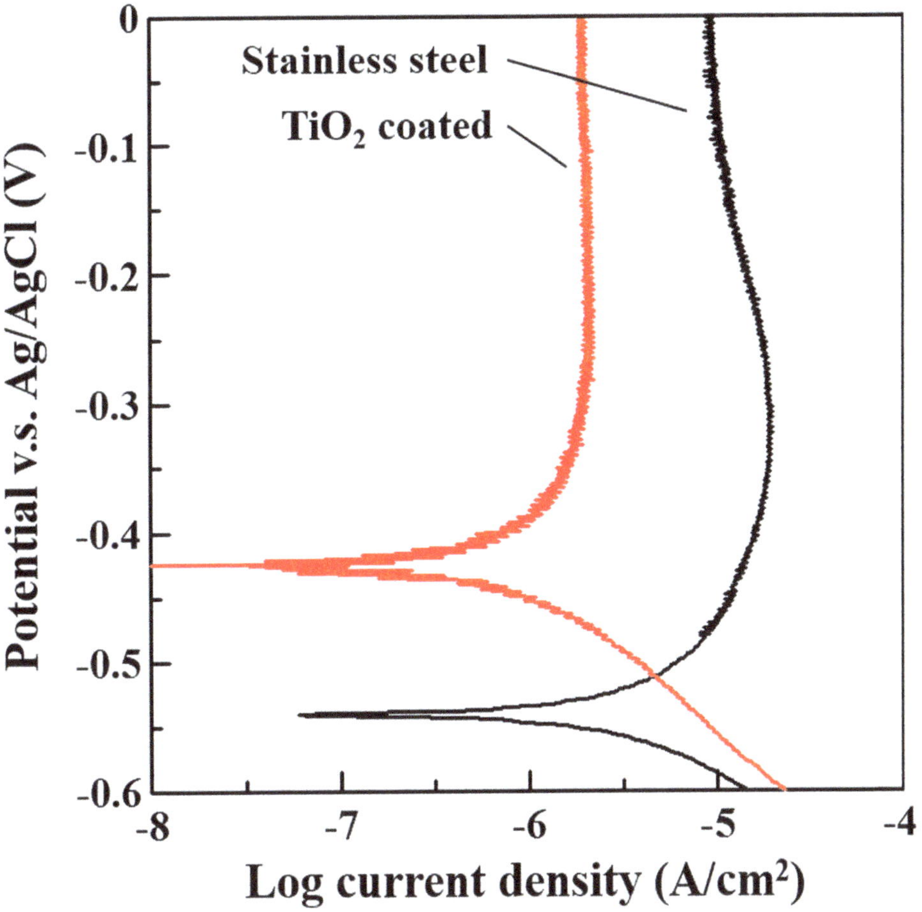

Figure 5. Polarization curves of bare stainless steel and TiO_2-coated stainless steel specimens in a 3.5 wt.% sodium chloride solution [20].

specimens. Once the specimens had been coated with TiO_2, the E_{corr} and I_{corr} of the specimens were −0.42 V and 1.0 × 10^{-8} A/cm^2, respectively. Notably, TiO_2 is an inorganic compound, and its inertness in corrosive environments (e.g., a sodium chloride solution) helps reduce the tendency and rate of substrate dissolution and species coating in an electrolyte. This increases the corrosive potential and decreases the corrosive current, as noted herein.

In summary, the research results show that A-TiO_2 adds effective antimicrobial characteristics to stainless steel. The key to providing efficient antimicrobial efficacy lies in the photocatalytic performance of the coating, which originates from the anatase phase. Furthermore, based on the TEM observation results, the antimicrobial mechanisms that inhibit *S. aureus* and *E. coli* bacteria under the photocatalytic action of A-TiO_2 are different; specifically, the antimicrobial efficacy of A-TiO_2 against *E. coli* is more thorough. The A-TiO_2 coating also reduces the overall rate of corrosion and increases the corrosion barrier, compared with the features of bare stainless steel.

3. Bioactive titanium dioxide coating on polyetheretherketone for spinal implant application

3.1. Background

Orthopedic implants have become one of the most highly developed fields in hard-tissue replacement. Polyetheretherketone (PEEK) polymer, with its high chemical resistance, radiolucency to X-ray scanning, and low elastic modulus similar to human cancellous bone, has become a highly preferred biomaterial, providing a promising alternative to metallic implants [21]. In particular, the elastic modulus can avoid the stress shielding effect, and prevent compression fractures and osteopenia syndrome; the X-ray radiolucency characteristic does not present a medical image shielding problem. PEEK can also be sterilized and shaped by machining to fit the contour of bones [22]. Consequently, PEEK has been widely used for load-bearing orthopedic applications, including dental implants, screws, and spinal interbody fusion cages [23, 24].

Despite these excellent properties, PEEK is still categorized as a bioinert material because of its hydrophobic feature and inertness with the surrounding tissue [21]. To overcome this problem, two primary strategies, bulk modification and surface modification, have been proposed to enhance the bone fusion ability of the PEEK. Bulk modification incorporates various bioactive materials, such as hydroxyapatite (HA) [25], strontium-containing hydroxyapatite [26], β-tricalcium phosphate [27], or TiO_2 [28], into the PEEK matrix to form PEEK-based biocomposites. However, their tensile strength and toughness decrease as more of the bioactive materials are added, resulting in a substantial increase in the elastic modulus of these biomedical composites; the biomechanical property of these PEEK-based biocomposites is therefore no longer similar to that of human cancellous bone [21]. Conversely, surface modification only alters the surface properties of a material, without adversely affecting its bulk properties. In other words, surface modification is a more suitable approach for adapting PEEK to be used as implant. Consequently, various surface modification approaches have been developed to promote the hydrophilic and biological characteristics of PEEK, such as using plasma treatment to change the surface chemistry [29], using chemical treatment to graft functional groups [30], and using laser treatment to roughen the surface [31]. Moreover, adding a functional coating to PEEK to create a bioactive surface is a more effective method for enhancing osseointegration performance [32–38]. Functional coating materials include HA [32], titanium [33, 34], TiO_2 [35–37], and diamond-like carbon [38]. To date, by taking the advantage of good biocompatibility of titanium with human body, very thick titanium produced over PEEK surface via vacuum plasma spray for spinal implant has been clinically available.

It has been well established that under humid conditions, the surface of TiO_2 generates hydroxyl groups (—OH^-), followed by the conjunction with calcium ions (Ca^{2+}) and phosphate groups (PO_4^{3-}) from physiological fluid. Therefore, bone-like apatite compounds can be formed on the TiO_2 surface to induce osteoblast cell adhesion and proliferation [39, 40]. Based on the results, TiO_2 has been reported to exhibit excellent biocompatibility and further classified as a bioactive material [39, 40]. Furthermore, TiO_2 demonstrated excellent osseointegration ability, according to the animal experiment study [41]. These biological characteristics render TiO_2 film an even more promising material for the successful modification of PEEK surfaces, in comparison with regular titanium film.

In this research, the AIP technique was used to deposit TiO_2 films with controllable A-TiO_2 and rutile (R-TiO_2) phases onto PEEK substrates. The investigation focused on determining the effects of introducing a TiO_2 coating on the *in vitro* and *in vivo* characteristics of TiO_2-coated PEEK specimens, and evaluating the ability of the modified PEEK in a clinical application to shorten the osseointegration period for spinal implants and bone tissues.

3.2. Preparation of biocompatible TiO_2 films

The detailed AIP-TiO_2 deposition work is described in Section 2.2. The deposition conditions used in this section are listed in **Table 1**; target current and substrate bias were systematically manipulated to achieve specific ratios of A-TiO_2 and R-TiO_2 in the deposited films, characterized by a fixed 100% oxygen pressure of 0.5 Pa and a cathode target voltage of 20 V.

Based on the microstructure characteristics results [12], the AIP process can successfully fabricate TiO_2 films of varying A-TiO_2 and R-TiO_2 composition when appropriate coating parameters are used. Specifically, the A-TiO_2 phase in the deposited films ranged from 9.1% to 92.7% (**Table 1**).

Sample code	Target current (A)	Substrate bias (V)	A-TiO_2 content (%)
60A0V	60	0	92.7
90A0V	90	0	76.8
90A20V	90	–20	46.6
90A25V	90	–25	21.5
90A30V	90	–30	9.1

Table 1. Deposition conditions and the proportions of A-TiO_2 phases for TiO_2 coatings.

A low target current promotes the growth of A-TiO_2, whereas a high substrate bias induces the formation of R-TiO_2. The mechanism behind this outcome was previously investigated [12].

3.3. *In vitro* characteristics of TiO_2-coated PEEK

First, the MC3T3-E1 osteoblast cell line was used in the osteoblast compatibility test to assess the cell adhesion test, cell proliferation test, cell differentiation test, and osteogenesis performance [namely quantification of osteopontin (OPN), osteocalcin (OCN), and calcium content]. Next, the cell morphology that had attached to the PEEK and TiO_2-coated PEEK specimens was observed using field emission scanning electron microscopy (FESEM; Hitachi S-4800).

Figure 6 shows the osteoblast cell adhesion ability, cell proliferation ability, cell differentiation ability, and osteogenesis performance on the PEEK and TiO_2-coated PEEK specimens at various deposition conditions [36]. Notably, the osteoblast cell adhesion, proliferation, and differentiation abilities on TiO_2-coated PEEK specimens were superior to the bare PEEK specimens for all of the deposition conditions. This indicates that all of the obtained TiO_2 coatings possessed cell induction capabilities, which led to accelerated cell adhesion and growth and increased cell proliferation and maturity. These three indicators confirmed the osteoblast compatibility of the TiO_2-coatings deposited on PEEK. Furthermore, the osteogenesis performance (revealed by OPN, OCN, and calcium content as shown in **Figure 6(d)–(f)** [36], respectively) demonstrated that TiO_2 coatings also significantly increased the osteogenesis performance. This suggests that TiO_2 coatings enhance extracellular bone matrix growth. **Figure 6** [36] also shows that the specimen 90A30V, which was the richest in R-TiO_2 phase, exhibited the most osteoblast compatibility.

Figure 7 shows the morphologies of the osteoblast cells after they were cultured for 0.5 and 48 h on PEEK and TiO_2-coated PEEK specimens at different deposition conditions [36].

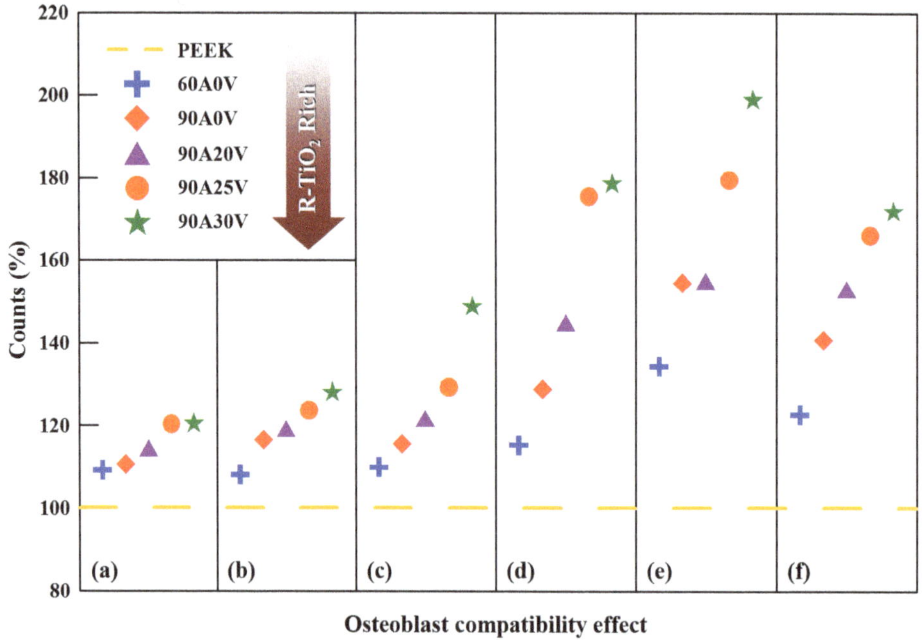

Figure 6. (a) Cell adhesion ability, (b) cell proliferation ability, (c) cell differentiation ability, (d) OPN, (e) OCN, and (f) calcium content of the osteoblast inoculated on bare PEEK and TiO_2-coated PEEK specimens with various deposition conditions [36].

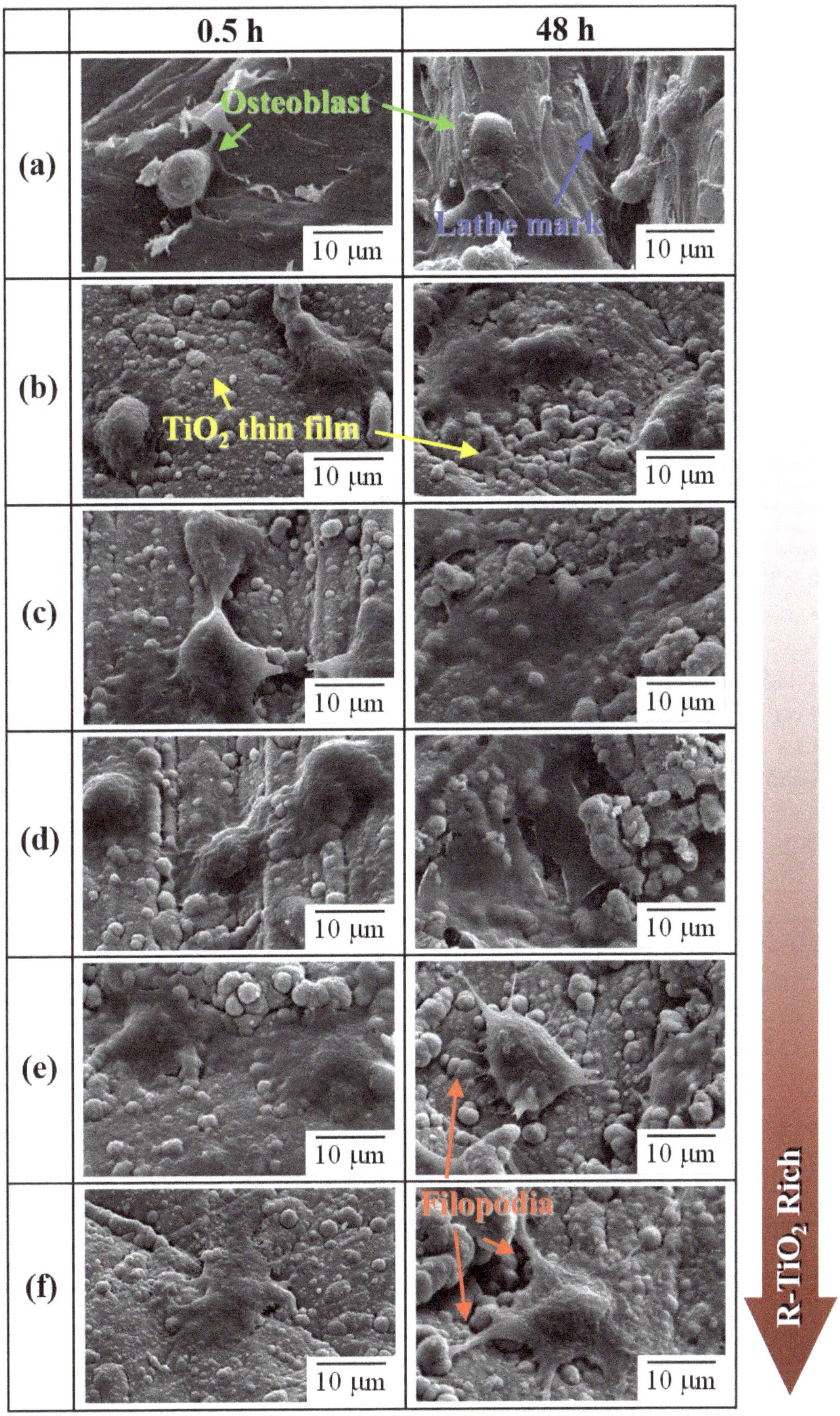

Figure 7. Morphologies of the osteoblasts cultured for 0.5 and 48 h on (a) the bare PEEK specimens, and the TiO_2-coated PEEK specimens at different deposition conditions: (b) 60A0V, (c) 90A0V, (d) 90A20V, (e) 90A25V, and (f) 90A30V [36].

Specifically, the morphology of osteoblast cells on the bare PEEK specimens remained spherical without the appearance of filopodium, suggesting the poor adhesion to the specimen. By comparison, osteoblasts on the TiO_2-coated PEEK specimens with the culturing time of 0.5 h showed a very comfortable adhesion features, that is, the filopodia extension and well-developed lamellipodia on the cells; this was particularly notable on the films with high ratios of R-TiO_2 to A-TiO_2. Similar results were observed in the cells cultured for 48 h. Overall, these results further confirm that a deposited film with high R-TiO_2 content has superior osteoblast growth.

Furthermore, bare PEEK, and TiO_2-coated PEEK specimens were then immersed in a simulated body fluid (SBF) for 1, 3, 7, 14, and 28 days, to investigate the effect of TiO_2 coating on the ability to induce HA formation. The TiO_2 coatings that possessed A-TiO_2 and R-TiO_2 under the deposition conditions of 60A0V and 90A30V, respectively, were examined. This biomimetic immersion test is a valuable approach for evaluating bioactivity of a candidate bone implant material prior to an *in vivo* test [42].

Figure 8 illustrates the X-ray diffraction (XRD) patterns of bare PEEK, A-TiO_2-coated PEEK, and R-TiO_2-coated PEEK specimens after immersion in the SBF for a varying number of days [43]. During the early immersion period, the diffraction peaks that are ascribed to PEEK showed no observable change, indicating that the growing layer was undetectable in all of the specimens. After 28 days of immersion, weak and broadened diffraction peaks that are ascribed to HA were found, as shown in **Figure 8(a)** [43]. This implies that a very poor crystalline or even amorphous calcium phosphate layer had formed on the PEEK specimens. By contrast, after only 7 days and 3 days of immersion in the SBF solution, diffraction peaks that are ascribed to HA could be observed in A-TiO_2- and R-TiO_2-coated PEEK specimens, respectively. Over time, the intensity of these diffraction peaks increased significantly, as shown in **Figure 8(b)** and **(c)** [43], suggesting that additional crystalline HA was formed on them.

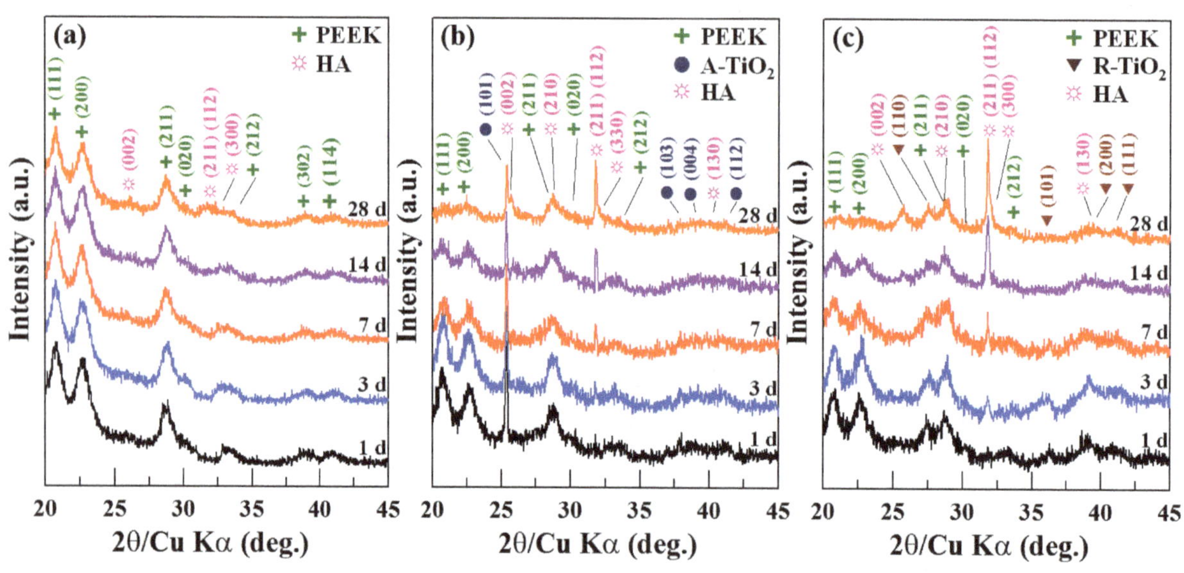

Figure 8. XRD patterns of the (a) bare PEEK, (b) A-TiO_2-coated PEEK, and (c) R-TiO_2-coated PEEK specimens immersed in a SBF for 1, 3, 7, 14, and 28 days [43].

Overall, these results suggest that HA growth in a SBF solution can be enhanced by adopting TiO_2 coatings, and that the R-TiO_2 coating seems to exhibit a superior capability to induce HA formation. Therefore, the results of the biomimetic immersion tests agree well with the finding of *in vitro* characteristics from osteoblast compatibility tests.

3.4. *In vivo* characteristics of TiO_2-coated PEEK

Bullet-shaped PEEK implants with a diameter of ϕ 4.0 mm × L 6.0 mm were used in an animal experiment. Bare PEEK, A-TiO_2-coated PEEK, and R-TiO_2-coated PEEK implants were inserted into the femurs of New Zealand white male rabbits to evaluate the *in vivo* osseointegration capacity through the push-out test and histological observation.

The push-out test can precisely quantify the degree of fixation between an implant and bone tissues [44]. **Figure 9** shows the push-out test results for the three implants after 4, 8, and 12 weeks [37]. Notably, the shear strength between the bone tissues and the implant increased as implantation time increased; at 12 weeks, the shear strength of the bare, A-TiO_2-coated, and R-TiO_2-coated PEEK implants was 2.54 MPa, 3.02 MPa, and 6.51 MPa, respectively. It was thus concluded that the bare PEEK implant had the poorest shear strength, but this could be enhanced by adding a TiO_2 coating. Overall, the R-TiO_2 coating had the optimal fixation.

To identify the failure mode between the implant and bone tissues after the push-out test, FESEM was adopted to observe the fracture morphology of the implant surface at 12 weeks, as shown in **Figure 10** [37]. It was noted that new bone tissue had fully peeled off the surface

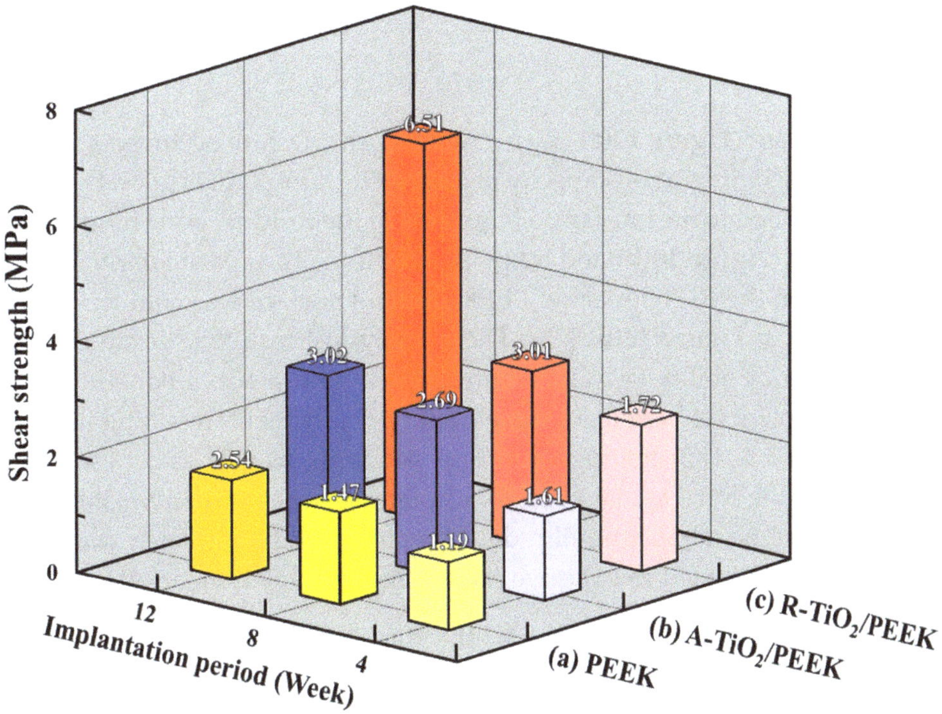

Figure 9. Shear strength between bone tissues and mplant for the (a) bare PEEK implant, (b) A-TiO_2-coated PEEK implant, and (c) R-TiO_2-coated PEEK implant at 4, 8, and 12 weeks after implantation [37].

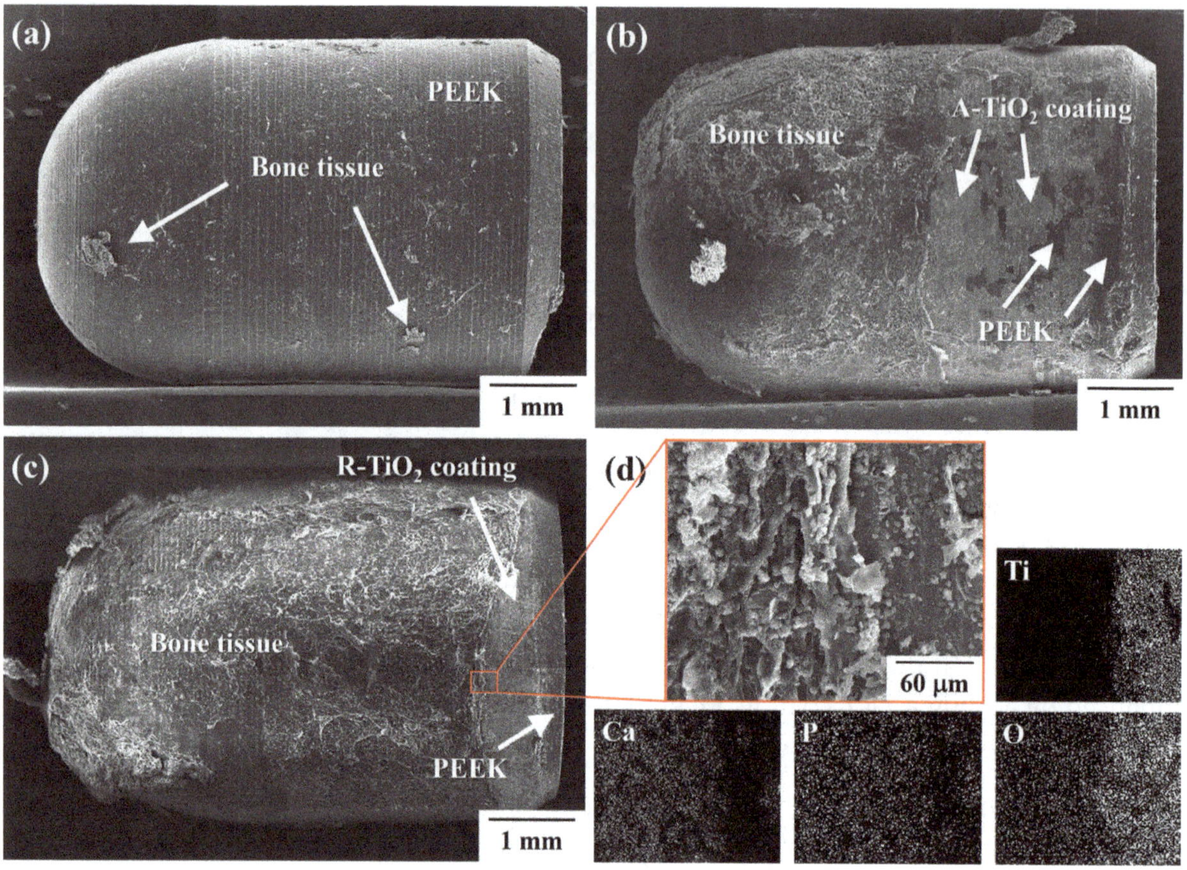

Figure 10. Fracture morphology of the (a) bare PEEK implant, (b) A-TiO_2-coated PEEK implant, and (c) R-TiO_2-coated PEEK implant with (d) the composition analysis of its bone tissues and implant interface after the push-out test conducted at 12 weeks [37].

of the bare PEEK implant (**Figure 10(a)** [37]), indicating that failure occurred at the bone/PEEK interface. Thus, the osseointegration capacity of a bare PEEK implant is poor. By contrast, when a TiO_2 coating was applied to the implant, a large area of the residual bone tissue adhered to the surface of the implant (**Figure 10(b)** and **(c)** [37]). Additionally, a particularly large amount of residual bone tissue on the R-TiO_2-coated PEEK implant surface was confirmed by elemental mapping, as revealed in **Figure 10(d)** [37]. These analytical results indicate that TiO_2-coated implants have a superior ability to induce bone growth and achieve bone ingrowth. The A-TiO_2-coated PEEK implants experienced some coating detachment, resulting in a mixed adhesive failure between the A-TiO_2 coating and PEEK substrate, as well as cohesive failure of the bone itself. However, the R-TiO_2-coated PEEK implant surfaces were almost completely covered with new bone tissue, almost no film detachment from the implants was observed, and thus, the failure can be regarded as cohesive failure by the bone tissue itself.

Figure 11 depicts the histological sections of the three implants at 4, 8, and 12 weeks after implantation [37]. Notably, new bone tissue that was generated by bone remodeling had formed mature lamellar bone, and directly connected to the TiO_2-coated PEEK implants after 4 weeks, indicating excellent osseointegration performance. Thus, it was concluded that

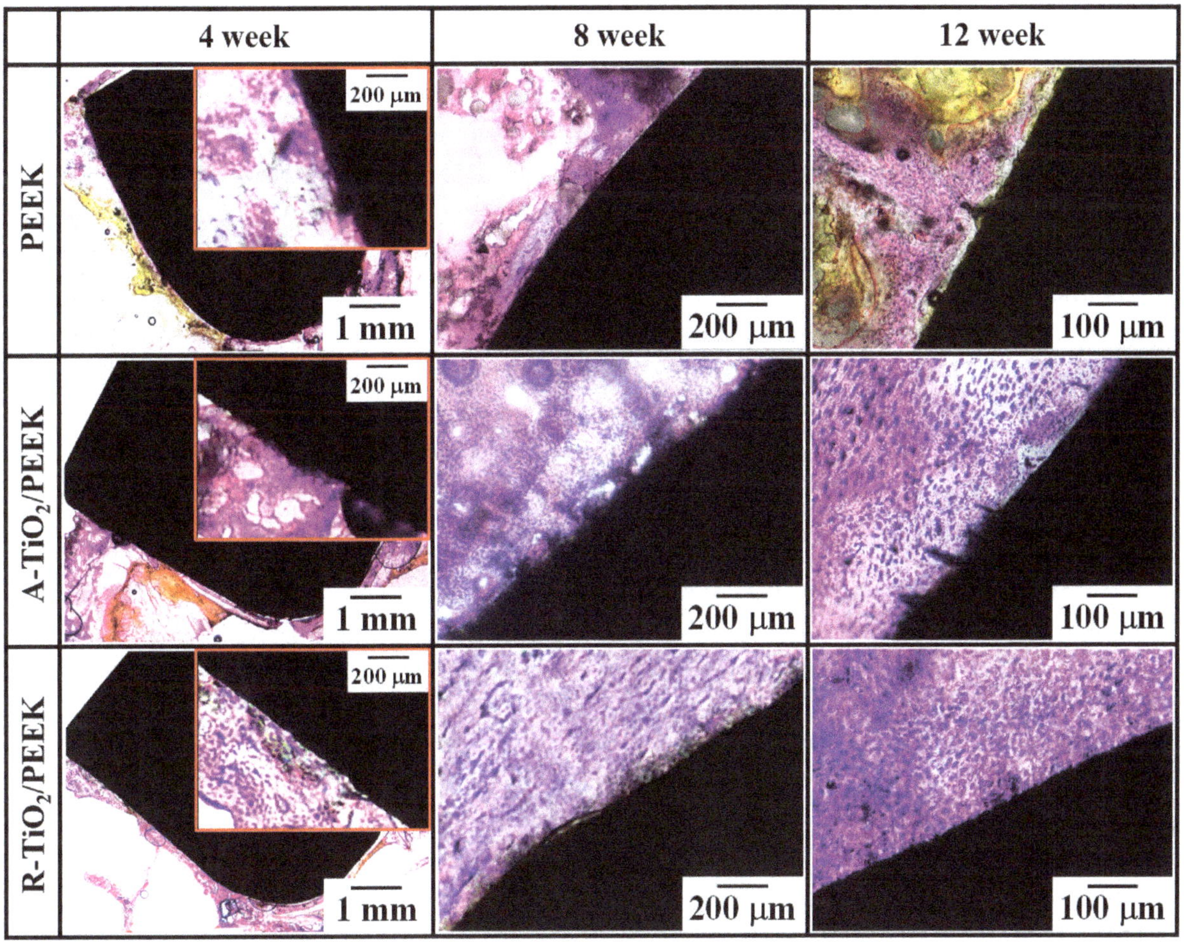

Figure 11. Histological sections of the bare PEEK implant, A-TiO_2-coated PEEK implant, and R-TiO_2-coated PEEK implant at 4, 8, and 12 weeks after implantation [37].

TiO_2 coating exhibits strong osteoblast compatibility and rapidly activates bone remodeling. Subsequently, the coating induced adhesion and proliferation of osteoblasts on the implant surface, and differentiation into osteocytes for the production of new bone tissue and later bone bonding. Conversely, new lamellar bone on the surface of the bare PEEK implants was not completely mature and not fully bonded with the implant.

The response of the TiO_2-coated PEEK implants in the marrow cavity (located far from the cortical bone) at 4 weeks indicated that regenerated bone tissues grew onto the implant surfaces; moreover, this new bone is the result of bone tissue repair, which proliferates from the endosteum of cortical bone. Due to the osteoconductive effect, the new bone tissues grew inward to the implant surfaces in the marrow [45]. These findings indicate that TiO_2 coatings have excellent osteoconductivity and promote new bone growth on the TiO_2-coated PEEK implant surfaces, with connections to cortical bone. By contrast, the surfaces of the bare PEEK implant were covered with fibrous tissue, implying that bone bonding did not occur between the implant and the cortical bone. Fibrous tissue growth is likely caused by micro movement in the implant and poor stability during the early implantation period [46].

When the implant period was extended to 8 weeks, immature osteogenesis was observed in the cortical bone around the bare PEEK implant, and new bone tissue was maturing after 12 weeks. However, fibrous tissue was still identified at the interface between the implants and bone tissues, indicating that the osseointegration capacity of bare PEEK implants is very limited, even when the implantation period is extended. By contrast, 8 weeks after the implantation of the TiO_2-coated PEEK implants, histological sections in the marrow cavity revealed that the new bone tissue was maturing and osteocytes covered the their surface. In other words, the osteoconductive effect of TiO_2 coating triggers quick bone remodeling. The new bone was fully mature and closely integrated with the TiO_2 coating in the cavity after 12 weeks (**Figure 11** [37]). However, a comparison of the TiO_2 coatings with different phase structures indicated that the degree of bone bonding between new bone and the R-TiO_2-coated PEEK implant was significantly better than that between new bone and the A-TiO_2-coated PEEK implant. In addition, some gaps existed between the A-TiO_2 coating and the new bone in some areas; detachment of the A-TiO_2 coating was also noted.

In summary, the *in vitro* and *in vivo* characteristics can be improved by TiO_2 coating because of its bioactivity; R-TiO_2 coatings perform particularly well, promoting biomimetic HA growth, osteoblast compatibility, and osseointegration. These phenomena are attributable to the abundance of negatively charged hydroxyl groups on the R-TiO_2 surface [35–37].

4. Conclusions

In this chapter, TiO_2 coatings prepared using the AIP technique to alter the surface properties of biomaterials were described. Specifically, it was found that introducing TiO_2 coating to stainless steel and PEEK specimens adds various anticorrosive, antimicrobial, and bioactive surface properties to the materials, which were systematically reviewed herein. The following conclusions can be drawn:

1. Owing to the efficient photocatalytic performance of its anatase phase structure, A-TiO_2-coated stainless steel exhibits excellent antimicrobial efficacy against *S. aureus* and *E. coli* bacteria. The material could possibly serve as a new antimicrobial treatment for surgical instruments and medical implements to reduce the risk of hospital-acquired infections.
2. The high corrosion resistance of TiO_2 coatings in a 3.5 wt% sodium chloride solution was postulated as a direct consequence of its ceramic nature, suggesting that TiO_2 is electrochemically inert in the human body environment.
3. Based on the *in vitro* and *in vivo* tests, the bioactivity and osseointegration of all TiO_2 coatings were far superior to bioinert PEEK; moreover, R-TiO_2 coatings exhibited greater performance than A-TiO_2 coatings because of the abundance of negatively charged hydroxyl groups on its surface. Consequently, TiO_2-coated PEEK specimens are suggested for use in clinical applications.
4. Overall, the aforementioned results prove that TiO_2 coatings are highly suitable for surface modifications of biomedical materials.

Author details

Hsi-Kai Tsou[1,2*] and Ping-Yen Hsieh[3]

*Address all correspondence to: tsouhsikai@gmail.com

1 Functional Neurosurgery Division, Neurological Institute, Taichung Veterans General Hospital, Taichung City, Taiwan, ROC

2 Department of Rehabilitation, Jen-Teh Junior College of Medicine, Nursing and Management, Miaoli County, Taiwan, ROC

3 Department of Materials Science and Engineering, National Tsing Hua University, Hsinchu City, Taiwan, ROC

References

[1] Gottenbos B, Van der Mei HC, Klatter F, Grijpma DW, Feijen J, Nieuwenhuis P, Busscher HJ. Positively charged biomaterials exert antimicrobial effects on gram-negative bacilli in rats. Biomaterials. 2003;**24**:2707-2710

[2] Kourai H. Surface science and microbiology: Antimicrobial finishings. Journal of the Surface Science Society of Japan. 2001;**22**:663-670

[3] Morihiro H, Katsuhisa M, Sadao H, Michiyuki K, Seiichis I. Improving method of antibacterial performance of copper-containing stainless steel. Japan Patent. 1996;**08**:229107

[4] Chungtsun TH, Tachiupao C, Kungnan KC. Antibacterial characteristic and material quality of ferrite antibacterial stainless steel "NSSAM-1". Nisshin Steel Technical Report. 1997;**76**:48-55

[5] Chungtsun TH, Tachiupao CJ, Shanpen CJ. Corrosion quality of galvanized stainless steel. Nisshin Steel Technical Report. 1998;**77**:69-81

[6] Shihyeh HH, Yutienchuan CK, Panching CN, Yuen KH. Antimicrobial pre-painted steel sheet. Nisshin Steel Technical Report. 1998;**78**:90-96

[7] Kim TN, Feng QL, Kim JO, Wu J, Wang H, Chen GC, Cui FZ. Antimicrobial effects of metal ions (Ag+, Cu2+, Zn2+) in hydroxyapatite. Journal of Materials Science: Materials in Medicine. 1998;**9**:129-134.

[8] Fujishima A, Honda K. Electrochemical photolysis of water at a semiconductor electrode. Nature. 1972;**238**:37-38

[9] Cheng Q, Li C, Pavlinek V, Saha P, Wang H. Surface-modified antibacterial TiO2/Ag+ nanoparticles: Preparation and properties. Applied Surface Science. 2006;**252**:4154-4160

[10] Sawunyama P, Yasumori A, Okada K. The nature of multilayered TiO_2-based photocatalytic films prepared by a sol-gel process. Materials Research Bulletin. 1998;**33**:795-801

[11] Matsunaga T, Tomoda R, Nakajima T, Wake H. Photoelectrochemical sterilization of microbial cells by semiconductor powders. FEMS Microbiology Letters. 1985;**29**:211-214

[12] Chung CJ, Tsou HK, Chen HL, Hsieh PY, He JL. Low temperature preparation of phase-tunable and antimicrobial titanium dioxide coating on biomedical polymer implants for reducing implant-related infections. Surface and Coatings Technology. 2011;**205**:5035-5039

[13] Chung CJ, Lin HI, Tsou HK, Shi ZY, He JL. An antimicrobial TiO_2 coating for reducing hospital-acquired infection. Journal of Biomedical Materials Research Part B: Applied Biomaterials. 2008;**85B**:220-224

[14] Chung CJ, Lin HI, Chou CM, Hsieh PY, Hsiao CH, Shi ZY, He JL. Inactivation of *Staphylococcus aureus* and *Escherichia coli* under various light sources on photocatalytic titanium dioxide thin film. Surface and Coatings Technology. 2009;**203**:1081-1085

[15] JIS Z2801:2000. Antimicrobial products—Test for antimicrobial activity and efficacy. Japanese Industrial Standard. 2001

[16] Huang Z, Maness PC, Blake DM, Wolfrum EJ, Smolinski SL. Bactericidal mode of titanium dioxide photocatalysis. Journal of Photochemistry and Photobiology A: Chemistry. 2000;**130**:163-171

[17] Maness PC, Smolinski S, Blake DM, Huang Z, Wolfrum EJ, Jacoby WA. Bactericidal activity of photocatalytic TiO_2 reaction: Toward an understanding of its killing mechanism. Applied and Environmental Microbiology. 1999;**65**:4094-4098

[18] Feng QL, Wu J, Chen GQ, Cui FZ, Kim TN, Kim JO. A mechanistic study of the antibacterial effect of silver ions on *Escherichia coli* and *Staphylococcus aureus*. Journal of Biomedical Materials Research. 2000;**52**:662-668

[19] ASTM G44-99. Standard practice for exposure of metals and alloys by alternate immersion in neutral 3.5% sodium chloride solution. American Society for Testing and Materials. 2005

[20] Chung CJ, Hsieh PY, Hsiao CH, Lin HI, Leyland A, Matthews A, He JL. Mutifunctional arc ion plated TiO_2 photocatalytic coatings with improved wear and corrosion protection. Surface and Coatings Technology. 2009;**203**:1689-1693

[21] Kurtz SM, Devine JN. PEEK biomaterials in trauma, orthopedic, and spinal implants. Biomaterials. 2007;**28**:4845-4869

[22] Barton AJ, Sagers RD, Pitt WG. Bacterial adhesion to orthopedic implant polymers. Journal of Biomedical Materials Research. 1996;**30**:403-410

[23] Williams DF, McNamara A, Turner RM. Potential of polyetheretherketone (PEEK) and carbon-fibre-reinforced PEEK in medical applications. Journal of Materials Science Letters. 1987;**6**:188-190.

[24] Schwitalla A, Müller WD. PEEK dental implants: A review of the literature. Journal of Oral Implantology. 2013;**39**:743-749

[25] Yu S, Hariram KP, Kumar R, Cheang P, Aik KK. In vitro apatite formation and its growth kinetics on hydroxyapatite/polyetheretherketon biocomposites. Biomaterials. 2005;**26**:2343-2352

[26] Wong KL, Wong CT, Liu WC, Pan HB, Fong MK, Lam WM, Cheung WL, Tang WM, Chiu KY, Luk KDK, Lu WW. Mechanical properties and in vitro response of strontium-containing hydroxyapatite/polyetheretherketone composites. Biomaterials. 2009;**30**:3810-3817

[27] Petrovic L, Pohle D, Münstedt H, Rechtenwald T, Schlegel KA, Rupprecht S. Effect of βTCP filled polyetheretherketone on osteoblast cell proliferation in vitro. Journal of Biomedical Science. 2006;**13**:41-46

[28] Wu X, Liu X, Wei J, Ma J, Deng F, Wei S. Nano-TiO_2/PEEK bioactive composite as a bone substitute material: In vitro and in vivo studies. International Journal of Nanomedicine. 2012;**7**:1215-1225

[29] Ha SW, Kirch M, Birchler F, Eckert KL, Mayerv J, Wintermantel E, Sittig C, Pfund-Klingenfuss I, Textor M, Spencer ND, Guecheva M, Vonmont H. Surface activation of polyetheretherketone (PEEK) and formation of calcium phosphate coatings by precipitation. Journal of Materials Science: Materials in Medicine. 1997;**8**:683-690

[30] Pino M, Stingelin N, Tanner KE. Nucleation and growth of apatite on NaOH-treated PEEK, HDPE and UHMWPE for artificial cornea materials. Acta Biomaterialia. 2008;**4**: 1827-1836

[31] Riveiroa A, Soto R, Comesaña R, Boutinguiza M, del Val J, Quintero F, Lusquiños F, Pou J. Laser surface modification of PEEK. Applied Surface Science. 2012;**258**:9437-9442

[32] Wu GM, Hsiao WD, Kung SF. Investigation of hydroxyapatite coated polyether ether ketone composites by gas plasma sprays. Surface and Coatings Technology. 2009;**203**:2755-2758

[33] Han CM, Lee EJ, Kim HE, Koh YH, Kim KN, Ha Y, Kuh SU. The electron beam deposition of titanium on polyetheretherketone (PEEK) and the resulting enhanced biological properties. Biomaterials. 2010;**31**:3465-3470

[34] Walsh WR, Bertollo N, Christou C, Schaffner D, Mobbs RJ. Plasma-sprayed titanium coating to polyetheretherketone improves the bone-implant interface. The Spine Journal. 2015;**15**:1041-1049

[35] Tsou HK, Hsieh PY, Chung CJ, Tang CH, Shyr TW, He JL. Low-temperature deposition of anatase TiO_2 on medical grade polyetheretherketone to assist osseous integration. Surface and Coatings Technology. 2009;**204**:1121-1125

[36] Tsou HK, Hsieh PY, Chi MH, Chung CJ, He JL. Improved osteoblast compatibility of medical-grade polyetheretherketone using arc ion plated rutile/anatase titanium dioxide films for spinal implants. Journal of Biomedical Materials Research Part A. 2012;**100A**:2787-2792

[37] Tsou HK, Chi MH, Hung YW, Chung CJ, He JL. In vivo osseointegration performance of titanium dioxide coating modified polyetheretherketone using arc ion plating for spinal implant application. BioMed Research International. 2015;**2015**:328943

[38] Wang H, Xu M, Zhang W, Kwok DTK, Jiang J, Wu Z, Chu PK. Mechanical and biological characteristics of diamond-like carbon coated poly aryl-etherether-ketone. Biomaterials. 2010;**31**:8181-8187

[39] Wu JM, Xiao F, Hayakawa S, Tsuru K, Takemoto S, Osaka A. Journal of Materials Science: Materials in Medicineacidic titanium tetrafluoride solution. Journal of Materials Science: Materials in Medicine. 2003;**14**:1027-1032

[40] Lindberg F, Heinrichs J, Ericson F, Thomsen P, Engqvist H. Hydroxylapatite growth on single-crystal rutile substrates. Biomaterials. 2008;**29**:3317-3323

[41] Erli HJ, Rüger M, Ragoβ C, Jahnen-Dechent W, Hollander DA, Paar OO, von Walter M. The effect of surface modification of a porous TiO_2/perlite composite on the ingrowth of bone tissue in vivo. Biomaterials. 2006;**27**:1270-1276

[42] Kokubo T, Kim HM, Kawashita M. Novel bioactive materials with different mechanical properties. Biomaterials. 2003;**24**:2161-2175

[43] Chi MH, Tsou HK, Chung CJ, He JL. Biomimetic hydroxyapatite grown on biomedical polymer coated with titanium dioxide interlayer to assist osteocompatible performance. Thin Solid Film. 2013;**549**:98-102

[44] Dhert WJA, Jansen JA. The validity of a single pushout test. In: An YH, Draughn RA, editors. Mechanical testing of bone and the bone–implant interface. New York: CRC Press; 2000. pp. 477-487

[45] Dua C, Meijer GJ, van de Valk C, Haan RE, Bezemer JM, Hesseling SC, Cui FZ, de Groot K, Layrolle P. Bone growth in biomimetic apatite coated porous Polyactive® 1000PEGT70PBT30 implants. Biomaterials. 2002;**23**:4649-4656

[46] Søballe K, Hansen ES, Rasmussen HB, Jørgensen PH, Bünger C. Tissue ingrowth into titanium and hydroxyapatite-coated implants during stable and unstable mechanical conditions. Journal of Orthopaedic Research. 1992;**10**:285-299

3

Photocatalytic Treatment Techniques using Titanium Dioxide Nanoparticles for Antibiotic Removal from Water

António Armando Lima Sampaio Duarte and
Maria Teresa Pessoa Amorim

Abstract

The increasing of emerging micropollutants presence in drinking water sources has brought new challenges to existing water treatment systems (WTS), highlighting the need of innovative and low-cost technological solutions. Recent advances in nanotechnology enable highly efficient and multifunctional processes, providing sustainable alternatives to current water treatment practices. This chapter presents the results of several pilot-scale studies developed to assess the effects of TiO_2 nanoparticles on antibiotic removal efficiency, using different low-cost photocatalytic reactors. The characterization of its photo-oxidation kinetics also performed considering different test scenarios in order to assess the effects of the major abiotic parameters on oxytetracycline (OTC) removal efficiency, which achieved the maximum values of 96% and 98% using the photocatalysis with TiO_2 and the photocatalytic filtration, respectively. It must be highlighted the surprising regeneration ability showed by the photocatalytic porous medium, developed at a lab-scale, which can completely recover its oxidative properties after few hours of simple sun exposure.

Keywords: heterogeneous photocatalysis, photo-oxidation kinetics, TiO_2 nanoparticles, photocatalytic filtration, antibiotic removal, safe drinking water

1. Introduction

Reliable access to clean and safe water remains a major worldwide challenge for the twenty-first century, in a global climate change context. In recent years, the classic problems associated with the presence in the ecosystems of priority pollutants have been extended to the detection of increasing amounts of micropollutants commonly called emerging. These, due to their toxicity and persistence in the environment (water column and sediments), have brought new challenges to existing water treatment systems (WTS) aiming to protect public health and the preservation of drinking water sources.

The Directive 2008/105/EC (PSD) lays down environmental quality standards (EQS) and presents the List of Priority Substances as afforded on the Article 16 and Annex X of the Water Framework Directive 2000/60/EC (WFD). However, the pharmaceuticals are not yet included among those compounds to be monitored, despite the increase in its occurrence reported in many European countries [1]. For urban water monitoring, possible priority pharmaceutical compounds (PhCs) should be the mainly analgesics, antidepressants, antibiotics, antineoplastics [2], synthetic estrogens, and hormones [3]. The inclusion of target PhCs in the EU List of Priority Substances implies the definition of their corresponding EQSs and the necessity to subject to monitoring EU aquatic ecosystems.

Recent advances in nanotechnology offer opportunities to develop next generation of WTS, as sustainable and safe alternative to current water treatment practices relied on centralized systems. The highly efficient and multifunctional processes, enabled by nanotechnological solutions, can also provide new capabilities allowing economic utilization of unconventional water sources on water-stressed regions [4]. Future water treatment systems in developing countries will most likely opt for nanotechnology-based water monitoring, treatment and reuse systems that can efficiently immobilize a wide variety of water emergent pollutants (for which existing technologies are inefficient or ineffective) coupled with affordability and ease of operation [5].

Advanced oxidation processes (AOPs) have been widelly studied because of their potential as a complementary or alternative process to conventional wastewater treatment. These AOPs have proven to be particularly efective in the degradation of many toxic pollutants [6–8] when nanomaterials are applied as photocatalyst. Photocatalytic oxidation with TiO_2 has been used in the removal of micropollutants (like antibiotics) and microbial pathogens from waters, as a useful pre-treatment and/or a polishing step to oxidize hazardous and recalcitrant organic compounds.

This chapter presents the development and results of several pilot-scale studies aiming to assess the effects of TiO_2 nanoparticles on antibiotic removal efficiency and to define its photo-oxidation kinetics, using different low-cost photocatalytic water treatment systems.

The antibiotic tested in this work was oxytetracycline (OTC) is a widely used broad spectrum antibiotic, especially employed in veterinary medicine [9, 10] and for human therapy [11]. It can be found not only in raw and treated wastewaters but also in surface water sources [12]. The catalyst used is Degussa (*Evonik*) P25 TiO_2, which was applied as suspended and immobilized nanoparticles exposed to UV and solar radiation in two photocatalytic reactors: water columns and columns filters with a granular porous medium coated by immobilized TiO_2 nanoparticles using a sol-gel method.

For both photo-oxidation reactors, different test scenarios are defined in order to assess the effect on OTC removal efficiency of the major abiotic parameters, such as hydraulic conditions, OTC initial concentration, pH, cumulate solar energy, and media granulometry.

The experimental results were very promising, because removal efficiencies in both reactors achieved the maximum value of 96% for water columns with suspended TiO_2 nanoparticles

[13] and 98% for the photocatalytic filtration performed by the porous medium coated with TiO_2 [14].

It must be also highlighted that the surprising regeneration ability showed by the developed photocatalytic porous media can completely recover its oxidative properties after a simple sun exposure [15], allowing a truly sustainable use of the developed photocatalytic filter.

2. Urban water cycle sustainability: new challenges to ensure safe water

Urban water cycle management involves the fields of water supply, urban drainage, wastewater treatment, reutilization, and sludge handling with a river basin scale approach.

Conventional approaches to urban water management for providing water supply and sanitation services are often costly, inefficient, and not integrated. Hence, there is a need for finding new ways for improving and assess the urban water systems to enable better sustainability of these systems [16] to face new challenges in a climate change context.

In an urban water systems context, life cycle assessment (LCA) can provide a pertinent holistic approach supporting the critical processes identification and the potential improvements of these systems, including the water and wastewater treatment facilities, as well as, its interactions with source or receiving waters. Several researchers used LCA approach for comparing water treatment technologies sustainability [17, 18], as well as the major environmental impact changes resulting from centralized wastewater treatment systems commutation to decentralized ones [19].

This kind of approaches allowed to identify new threats for the urban water cycle sustainability, concerning with the obligation to ensure safe drinking water in order to safeguard public health and urban aquatic ecosystems.

2.1. Occurrence of emerging micropollutants in urban water systems

Aquatic ecosystem pollution is particularly problematic due to the cumulative effect of pollutants on aquatic organisms during its life cycle. This cumulative effect can occur so slowly that major impacts may remain undetectable until the hatching of irreversible ecosystem changes [20]. The hydrodynamics and the longitudinal dispersion patterns presented by receiving water systems have a decisive role in its ability to self-regenerate [21] and to wash-out inflow pollutants like nutrients and xenobiotics [22].

During the last decades, the impact of chemical pollution has focused almost exclusively on the conventional priority pollutants, especially those acutely toxic/carcinogenic pesticides displaying persistence in the environment.

At the same time (but receiving much less attention), the anthropogenic activities increased the diversity and load discharge of another groups of bioactive hazardous chemicals into urban water systems (**Figure 1**), namely:

- Contaminants of emerging concern (CECs), such as pharmaceutical compounds (PhCs), diagnostic agents, steroids, phthalates, and disinfectants.

- Endocrine disrupting compounds (EDCs), like natural and synthetic estrogenic or androgenic chemicals.
- Personal care products (PCPs), such as fragrances, sun-screen agents, and cosmetics.

The widespread use of antibiotics as a therapy for bacterial infections in humans and animals (even for promoting it growth) has led to the concentration increase of antibiotic-resistant bacteria (ARB) in surface waters and urban waterways [23–25], used for domestic sewage, hospital wastewater, and livestock feeding operations drainage. As opposed to the conventional persistent priority pollutants, PhCs need not be (necessarily) "persistent" if they are continually introduced to surface waters, even at very low concentrations.

The use of conventional water treatment technologies against these emerging contaminants is limited due to their ineffectiveness and incomplete biodegradation of the waste products as outlined in the applicable EU directives.

The presence of PhCs, PCPs, and EDCs in drinking water indicates that conventional and most commonly used water treatment technologies may not be enough to completely eliminate these compounds from source waters [26], which can be polluted because existing Wastewater Treatment Plants (WWTPs) were usually not designed to remove antibiotics present at trace levels, implying the need for its urgent improvement. Indeed, if urban WWTPs play a vital role in minimizing the discharge of many water pollutants, including antibiotics [27] and pathogenic microorganisms [28] to the aquatic ecosystems, they are also potential breeding grounds and point sources for environmental dissemination of antibiotic resistance [29].

Indeed, the very high bacterial density into biological reactors (e.g., activated sludge) promotes selective elimination and/or changes in the proportions of phenotypes within effluent bacterial populations turning WWTPs into important reservoirs of enteric bacteria which

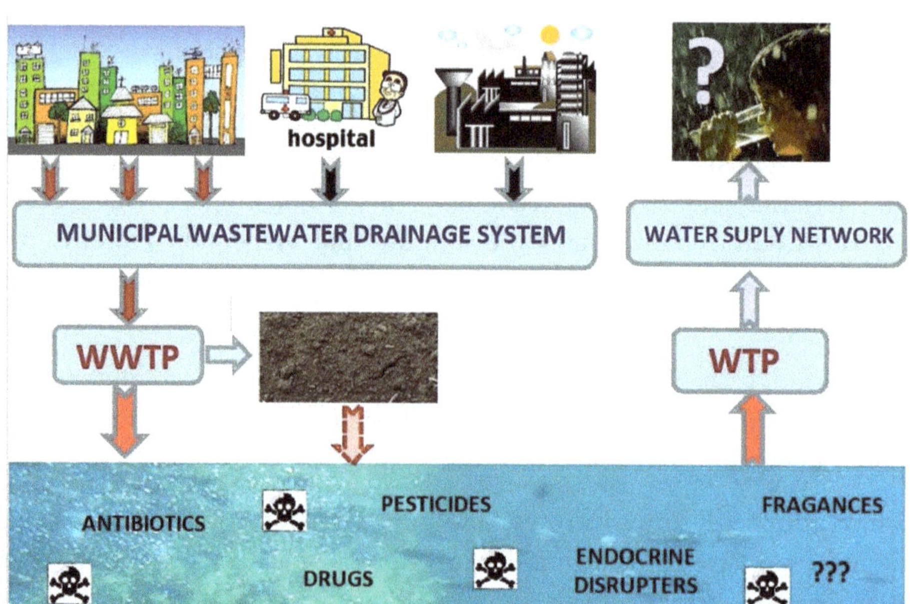

Figure 1. Threats to urban water cycle sustainability due to xenobiotic load increase.

carry potentially transferable resistance genes. For these reasons, higher frequency of multiple resistant coliform bacteria in treated sewage than in raw sewage [30, 31] for most antibiotics, especially for ciprofloxacin and tetracycline, have been found.

Human health risk characterization related to the pharmaceutical water ingestion exposure can be performed by the assessment of risk quotients (RQs). This risk index can be estimated dividing the maximum concentration of a pharmaceutical (MPC) found in the water matrix by the respective Drinking Water Equivalent Level (DWEL), which can be obtained as an exposure criteria based on other related parameters, such as acceptable daily intake; body weight, hazard quotient, and drinking water daily ingestion; gastrointestinal absorption rate; and frequency of exposure. So, a RQ value higher than 1 leads to a risk concern related to inadvertent exposure through drinking water, and measures must be considered in order to prevent public health.

A recent monitoring program performed along Lisbon's drinking water supply system [32] showed that appreciable risks to the consumer's health arising from exposure to trace levels of pharmaceuticals in drinking water were yet extremely unlikely, because all risk quotient (RQ) values were less than 0.001. Therefore, a high environmental risk was detected for *Erythromycin* (RQ = 1.55), the urgency of the study and development of new low-cost technologies for an effective removal of the most prevalent antibiotics in WTP raw waters.

2.2. Advanced oxidation processes: the role of photocatalysis as a *low-cost* alternative technology

Nanotechnology offers significant opportunities to revolutionize approaches toward drinking water treatment by enhancing the multifunctionality and versatility of treatment systems, while reducing reliance on stoichiometric chemical addition, shrinking large facilities with relatively long hydraulic contact times, and minimizing energy intensive processes [33]. So, it can provide low-cost, safe, and efficient water treatment systems with minimal energy requirements contributing for a more sustainable urban water cycle.

Nanomaterials properties have been explored for applications in water and wastewater treatment, due to its advantages related to the high specific surface area, fast dissolution, high reactivity, and strong sorption. Micropollutants' removal ability of new materials, such as carbon nanotubes, nanofibers, nanoscale metal oxide, nano-zeolites, and magnetic nanoparticles, is being tested and assessed when used in selected treatment unit processes, like adsorption, photocatalysis, membrane filtration, and disinfection.

Different advanced water treatment techniques for antibiotic removal have been studied, especially focus on membrane filtration, activated carbon adsorption, and advanced oxidation processes (AOPs). AOPs are recommended when water pollutants (such pharmaceuticals) have a high chemical stability and/or low degradability, allowing a more useful and cost-efficient combination with biological processes, namely in wastewater treatment [34].

The efficacy of AOPs depends on the generation of very reactive and nonselective free radicals—such as hydroxyl radicals ($^{\bullet}OH$), superoxide radical (O^{2-}), hydroperoxyl radical ($HO_2^{\bullet}$), and alkoxy radical (RO•)—involving chemical (e.g., O_3, O_3/H_2O_2), photochemical (UV/O_3, UV/H_2O_2), or photocatalytic (TiO_2/UV, ZnO/UV) oxidation processes. In recent years, semiconductor photocatalytic process has shown a great potential as a low-cost, environmental friendly, and sustainable treatment technology to align with the "zero" waste scheme in the water/wastewater industry. The ability of this advanced oxidation technology has been widely demonstrated to remove persistent organic compounds and microorganisms in water [35] and some hazardous inorganic micropollutants (e.g., arsenic, heavy metals, uranium).

Recent research works were mainly focused on AOPs assisted by solar radiation (a clean and renewable energy source), such as heterogeneous photocatalysis, in order to develop more sustainable and *low-cost* processes. The photocatalytic reactors can be divided into two main groups: with suspended nanoparticles (e.g., TiO_2, ZnO) in the reaction mixture (water and wastewater) and with immobilized nanoparticles on a carrier material (e.g., glass, quartz, stainless steel, zeolites).

When the catalyst is in suspension, the active surface is greater. However, its particles have to be removed from the treated water after the detoxification, and the manipulation of powdered semiconductors are difficult. To ensure complete rejection of TiO_2 nanoparticles, an extensive and relatively costly installation technology is necessary, including pumps. Very promising techniques for solving problems concerning separation of the photocatalyst as well as products and by-products of photo-degradation from the reaction mixture are the use of photocatalytic membrane reactors (PMRs) and the introduction of a magnetic into the nanocomposite [36]. However, the energy costs evolved in membrane processes can compromise the economic sustainability of the water treatment utilities, namely in medium and small water supply systems.

A solution for avoiding the contamination with the photocatalytic nanoparticles is their immobilization on the surface of specified materials by use of suitable coating techniques, as a wet chemical process. Quartz has been found to be the best support for titanium dioxide, because it is the most neutral and stable one at high temperatures. As a consequence, it has been chosen as the ideal support for new experiments with TiO_2 in the photodegradation of organic micropollutants in water [37].

3. Experimental methodology

During this research work, a set of experiments under different test scenarios were performed in order to assess the antibiotic removal efficiency and to characterize its photo-oxidation kinetics, using two different lab-scale photoreactors. In the first one (PR1), the heterogeneous photocatalysis was performed using suspended TiO_2 nanoparticles as catalyst to remove the antibiotic from water. In the second (PR2) one, a photocatalytic filtration was performed using a granular porous medium coated by immobilized TiO_2 nanoparticles.

In these experiments, the antibiotic used to prepare all synthetic solutions was the oxytetracycline hydrochloride (MW = 496.89, CAS# 2058-46-0), supplied by Sigma-Aldrich with a purity higher than 95%. The OTC concentrations were always measured using a UV-VIS spectrophotometer, Shimadzu UV-1800, at 354-nm wavelength. Titanium dioxide (TiO_2) used was Degussa (Aeroxide®) P-25 (80% anatase and 20% rutile).

The intensity of solar radiation is measured by a global UV radiometer (OHM – HD 9021), which was placed next to the solar reactors, in order to provide data in terms of incident solar radiation intensity (W/m^2). A reagent kit for rapid analysis of the amount of iron (Aquaquant®, E. Merck Darmstadt Germany) was also used.

Test scenarios were defined aiming to assess the effect on OTC removal efficiency of some abiotic parameters (e.g., OTC initial concentration, pH, hydraulic conditions, UV radiation source, and water matrices).

3.1. Photo-oxidation experiments using suspended TiO_2

In reactor PR1, photo-oxidation experiments were performed, with and without suspended TiO_2 nanoparticles, using two different UV radiation sources: solar radiation and UV lamp reactor (**Figure 2**).

For the OTC photo-degradation under solar radiation, bottles of colorless polyester with a capacity of 1.5 L were used as reactor. These water bottles were placed vertically, being shaken manually every 10 minutes to prevent the deposition of TiO_2 at the bottom. The sun exposure time was 210 minutes for all photodegradation tests.

The UV reactor (*Heraeus Noblelight, System 2*) used in photodegradation assays consists of an UV immersion lamp TQ 150, an immersion tube, a cooling tube, and a reactor vessel. The UV

Figure 2. UV radiation sources used in OTC degradation experiments: solar (polyester bottles); UV reactor *Heraeus Noblelight*.

immersion lamp is a medium-pressure mercury vapor lamp with a broad emission spectrum in the UV range above 190 nm and lamp output of 150 W. The reactor vessel has a capacity of 0.8 L and three openings (one central and two sideways). In the central opening, the UV lamp tube is inserted, and only a side opening is used to carry out the extraction of the samples during the tests. The container is placed on a magnetic stirrer that was in operation throughout the test. The UV lamp exposure time was 60 minutes for all tests.

Equation (1) allows the calculation of the amount of accumulated UV energy ($Q_{450\text{-}950n}$) received on any surface in the same position with regard to the sun, per unit of volume of water inside the reactor, in the time interval Δt.

$$Q_{450-950_n} = Q_{450-950_{n-1}} + \Delta t_n \times \overline{450-950} \times \frac{A_r}{V_t} \;;\; \Delta t_n = t_n - t_{n-1} \tag{1}$$

Where t_n is the experimental time of each sample (s); V_t is the total reactor volume (L); A_r is the exposed surface area (m^2) of the reactor; and $\overline{450-950}$ is the average solar radiation $(W/m2)$ measured during the period Δt_n (s).

Photolytic and photocatalytic experiments were carried out under static hydraulic conditions using 20 mg/L of OTC, as initial pollutant concentration, in all tests. For photocatalysis, the chosen initial suspended catalyst concentrations were 50 and 25 mg/L of TiO_2, in order to assess the effect of doubling the value of this parameter on OTC removal efficiency.

In order to assess the photocatalysis ability as post-treatment unit in WTPs for antibiotic removal, OTC solutions were prepared using two different water matrices (distilled and tap water) in order to assess the potential influence of other water supply constituents on OTC removal efficiency. The pH values measured in all experiments ranged between 4.3–4.9, for distilled water, and 6.6–7.3, for tap water.

To evaluate the influence of radiation in OTC degradation, at any given irradiation time interval, the dispersion was sampled (5 mL), filtered through a Millipore filter (pore size of 0.22 μm) to separate the TiO_2 particles, and the absorption was monitored to obtain OTC concentration.

Table 1 summarizes the different assay conditions (scenarios) under which the OTC photodegradation tests, using suspended TiO_2, were performed (reactor PR1).

Most of the studies carried out on heterogeneous photocatalysis with TiO_2 have shown that the kinetics underlying the photo-oxidation of emerging pollutants can be represented by Eq. (2), according to the *Langmuir-Hinshelwood* model [38, 39].

$$r_0 = -\frac{dC}{dt} = \frac{k \times K \times C_0}{1 + K \times C_0} \tag{2}$$

Where r_0 is the initial rate of photo-oxidation (ppm minutes^{-1}); C_0 is the initial pollutant concentration (ppm); k is the reaction rate constant (ppm minutes^{-1}); and K is the pollutant adsorption coefficient (L/mg) measured during the period Δt_n (s).

Scenario	UV radiation	Water matrix	$[TiO_2]_0$ (ppm)
S1	Solar	Distilled	50
S2			25
S3		Tap	50
S4			25
S5		Distilled	–
S6		Tap	–
S7	UV lamp	Distilled	50
S8			25
S9		Tap	50
S10			25
S11		Distilled	–
S12		Tap	–

Table 1. Scenario analysis for OTC photo-oxidation in reactor PR1.

Considering that "$K \times C_0$" product can be a value quite low for photo-oxidation processes, which can be described by a pseudo-first order decay kinetics [35], the final pollutant concentration (C_t) is given by Eq. (3).

$$C_t = C_0 \times e^{-K_{aap} \times t} \tag{3}$$

Where K_{aap} is the apparent velocity reaction constant (minutes^{-1}).

So, the initial rate of photo-oxidation can be obtained by Eq. (4) when the pollutants present vestigial concentrations.

$$r_0 = K_{aap} \times C_0 \tag{4}$$

3.2. Photocatalytic filtration experiments using immobilized TiO_2

The lab-scale reactive filter applied on photocatalytic oxidation of OTC consists of two borosilicate glass cylinder (DURAN®) with 750 mm length, 70 mm external diameter, and 62 mm inner diameter. The filtration columns, with this quartz porous medium coated with $TiO_{2,}$ are assembly as showed in **Figure 3**, and the OTC solution was feed to the columns by a peristaltic pump (Watson-Marlow 503U).

The porous bed consists of a quartz extracted from a quarry located in Ponte da Barca (Portugal), which was characterized by X-ray diffraction (XRD) (**Figure 4**).

The quartz was crushed and sieved in order to reduce its grains size till the desired granulometry, as well as, to facilitate the removal of the usual impurities. After sieving out, a grain

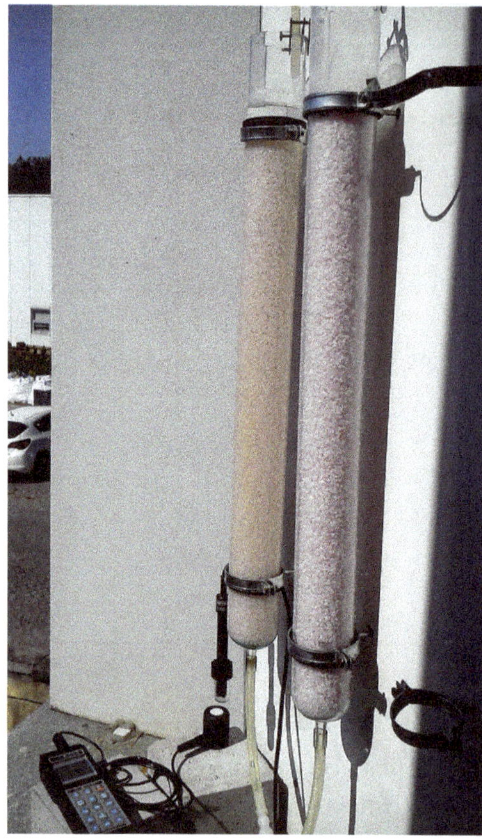

Figure 3. Filtration columns with a quartz porous media for OTC photo-oxidation.

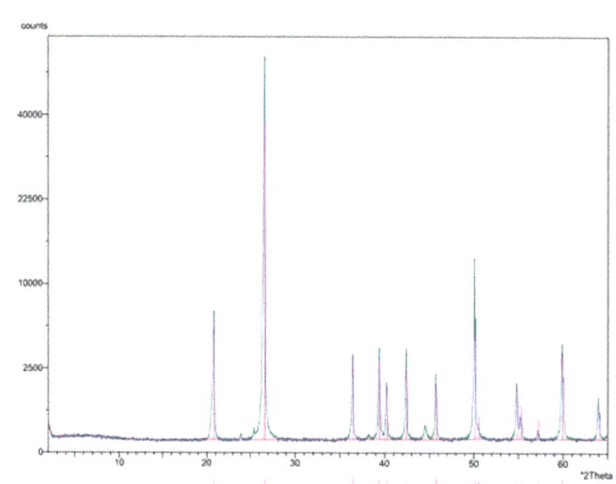

Figure 4. Characterization of a quartz sample by X-ray diffraction.

size distribution between 2.36 and 4.75 mm was dipped coated with TiO_2, also from Degussa (Aeroxide®), using the method described by Jeong et al. [40].

Prior to the start of the photocatalytic filtration tests, a study was carried out to optimize the hydraulic operation of the filter (e.g., flow rates ranges, head losses, hydraulic retention times) in order to select the most suitable flow rates: for photocatalysis, experiments were defined 4, 6 and 12 L/h; for adsorption, tests were defined 2, 4 and 6 L/h.

The selected range of flow rates for photocatalysis allows to simulate filtration (loading) rates similar to those occurring in WTP rapid and high rate filters (real scale hydraulic conditions) and also leads to OTC contact times with the TiO_2 that can provide an efficient photodegradation.

The hydraulic tests were performed both in open and closed (looped) circuit. An open circuit operation (without filtered water recycling) allows to maintain the initial OTC concentration constant and thus to evaluate the maximum capacity of retaining pollutant mass corresponding to the occurrence of porous medium saturation. A closed circuit operation allows to perform the number of loops (cycles) necessary to obtain the desired OTC contact time with the porous medium coated with TiO_2 nanoparticles.

The photocatalytic filtration tests of OTC solutions were performed in looped circuit during 270 minutes, considering different flow rates, initial OTC concentration (20 and 40 ppm), and aeration conditions. Final OTC concentrations were obtained by absorbance measurement using an UV-VIS spectrophotometer (Shimadzu UV—1800) at 354 nm wavelength. The effect of the aeration on the photo-degradation efficiency of OTC feed solution was also evaluated.

Table 2 summarizes the different test conditions (scenarios) under which the photocatalytic filtration was performed (reactor PR2)

Adsorption test was carried out under similar hydraulic conditions and the same duration of photodegradation tests, passing the OTC solution through the filter, first with quartz and after with quartz coated with TiO_2, in darkness to avoid any photodegradation contribute on final OTC removal.

3.3. Acute toxicity test

In order to assess the toxicity of OTC and oxidation by-products, it was used a simple toxicity test, not normalized but standardized by the international organization WaterTox Network [41]. In this toxicity assay, lettuce seeds (*Lactuca sativa*) are used.

Scenario	$[OTC]_0$ (ppm)	Flow rate (L/h)	Filter aeration
F1	20	4	No
F2	40	6	
F3	20		
F4		12	
F5			Yes

Table 2. Scenario analysis for OTC photocatalysis in reactor PR2.

Each of the lettuce seed root growth inhibition test was performed with 20 seeds in a Petri dish, containing a filter paper embedded in 2 mL of each sample dilution (100, 75, 50, and 25%). Root lengths were measured after 72 hours of incubation (**Figure 5**), and the average lethal concentration (LC50) was calculated as stated by Dutkka [42].

The samples used consisted of the oxytetracycline before and after photocatalytic treatment and, as negative control, distilled water. The tests were always carried out in triplicate.

Figure 5. Preparation and final result of the acute toxicity bioassay using *L. sativa*.

4. Results and discussion

4.1. Photo-oxidation experiments (reactor PR1)

Figure 6 shows the degradation kinetics of OTC photocatalysis (scenarios S1–S4) and photolysis (scenarios S5 and S6) performed in two different aqueous matrices (distilled and tap water), always with an initial concentration of 20 mg/L and exposed to solar radiation (a free and renewable energy source) during 210 minutes (experimental).

For OTC degradation using solar radiation exposure, the maximum average value of 88% was reached for the scenarios S1 and S3 (different water matrix), which correspond to the highest TiO_2 concentration.

The constituents present in the tap water, namely the iron, showed to have a significant effect on the OTC degradation efficiency, with special emphasis in photolysis experiments (almost quintupled), while in photocatalysis, this increase was only about 20%, under similar conditions of accumulated UV energy. Indeed, auxiliary control testing of tap water quality parameters detected the presence of iron concentrations in the range of 0.08–0.1 mg/L.

In order to assess a potential efficiency increase in OTC removal, due to an alternative UV radiation source (although with energy costs), those two photo-oxidative processes were also performed for the same aqueous matrices and OTC initial concentration but using the described UV lamp reactor with an exposure time of 60 minutes (scenarios S7–S12). The obtained results are depicted in **Figure 7**.

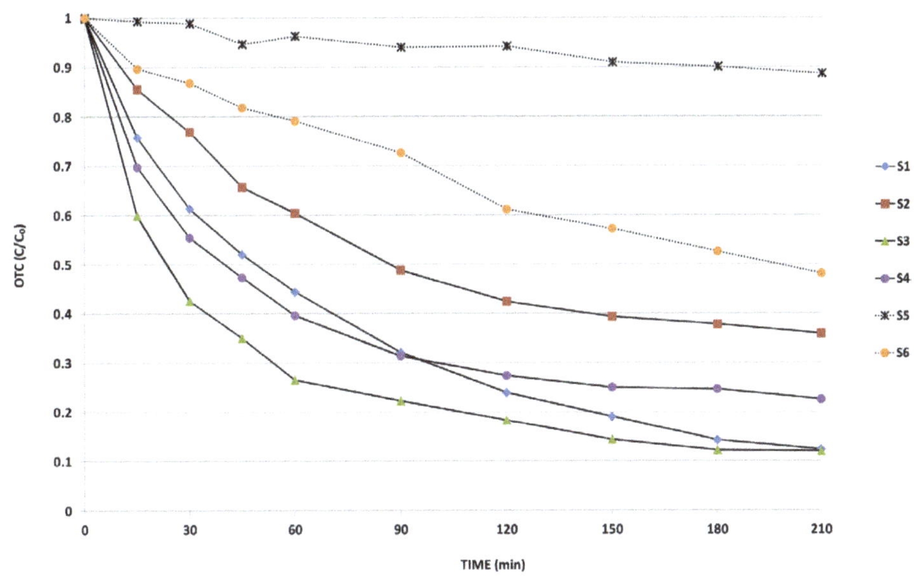

Figure 6. OTC photo-oxidation efficiency with solar radiation.

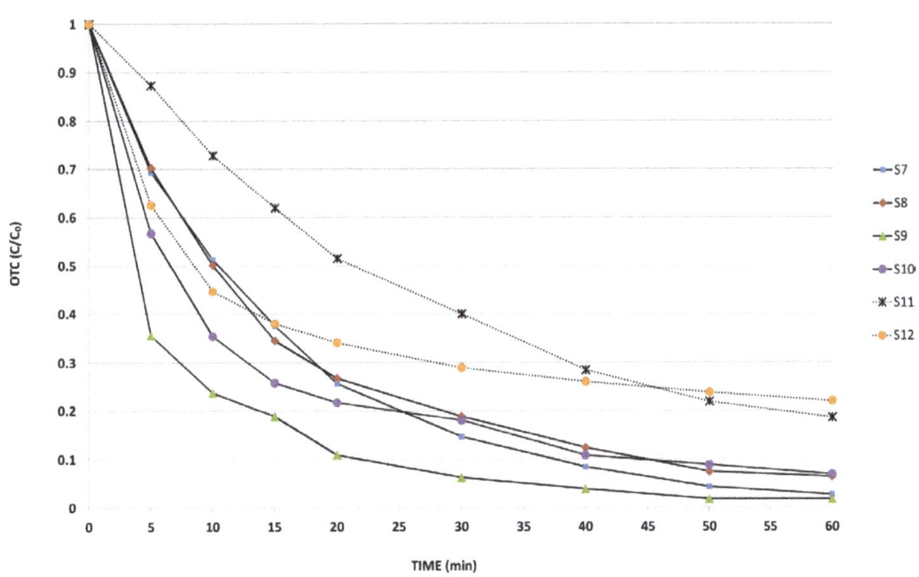

Figure 7. OTC photo-oxidation efficiency with UV lamp reactor.

For OTC degradation using UV reactor exposure, the same behavior was observed. The maximum efficiency (near 96%) was reached for the scenarios S7 and S9 (different water matrix), which correspond to the highest TiO_2 concentration. As depicted in **Figure 7**, the use of those two different aqueous matrices had a negligible effect on final OTC removal efficiency.

In this case (UV lamp reactor), the efficiency gains on OTC removal, related to the catalyst action, are much less significant than in the case of the solar radiation tests. Due to this finding the benefit of the use of photocatalysis would not be sufficiently attractive given the costs inherent to the necessary removal process of suspended TiO_2 nanoparticles.

Table 3 summarizes the major experimental results obtained for OTC removal using suspended TiO_2, namely the maximum average efficiencies, some photo-oxidation kinetic parameters, and the coefficient of determination (R^2) observed in the adjustment of the *Langmuir-Hinshelwood* model to the experimental data set obtained for each assay.

The obtained R^2 values (**Table 3**) allow to conclude that the *Langmuir-Hinshelwood* model adapts adequately to the kinetic behavior observed in the OTC photo-oxidation for any of those experimental scenarios tested and analyzed in this study.

For both water matrices solutions and in the scenarios using 50 mg/L of TiO_2, OTC removal efficiencies may achieve values higher than 88% if the accumulated solar energy quantity is higher than 113 kJ/L.

Comparing the results obtained using these two different UV radiation sources, the photocatalysis using TiO_2 with solar radiation seems to be a sustainable alternative for antibiotic removal in WTPs due to its minor energy costs and high efficiency removal, even requiring more exposure/retention time and achieving lower efficiencies, when compared with the ones observed in UV reactor tests.

4.2. Photocatalytic filtration experiments (reactor PR2)

The results of OTC removal efficiency by photocatalytic filtration performed in the reactor PR2 are depicted in **Figure 8**, considering the experimental scenarios F1–F5, which were defined aiming to assess the influence of different filtration fluxes, OTC initial concentration, and the OTC solution aeration in the feed tank.

The results showed that slower flux resulted in better OTC removal efficiency at the beginning of the experiment, due to longer retention times in the filter (curves F1, F3 and F4). Aeration is important for the oxidation reaction in photocatalytic processes. This process requires dissolved oxygen to act as an oxidant and to slow down the electron-hole recombination reaction. The curve F5 for the experiment with aeration shows the highest value for the initial photo-degradation rate.

In **Figure 8**, it can be seen that the experiments F1, F4, and F5 with 4 and 12 L/h had higher initial degradation rates, and these tests removed more than 96% of OCT by 270 minutes of solar irradiation time. The highest OTC removal efficiency obtained for photocatalytic filtration,

Parameter	S1	S2	S3	S4	S5	S6	S7	S8	S9	S10	S11	S12
C_0 (ppm)	19.25	20.59	18.69	20.83	15.39	20.69	19.57	20.62	12.13	18.13	20.69	21.04
C_f (ppm)	2.48	4.80	2.48	7.60	13.67	10.06	0.86	2.16	0.48	1.43	4.02	3.22
K_{aap} (minutes^{-1})	0.013	0.009	0.011	0.006	0.001	0.004	0.061	0.047	0.078	0.052	0.030	0.040
R^2	0.844	0.909	0.974	0.987	0.931	0.982	0.899	0.866	0.816	0.807	0.992	0.321
r_0 (ppm·minutes^{-1})	0.25	0.19	0.21	0.12	0.01	0.08	1.19	0.97	0.95	0.94	0.62	0.84
OTC removal (%)	87	77	87	64	11	51	96	90	96	92	81	85

Table 3. Results synthesis of OTC photo-oxidation experiments in reactor PR1.

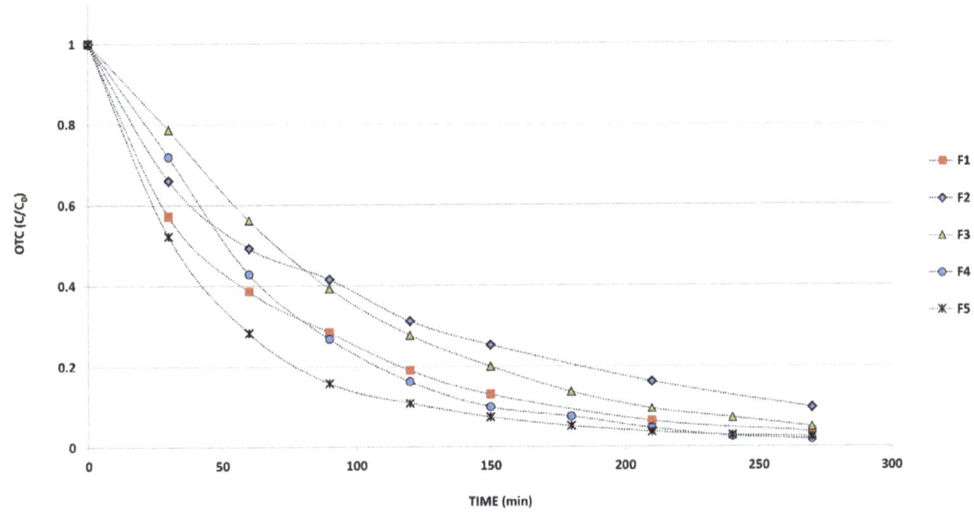

Figure 8. OTC removal efficiency using photocatalytic filtration with TiO_2 (PR2).

using a quartz porous medium coated with TiO_2, was 98% achieved for scenarios F4 and F5, which correspond to the higher flow rates tested (without and with filter aeration).

Table 4 summarizes the major experimental results of OTC removal experiments using photocatalytic filtration with a porous medium coated with TiO_2, namely the maximum average efficiencies, some photo-oxidation kinetic parameters, and the coefficient of determination (R^2) observed in the adjustment of the *Langmuir-Hinshelwood* model to the experimental data sets.

The results presented were obtained on different days with variations in the amount of accumulated energy from solar radiation received on the surface of the porous medium.

The calculated R^2 values (**Table 4**) allow to verify that the *Langmuir-Hinshelwood* model also adapts adequately to the kinetic behavior observed in the OTC photocatalytic filtration performed in this study for any of the analyzed experimental scenarios.

The effect of the flow rate variation on OTC adsorption was assessed using the reactor PR2 in darkness conditions and filtration with two different porous media (quartz without and with TiO_2 functionalization). The results of the OTC adsorption tests are depicted in **Figure 9**,

Parameter	F1	F2	F3	F4	F5
C_0 (ppm)	20.01	40.59	19.38	18.62	18.48
C_f (ppm)	0.74	3.94	0.95	0.33	0.47
K_{aap} (minutes^{-1})	0.011	0.009	0.007	0.010	0.013
R^2	0.941	0.956	0.999	0.999	0.958
r_0 (ppm·minutes^{-1})	0.44	0.36	0.14	0.38	0.51
OTC removal (%)	96.3	90.3	95.1	98.2	97.5

Table 4. Results synthesis of OTC photo-oxidation experiments in reactor PR2.

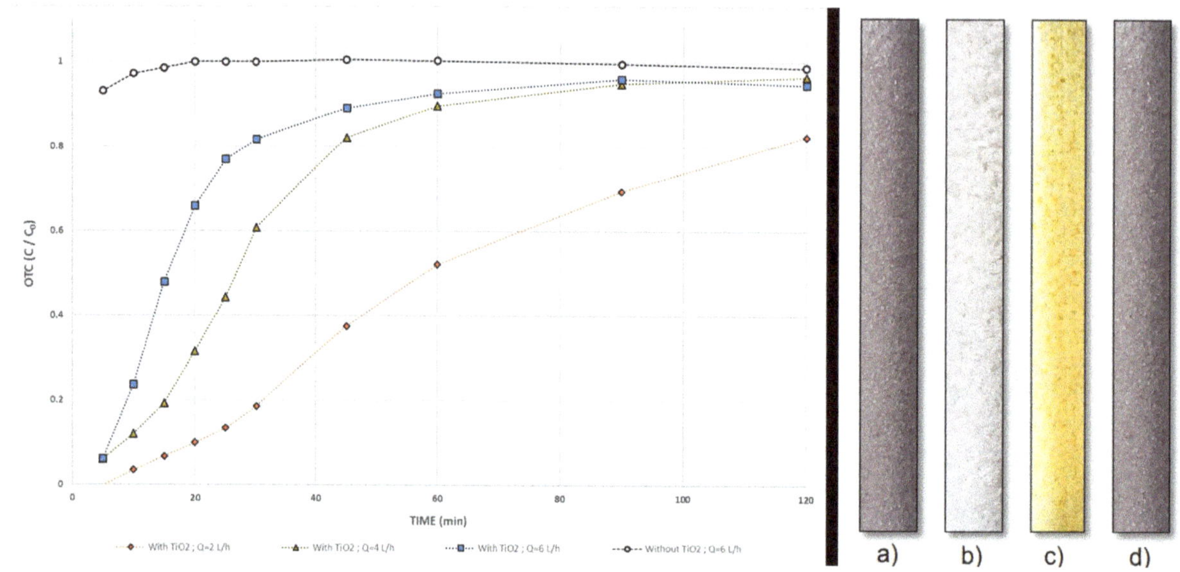

Figure 9. Results of the OTC adsorption tests and final look (color changes) of the porous medium.

as well as the final look (color changes) of the porous medium in the following four distinct situations:

a. Quartz without TiO_2

b. Quartz coated with TiO_2

c. Quartz coated with TiO_2 after saturation (OTC adsorption)

d. Quartz without TiO_2 after OTC adoption test.

In darkness and after 120 minutes, the quartz (without TiO_2) has a negligible OTC adsorption, but in the column filter with the coated quartz, the adsorption is function of the feed flow rate. With a flow rate of 6 L/h, the equilibrium concentration was reached within 90 minutes, and for 4L/h, the equilibrium concentration was only reached after 120 minutes.

Moreover, it was also observed a high regeneration ability by the photocatalytic porous medium, which can completely recover its oxidative properties after a simple solar radiation exposure of about 4 hours [15]. **Figure 10** presents the time evolution of saturation and regeneration processes observed in this photocatalytic filter.

As reported on item 3.3, the toxicity of the oxytetracycline both before and after the photocatalytic degradation (performed in each reactor – PR1 and PR2) was evaluated by using *L. sativa* seeds germination as a bioindicator.

The results of these toxicity tests toward lettuce seed growth showed a toxicity decrease after the photocatalytic OTC degradation, enabling the adoption of this emerging water treatment technique as an apparently safe alternative for the antibiotics removal challenge.

Figure 10. Saturation and regeneration processes evolution observed in the photocatalytic filter (PR2).

5. Conclusions

The chemical structure of OTC was effectively degraded by in both lab-scale photo-oxidation reactors, achieving very high OTC removal efficiencies for photocatalytic experiments (96–98%), even with small amounts of suspended and coated TiO_2 nanoparticles.

The results obtained in the photo-oxidation experiments using suspended TiO_2 indicate that:

- The photocatalytic tests were more effective than testing photolysis, which proves the high catalyzing power of TiO_2 particles described in the literature.
- OTC solutions exposed to UV-lamp radiation reached higher OTC removal efficiency (maximum about 96%) than those exposed to solar radiation (maximum about 88%). Nevertheless, the last one UV radiation source seems to be a more sustainable alternative for antibiotic removal in WTPs due to its minor energy costs and high efficiency removal.
- The overall efficiencies of the OTC degradation in distilled and tap waters are very close, namely when the iron concentration in water is low.
- The kinetics of OTC photo-oxidation reveals a faster degradation during the first 10–20 minutes.

The results obtained in the photocatalytic filtration experiments performed by a porous media coated with TiO_2 nanoparticles indicate that:

- The best OTC removal efficiency was 98%, achieved for an antibiotic initial concentration of 20 mg/L, a flow rate of 12 L/h in a looped hydraulic circuit, and for a cumulate solar energy near 805 kJ/L.
- Slower flux seems increase OTC removal efficiency at the beginning of the experiment, due to longer retention/contact time into the column filter.
- The experiment performed with aeration shows the highest value for the initial photo-oxidation rate, and one of the best final OTC removal efficiency.

- In darkness, the quartz (without TiO_2) has a negligible OTC adsorption, but in the column filter with the coated quartz the adsorption is relevant, and function of the feed flow rate increase.
- It must be highlighted the surprising regeneration ability showed by the developed photocatalytic porous media, which can completely recover its oxidative properties after a simple sun exposure for 4 hours, allowing sustainable use of the photocatalytic filter.

The *Langmuir-Hinshelwood* model was adequately adapted to the kinetic behavior observed in the OTC photo-oxidation processes in all of the analyzed scenarios, and for both used photocatalytic reactors.

The toxicity tests carried out showed that the use of heterogeneous photocatalysis with suspended TiO_2 does not induce the appearance of toxic by-products in the water, since the seeds of lettuce *L. sativa* always showed inhibition percentages lower than 22% after treatment.

Acknowledgements

This work was financed by FEDER funds through the Operational Competitiveness Programme—COMPETE and by national funds through FCT—Foundation for Science and Technology within the scope of the project POCI-01-0145-FEDER-007136.

Author details

António Armando Lima Sampaio Duarte[1]* and Maria Teresa Pessoa Amorim[2]

*Address all correspondence to: aduarte@civil.uminho.pt

1 Department of Civil Engineering, University of Minho, Portugal

2 Department of Textile Engineering, University of Minho, Portugal

References

[1] Verlicchi P, Al Aukidy M, Zambello E. Occurrence of pharmaceutical compounds in urban wastewater: Removal, mass load and environmental risk after a secondary treatment—A review. Science of the Total Environment. 2012;**429**(1):123-155. DOI: 10.1016/j.scitotenv.2012.04.028

[2] Ashton D, Hiltonb M, Thomas KV. Investigating the environmental transport of human pharmaceuticals to streams in the United Kingdom. Science of the Total Environment. 2004;**333**(1-3):167-184. DOI: 10.1016/j.scitotenv.2004.04.062

[3] Richardson SD, Ternes TA. Water analysis: Emerging contaminants and current issues. Analytical Chemistry. 2011;**83**(12):4614-4648. DOI: 10.1021/ac200915r

[4] Qu X, Alvarez PJJ, Li Q. Applications of nanotechnology in water and wastewaters. Water Research. 2013;**47**:3931-3946. DOI: 10.1016/j.watres.2012.09.058

[5] Baruah S, Khan, MN, Dutta J. Perspectives and applications of nanotechnology in water treatment. Environment Chemical Letters. 2016;**14**:1-14. DOI: 10.1007/s10311-015-0542-2

[6] Segneanu AE, Orbeci C, Lazau C, Sfirloaga P, Vlazan P, Bandas C, Grozescu I. Waste water treatment methods. In: Elshorbagy W, Chowdhury RK, editors. Water Treatment. Rijeka: InTech; 2013. pp. 53-71. DOI: 10.5772/53755

[7] Klavarioti M, Mantzavinos D, Kassinos D. Removal of residual pharmaceuticals from aqueous systems by advanced oxidation processes. Environment International. 2009; **35**(2):402-417.

[8] Schranka SG, Joséb HJ, Moreirab RFPM, Schröder HF. Applicability of Fenton and H_2O_2/UV reactions in the treatment of tannery wastewaters. Chemosphere. 2005;**60**(5):644-655. DOI: 10.1016/j.chemosphere.2005.01.033

[9] Pereira JHOS, Reis AC, Queirós D, Nunes OC, Borges MT, Vilar VJP, Boaventura RAR. Insights into solar TiO_2 – assisted photocatalytic oxidation of two antibiotics employed in aquatic animal production, oxolinic acid and oxytetracycline. Science of the Total Environment. 2013;**463-464**:274-283.

[10] Sarmah AK, Meyer MT, Boxall AB. A global perspective on the use, sales, exposure pathways, occurrence, fate and effects of veterinary antibiotics (VAs) in the environment. Chemosphere. 2006;**65**(5):725-759. DOI: 10.1016/j.chemosphere.2006.03.026

[11] Pena A, Paulo M, Silva LJG, Seifrtiva M, Lino CM, Solich P. Tetracycline antibiotics in hospital and municipal wastewaters: A pilot study in Portugal. Analytical and Bioanalytical Chemistry. 2010;**396**(8):2929-2936.

[12] Feitosa-Felizzola J, Chiron S. Occurrence and distribution of selected antibiotics in a small Mediterranean stream (Arc River, Southern France). Journal of Hydrology. 2009;**364**:50-57. DOI: 10.1016/j.jhydrol.2008.10.006

[13] Pereira MSTG, Duarte AALS, Amorim MTSP. Environmental risk assessment mitigation of effluent hospital discharge. A contribution for pharmaceutical compounds removal (in Portuguese) [thesis]. Braga, Portugal: University of Minho; 2013. p. 131

[14] Dias DNF, Duarte AALS, Amorim MTSP. Reactive filtration by photo-oxidative porous media (in Portuguese) [thesis]. Braga, Portugal: University of Minho; 2014. p. 121

[15] Guerra AFB, Duarte AALS, Amorim MTSP. Micropollutants photocatalytic removal by functionalized porous media (in Portuguese) [thesis]. Braga, Portugal: University of Minho; 2015. p. 127

[16] Porto M, Khatri K, Vairavamoorthy K. Challenges for urban water supply and sanitation in the developing countries. Proceedings of the International Symposium "Water for a Changing World Developing Local Knowledge and Capacity", Alaerts GJ, Dickinson NL (eds.), Delft, The Netherlands, June 13-15, 2007. DOI: 10.1201/9780203878057

[17] Houillon G, Jolliet O. Life cycle assessment of processes for the treatment of wastewater urban sludge: Energy and global warming analysis. Journal of Cleaner Production. 2005;**13**:287-299

[18] Friedrich E, Buckley CA. The use of life cycle assessment in the selection of water treatment processes. Pollution Research Group ed., School of Chemical Engineering, University of Natal, Durban, South Africa. 2001.

[19] Renzoni R, Germain A. Life cycle assessment of the anthropic water cycle – Part 1 waste water treatment plants. In: 9th SETAC LCA Case Studies Symposium; January; Noordwijkerhout, The Netherlands. 2001.

[20] Daughton CG, Ternes TA. Pharmaceuticals and personal care products in the environment: Agents of subtle change? Environmental Health Perspectives. 1999;**107**(6):907-938.

[21] Duarte AALS, Boaventura RAR. Pollutant dispersion modelling for Portuguese river water uses protection linked to tracer dye experimental data. WSEAS Transactions on Environment and Development. 2008;**4**(12):1047-1056.

[22] Duarte AALS, Pinho JLS, Pardal MAC, Neto JM, Vieira JMP, Santos FS. Hydrodynamic modelling for Mondego estuary water quality management. In: Pardal MAC, Marques JC, Graça MA, editors. Aquatic Ecology of the Mondego River Basin Global Importance of Local Experience. Coimbra, Portugal: University of Coimbra Press; 2002. pp. 29-42.

[23] Kappell AD, DeNies MS, Ahuja NH, Ledeboer NA, Newton RJ, Hristova KR. Detection of multi-drug resistant *Escherichia coli* in the urban waterways of Milwaukee, WI. Frontiers in Microbiology. 2015;**6**:336-348. DOI: 10.3389/fmicb.2015.00336

[24] Behera SK, Kim HW, Oh JE, Park HS. Occurance and removal of antibiotics, hormones and several other pharmaceuticals in wastewater treatment plants of largest industrial city of Korea. Science of the Total Environment. 2011;**409**:4351-4360.

[25] Jean J, Perrodin Y, Pivot C, Trepo D, Perraud M, Droguet J, et al. Identification and prioritization of bioaccumulable pharmaceutical substances discharged in hospital effluents. Journal of Environment. 2012;**103**:113-121.

[26] Padhye LP, Yao H, Kung FT, Huang CH. Year-long evaluation on the occurrence and fate of pharmaceuticals, personal care products, and endocrine disrupting chemicals in an urban drinking water treatment plant. Water Research. 2014;**51**:266-276. DOI: 10.1016/j.watres.2013.10.070

[27] Zhou LJ, Ying GG, Liu S, Zhao JL, Yang B, Chen ZF, Lai HJ. Occurrence and fate of eleven classes of antibiotics in two typical wastewater treatment plants in South China. Science of the Total Environment. 2013;**452**:365-376.

[28] Frigon D, Biswal BK, Mazza A, Masson L, Gehr R. Biological and physicochemical wastewater treatment processes reduce the prevalence of virulent *Escherichia coli*. Applied Environmental Microbiology. 2013;**79**(3):835-844.

[29] Pruden A, Larsson DG, Amezquita A, Collignon P, Brandt KK, Graham DW, et al. Management options for reducing the release of antibiotics and antibiotic resistance genes to the environment. Environmental Health Perspectives. 2013;**121**(8):878-885. DOI: 10.1289/ehp.1206446

[30] Silva MF, Moreira IV, Gonzalez-Pajuelo M, Nunes OC, Manaia CM. Antimicrobial resistance patterns in *Enterobacteriaceae* isolated from an urban wastewater treatment plant. FEMS Microbiol Ecology. 2007;**60**(1):166-176. DOI: 10.1111/j.1574-6941.2006.00268.x

[31] Figueira V, Serra E, Manaia, C. Differential patterns of antimicrobial resistance in population subsets of *Escherichia coli* isolated from waste- and surface waters. Science of the Total Environment. 2011;**409**(6):1017-1123. DOI: 10.1016/j.scitotenv.2010.12.011

[32] Gaffney VJ, Almeida CM, Rodrigues A, Ferreira E, Benoliel MJ, Cardoso VV. Occurrence of pharmaceuticals in a water supply system and related human health risk assessment. Water Research. 2015;**72**:199-208. DOI: 10.1016/j.watres.2014.10.027

[33] Westerhoff P, Alvarez PJ, Li Q, Zimmerman J. Overcoming implementation barriers for nanotechnology in drinking water treatment. Environmental Science: Nano. 2016;**3**(6): 1241-1253. DOI: 10.1039/C6EN00183A

[34] Marco A, Esplugas S, Saum G. How and why combine chemical and biological processes for wastewater treatment. Water Science and Technology. 1997;**35**:321-327

[35] Chong MN, Jin B, Chow CW, Saint C. Recent developments in photocatalytic water treatment technology: A review. Water Research. 2010; **44**(10):2997-3027.

[36] Gomez-Pastora J, Dominguez S, Bringas E,Rivero MJ, Ortiz I. Review and perspectives of the use of magnetic photocatalysis (MVPCs) in water treatment. Chemical Engineering Journal. 2017;**310**:407-427. DOI: 10.1016/j.cej.2016.04.140

[37] Fernandez A, Lassaletta G, Jimenez VM, Justo A, Gonzalez-Elipe AR, Herrmann JH, Tahiri H, Ait-Ichou Y. Preparation and characterization of TiO_2 photocatalysts supported on various rigid supports (glass, quartz and stainless steel). Comparative studies of photocatalytic activity in water purification. Applied Catalysis B: Environmental. 1995;**7**:49-63

[38] Rao N, Dubey A, Mohanty S, Khare P, Jain R, Kaul S. Photocatalytic degradation of 2-chlorophenol: A study of kinetics, intermediates and biodegradability. Journal of Hazardous Materials. 2003;**101**(3):301-314

[39] Xu Y, Langford C. UV- or visible-light-induced degradation of X3B on TiO_2 nanoparticles: The influence of adsorption. Langmuir. 2001;**17**(3):897-902

[40] Jeong SH, Kim JK, Kim BS, Shim SH, Lee BT. Characterization of SiO_2 and TiO_2 films prepared using rf magnetron sputtering and their application to anti-reflection coating. Vacuum. 2004;**76**(4):507-515. DOI: 10.1016/j.vacuum.2004.06.003

[41] Ronco A, Gagnom P, Diaz-Baez MC, Arkhipchuk V, Castillo G, Castillo LE, et al. Overview of results from the WaterTox intercalibration and environmental testing phase II program: Part 1, statistical analysis of blind sample testing. Environmental Toxicology. 2002;**17**(3):232-240.

[42] Dutkka B. Short-Term Root Elongation Toxicity Biossay. Methods for Toxicological Analysis of Waters, Wastewaters and Sediments. Ottawa, Canada: National Water Research Institute; 1989

4

Recent Overview on the Abatement of Pesticide Residues in Water by Photocatalytic Treatment using TiO_2

Nuria Vela, Gabriel Pérez-Lucas, José Fenoll and Simón Navarro

Abstract

The water bodies' pollution with phytosanitary products can pose a serious threat to aquatic ecosystems and drinking water resources. The usual appearance of pesticides in surface water, waste water and groundwater has driven the search for proper methods to remove persistent pesticides. Although typical biological treatments of water offer some advantages such as low cost and operability, many investigations referring to the removal of pesticides have suggested that in many cases they have low effectiveness due to the limited biodegradability of many agrochemicals. In recent years, research for new techniques for water detoxification to avoid these disadvantages has led to processes that involve light, which are called advanced oxidation processes (AOPs). Among the different semiconductor (SC) materials tested as potential photocatalysts, titanium dioxide (TiO_2) is the most popular because of its photochemical stability, commercial availability, non-toxic nature and low cost, high photoactivity, ease of preparation in the laboratory, possibility of doping with metals and non-metals and coating on solid support. Thus, in the present review, we provide an overview of the recent research being developed to photodegrade pesticide residues in water using TiO_2 as photocatalyst.

Keywords: titania, pesticides, water, photocatalytic degradation

1. Introduction

Pesticide may be defined as 'any substance or mixture of substances intended for preventing, destroying, or controlling any pest including vectors of human or animal diseases, unwanted species of plants or animals causing harm during, or otherwise interfering with, the production, processing, storage, or marketing of food, agricultural commodities, wood and wood products, or animal feedstuffs, or which may be administered to animals for the control of insects, arachnids or other pests in or on their bodies. The term includes chemicals used as growth regulators, defoliants, desiccants, fruit thinning agents, or agents for preventing the premature fall of fruits, and substances applied to crops either before or after harvest to prevent deterioration during storage or transport' [1]. There are many hundreds of them including herbicides, insecticides, fungicides, rodenticides, nematicides, plant growth regulators and others. Pesticide products contain both 'active' and 'inert' ingredients: active ingredients are the chemicals in pesticide products that kill, control or repel pests. All other ingredients are called 'inert ingredients,' which are important for product performance and usability.

Pesticides have been widely applied to protect agricultural crops since the 1940s, and their use increased steadily during the subsequent decades. The *Green Revolution* was the notable increase in cereal-grain production in many developing countries in the 1960s and 1970s. This tendency resulted from the introduction of hybrid strains of wheat, rice and corn and the adoption of modern agricultural technologies, including irrigation and large doses of agrochemicals, fertilisers and pesticides [2]. However, Rachel Carson (1907–1964) with the publishing of her sensational book *Silent Spring* in 1962 [3] warned of the dangers to all natural systems from the misuse of some pesticides such as DDT (1,1,1-trichloro-2,2-di(4-chlorophenyl) ethane). As a result, a wider audience was warned of the environmental effects of the widespread use of pesticides, and DDT was banned for agricultural use 10 years later in the USA, and the regulation of chemical pesticide use was strengthened. Currently, all pesticides are subject to strict registration as meaning 'the process whereby the responsible national government authority approves the sale and use of a pesticide following the evaluation of comprehensive scientific data demonstrating that the product is effective for the purposes intended and not unduly hazardous to human or animal health or the environment' [4]. In evaluating a pesticide registration application, a wide variety of potential human health and environmental effects associated with their use must be tested. Registrants must generate the necessary scientific information to address concerns corresponding to the identity, composition, potential adverse effects and environmental fate of each pesticide. These data allow evaluating whether a pesticide could harm certain nontarget organisms and endangered species.

Protection of crop losses/yield reduction and increase in food quality are key benefits associated with the use of pesticides in agriculture. However, most organic pesticides characterised as persistent in the environment can bioaccumulate through the food web and can be transported in long distances [5], as evidenced by the accumulation in regions where persistent pesticides have never been used [6]. Persistent organic pollutants (POPs) are *chemical compounds that persist in the environment and adversely affect human health and the environment around the world* [7]. Because they can be transported by wind and water, most POPs can affect human health and wildlife far from where they are applied. They have high persistence in the environment and can accumulate passing from one species to the next through the food chain. To treat this environmental concern, the USA joined forces with 90 other countries and the European Community to sign the agreement United Nations treaty in Stockholm (Sweden, May 2001). The *Stockholm Convention on Persistent Organic Pollutants*, approved by Council Decision 2006/507/EC [8], entered into force on 17 May 2004. The aim of the Convention was to protect human health and the environment from POPs. Under the treaty, known as the *Stockholm Convention*, countries agreed to reduce or eliminate the production, use and/or release of 12 key POPs ('the dirty dozen', mainly organochlorine (OC) insecticides) and specified under the Convention a scientific review process that has led to the addition of other POP chemicals of global concern. Currently, there are 30 substances catalogued as POPs including mainly pesticides, industrial chemicals and by-products.

In addition, many pesticides are endocrine-disrupting chemicals (EDCs), *compounds that alter the normal functioning of the endocrine system of both wildlife and humans increasing incidence of breast cancer, abnormal growth patterns and neurodevelopmental delays in children, as well as changes in immune function* [9–11]. Most of them are organochlorine (OC) pesticides that affect the reproductive function.

Worldwide consumption of pesticides for agricultural use is constantly increasing as increased human population and crop production, and it has undergone significant changes since the 1960s. Nowadays, the worldwide consumption of pesticides is about 2 million tonnes per year, which 45% is used by Europe alone, 25% is consumed in the USA and 30% in the rest of the world. The proportion of herbicides in pesticide consumption increased rapidly, from 20% in 1960 to 47.5% in 2015. However, the proportion of consumption of insecticides (29.5%) and fungicides/bactericides (17.5%) declined despite their sales increased with other accounts for 5.5% only [12]. The rapid increase of herbicide consumption enhanced agricultural intensification and productivity.

The application of chemical pesticides, in particular the organic-synthesised pesticides, has been a significant mark of human civilisation, which greatly protects and facilitates agricultural productivity. Worlwide, insect pests cause an estimated 14% loss, plant pathogens cause a 13% loss, and weeds causes a 13% loss [13]. Pesticides are so indispensable in agricultural production. About one-third of the agricultural products are produced by using pesticides [14]. Without pesticide application, the loss of fruits, vegetables and cereals from pest injury would reach 78, 54 and 32%, respectively [15]. Ideally, a pesticide must be lethal to the targeted pests, but not to nontarget species, including humans. Unfortunately, this is not the case, so the controversy of use and abuse of pesticides is obvious. Consequently, the risks of using pesticides are serious as well [13]. Most pesticides are not spontaneously generated, and they are toxic to humans and the environment in greater or lesser degree [16]. Pesticides and their degraded products can pass into the atmosphere, soils and water, resulting in the accumulation of toxic substances and thus threatening human health and the environment. In addition, accumulated application leads to loss of biodiversity. Because many pesticides are barely degradable, they persist in soil and pollute surface water and groundwater. Depending on their chemical properties, they can bioaccumulate in food chains and consequently affect human health.

For many pesticides, there is evidence of long-term ubiquity in the aquatic environment at the European Union (EU) level, and therefore they need special consideration as regards their impact on the presentation of chemical status under the European Water Framework Directive (EWFD) [17]. The EWFD establishing a framework for community action in the field of water policy lays down a strategy against the pollution of water. Directive 2008/105/EC [18], amended by Directive 2013/39/EU [19] on environmental quality standards in the field of water policy, lays down environmental quality standards in accordance with the EWFD, for the 33 priority substances. The EWFD also sets out general provisions for the protection and conservation of groundwater. The establishment of detailed quality criteria for the assessment of groundwater chemical status in Europe was laid down in the European Groundwater Directive [20]. For this reason, the EU established the following groundwater quality standards: 0.1 $\mu g\ L^{-1}$ for individual pesticide and 0.5 $\mu g\ L^{-1}$ for the sum of all individual pesticides to safeguard people from harmful effects.

During the 1990s of the last century, atrazine (herbicide) and endosulfan (insecticide) were found most often in surface waters in the USA and Australia due to their widespread use. In addition, although in lower proportions, other pesticides such as pronofos, dimethoate, chlordane, diuron, prometryn and/or fluometuron were detected [21]. Studies that are more recent also reported the presence of several pesticides in environmental waters (surface water, groundwater and seawater) close to agricultural lands over the world [22–34]. In addition, different studies have corroborated the presence of some pesticides in drinking water [35–38].

Pesticides are being continuously released into the aquatic environment through anthropogenic activities. Their detection in storm and wastewater effluent has been reported to be a major obstacle as regards wide-ranging acceptance of water recycling [39]. In addition, their variety, toxicity and persistence present a threat to humans through pollution of drinking water resources (e.g. surface water and groundwater). The frequent occurrence of pesticides in surface water and groundwater has prompted the search for suitable methods to destroy them. Although conventional biological treatments of water offer some advantages such as their low cost and easy operation, most studies concerning the treatment of pesticides have concluded that they are not very effective due to their low biodegradability [40, 41]. Other technologies such as adsorption or coagulation merely concentrate pesticides by transferring them in other phases but still remain and not being completely eliminated. To solve this problem, apart from reducing emissions, two main water strategies are followed: (i) chemical treatment of drinking water, surface water and groundwater and (ii) chemical treatment of wastewaters containing biocides and bio-recalcitrant pollutants as pesticides. Chemical treatments of polluted surface water, waste water and groundwater are part of a long-term strategy to improve the quality of water by removing toxic compounds of anthropogenic origin before returning the water to its natural cycle. The Directive 2013/39/EU [19] promotes the preventive action and the polluter pays principle, the identification of pollution causes, dealing with emissions of pollutants at the source, and finally the development of innovative wastewater treatment technologies, avoiding expensive solutions. Therefore, effective, low-cost and robust methods to decontaminate waters are needed, as long as they do not further stress the environment or endanger human health, particularly prior to direct or indirect reuse of reclaimed water. In this context, the development of Solar Chemistry Applications is of special relevance, especially photochemical processes where solar photons are absorbed by reactants and/or a catalyst causing a chemical reaction. Consequently, in recent years there has been growing interest in the use of advanced oxidation processes (AOPs) to remove pesticide residues as alternative to methods that are more conventional because they allow the abatement of them by mineralisation.

2. Advanced oxidation processes to remove pesticides from water

AOPs have been commonly defined as *near-ambient temperature treatment processes based on highly reactive radicals*, especially the hydroxyl radical ($^{\bullet}OH$). Other radicals and active oxygen species involved are superoxide radical anions ($O_2^{\bullet -}$), hydroperoxyl radicals ($HO_2^{\bullet -}$), triplet oxygen ($3O_2$) and organic peroxyl radicals (ROO^-). In all probability, the $^{\bullet}OH$ (E_0 = 2.8 V) is among the strongest oxidising species used in water treatment and confers the potential to greatly accelerate the rates of pesticide oxidation. Hydroxyl radicals can degrade indiscriminately micropollutants with reaction rate constants usually around 109 L mol^{-1} s^{-1} [42], yielding CO_2, H_2O and, eventually, inorganic ions as final products. After fluorine (E_0 = 3.1 V), $^{\bullet}OH$ is the strongest oxidant [43], and its production can be achieved by many pathways, which allows one to choose the appropriate AOP according to the specific characteristics of the target water/wastewater and treatment requirements. Regarding the methodology to generate hydroxyl radicals, AOPs can be divided into chemical, electrochemical, sono-chemical and photochemical processes. Typical AOPs can be also classified as homogeneous that occur in a single phase and heterogeneous processes because they make use of a heterogeneous catalyst like metal-supported catalysts, carbon materials or semiconductors such as TiO_2, WO_3, ZnO, CdS, SnO_2, ZnS and others [44]. **Figure 1** shows some homogeneous and heterogeneous processes [41, 45].

When the chemical process (mineralisation) destroys the contaminants and their reaction intermediate products (metabolites), critical secondary wastes are not generated and, thus, post-treatment or final disposal is not required [46]. However, if complete mineralisation is not achieved or the reaction period is too long, a final post-treatment may be necessary. A higher biodegradability and lower toxicity of the reaction by-products, in comparison with the parent compounds, are desirable benefits of applying AOPs to treat wastewaters. However, in some cases, these by-products are less biodegradable and/or more toxic than the parent compounds. For this reason, AOPs can be applied as post- or pretreatment of biological processes. The integration of different AOPs in a sequence of treatment processes is a common

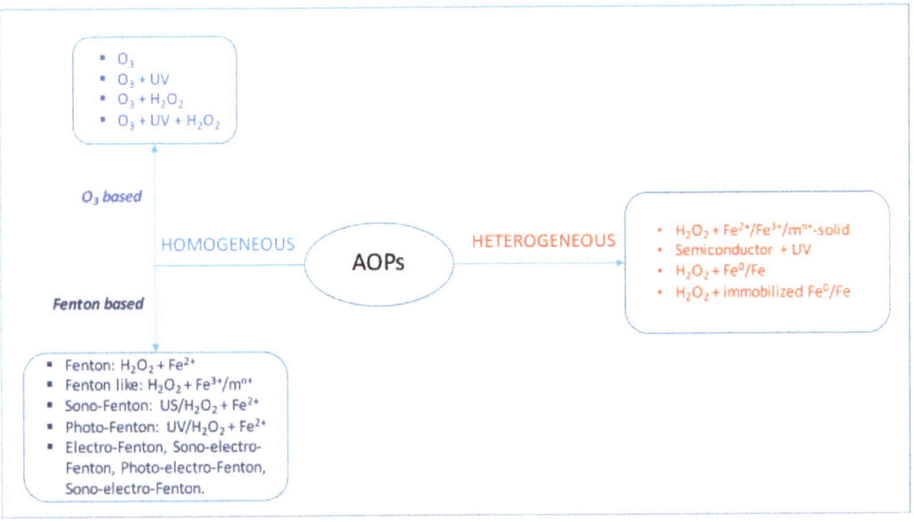

Figure 1. Scheme for conventional AOPs.

approach to achieve a biodegradable effluent, which can be further treated by a conventional biological process, reducing the residence time and reagent consumption in comparison with AOPs alone. **Figure 2** shows possible integration of AOPs in wastewater and drinking water treatment plants.

One drawback of these processes is the presence of scavengers in wastewaters because these species consume •OH, competing with pesticides. They can be organic matters (e.g. humic and/or fulvic acids, amino acids, proteins and carbohydrates) or inorganic ions ($CO_3^{=}$, HCO_3^-, $S^=$, Br^-, NO_3^- and others). Because most natural waters contain these scavengers, optimisation of AOPs must be performed bearing them in mind. The knowledge of their effect on the process efficiency is difficult because they have different reactivities, as well as due to the constant variations in the aqueous phase when the parent pollutants are continuously transformed into many different intermediates. Among these techniques, photocatalytic methods in the presence of artificial or solar light, like heterogeneous photocatalysis (HP), have been proven very effective for the degradation of a wide range of pesticides [48].

2.1. Basis of heterogeneous photocatalysis

Photocatalysis may be defined as the *acceleration of a photoreaction by the presence of a catalyst*. HP, *the use under irradiation of a stable solid semiconductor for stimulating a reaction at the solid/ solution interface*, is a technique of environmental interest for the treatment of pesticide-polluted water combining the low cost, the mild conditions and the possibility of using natural sunlight as the source of irradiation [39, 49]. Progress and challenges of HP can be reviewed in recent published papers [50, 51]. In brief, HP is based on the irradiation of semiconductor (SC) particles, usually suspended in aqueous solutions, with wavelength energy $h\nu \geq E_g$ (band-gap energy). Thus, an electron (e^-) is driven to the conduction band (cb), remaining a positive hole (h^+) in the valence band (vb). Both the e^- and h^+ migrate to the particle surface (**Figure 2**). The e^-_{cb} and the h^+_{vb} can recombine on the surface or in the bulk of the particle in

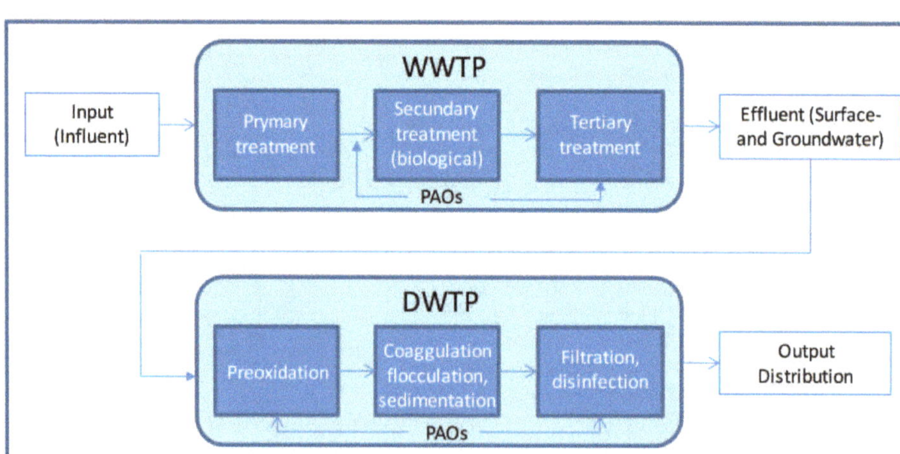

Figure 2. Possible implementation of AOPs in waste water and drinking water treatment plants (adapted from Petrovic et al. [47].

short time, and the energy dissipated as heat. In addition, they can be trapped on the surface reacting with donor (D) or acceptor (A) species adsorbed or close to the surface of the particle [52] as can be seen in **Figure 3**. The wavelength (λ) of radiation required to activate the catalyst must be equal or lower than the calculation by Planck's equation, $\lambda = hc/E_g$, where h is Planck's constant (6.626176×10^{-34} J s^{-1}), c is the speed of light and E_g is the semiconductor band-gap energy.

As a rule, TiO_2 is considered the best photocatalyst due to different qualities such as high photochemical stability, high efficiency, non-toxic nature and low cost, whose behaviour is very well documented in the recent literature [53, 54]. Excellent reviews have been published during the last years on the photoactivity of TiO_2 to purify pesticide-polluted waters [39, 48, 55–60].

2.2. Properties and characteristics of titanium dioxide photocatalysts

Titanium dioxide is by far, the most investigated photocatalyst to remove organic pollutants from water. The photocatalytic activity of TiO_2 slurries depends on physical properties of the catalyst (crystal and pore structure, surface area, porosity, band gap, particle size and surface hydroxyl density) [61, 62]. On the other hand, operating conditions such as light intensity and wavelength, initial concentration and type of pollutants, catalyst loading, oxygen content, interfering substances, presence of oxidants/electron acceptor, pH value and configuration of photoreactor have a key role [39, 58, 63–66]. Finally, the mode of TiO_2 application (suspended, immobilised or doped) is fundamental to rate the photocatalytic activity.

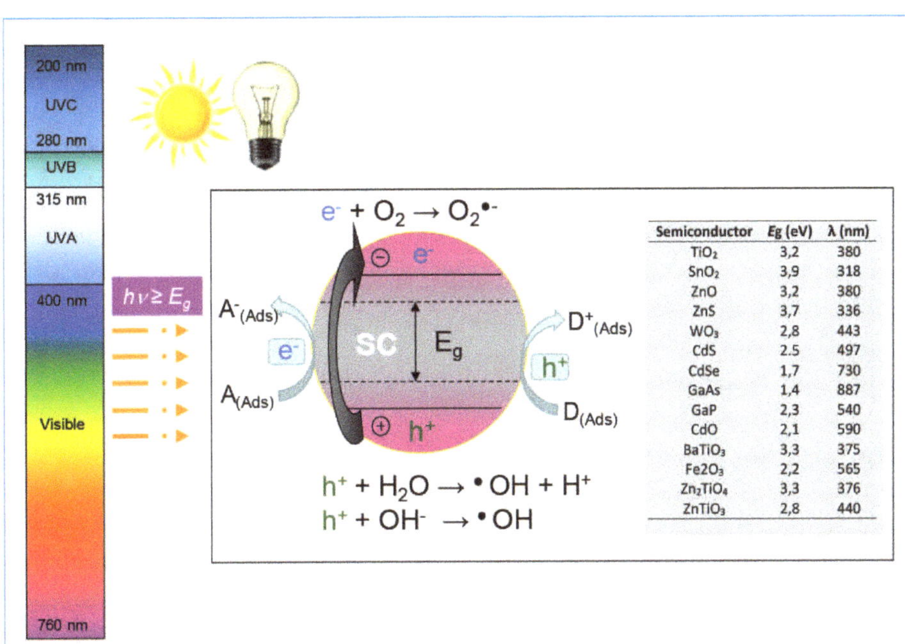

Figure 3. Scheme for the heterogeneous photocatalysis.

2.2.1. *Composition of TiO_2 and its types*

Titanium dioxide is known to occur in nature as anatase (At), brookite (Bk) and rutile (Rl) (**Figure 4**). Rl is usually considered to be the high-temperature and high-pressure phase relative to At, whereas Bk is often considered to be of secondary origin. Rl is the most common, most stable and chemically inert and can be excited by both visible and ultraviolet (UV) light (wavelengths smaller than 390 nm) [67]. At is only excited by UV light and can be transformed into Rl at high temperatures. Both Rl and At have a tetragonal ditetragonal dipyramidal crystal system but have different space group lattices. UV light does not excite Bk, but its orthorhombic crystal system can be transformed into Rl with the application of heat.

Generally, At exhibits higher photocatalytic activities than Rl. However, the reasons for the differences in photocatalytic activity between At and Rl are still being debated. Although At has lower absorbance ability to solar light than Rl because its band gap is larger (3.2 eV) than that of Rl (3.0 eV), the photocatalytic activity of At is higher than that of Rl. This can be explained because At has a higher surface adsorption capacity to hydroxyl groups and a lower charge carrier recombination rate than Rl [68]. Also, the lower photocatalytic activity of Rl is due to its larger grain size [69], lower specific surface areas and less capacity for surface adsorption. In addition, the lifetime of photo-generated e^- and h^+ in Rl is about an order of magnitude smaller than that of At. Consequently, the chance of participation of photoexcited e- and h+ of At in surface chemical reactions is greatly enhanced. According to Zhang et al. [70], Rl and Bk

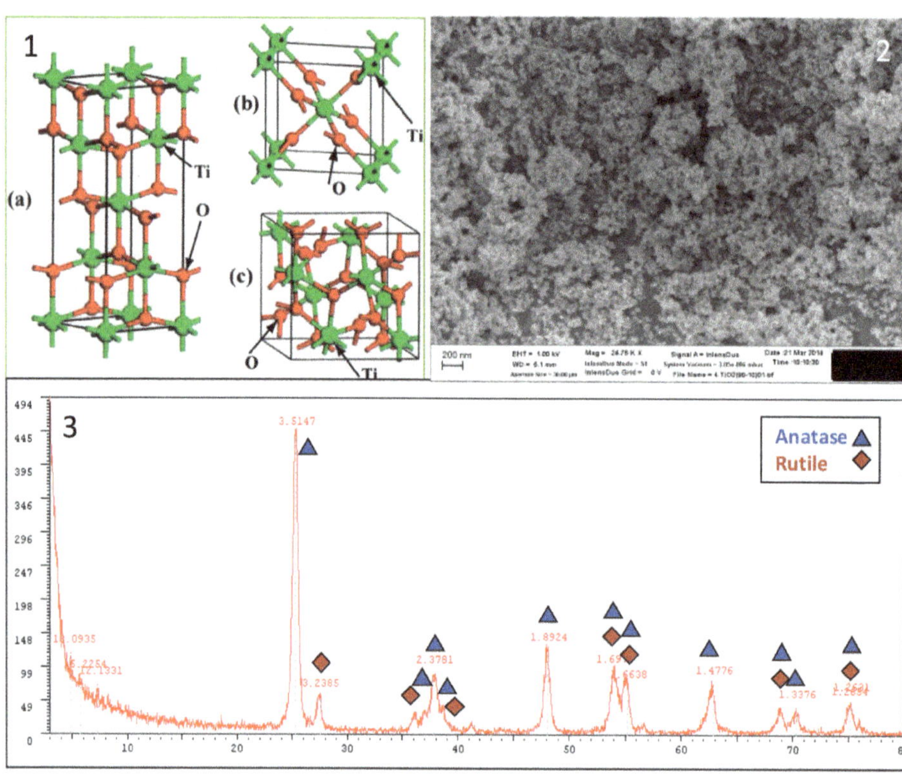

Figure 4. (1) The schematic conventional cells for anatase (a), rutile (b) and brookite (c) phases [73]; (2) SEM image of TiO_2 P25 and (3) XRD pattern of TiO_2 P25.

belong to the direct band-gap SC category, while At appears to be an indirect band-gap SC. As a consequence, At exhibits a longer lifetime of photoexcited e^- and h^+ than Rl and Bk because the direct transition of photo-generated electrons from the conduction band to valence band of At is not possible. Moreover, At has the lightest average effective mass of photo-generated e^- and h^+ than Rl and Bk suggesting the fastest migration of photo-generated e^- and h^+ from the interior to the surface of At. This results in the lowest recombination rate of photo-generated charge carriers within At. Therefore, it is not surprising that Rt and Bk show a smaller photocatalytic activity than At.

TiO_2 can use natural sunlight because it has an appropriate energetic separation between its valence and conduction bands which can be surpassed by the energy content of a solar photon (380 nm > λ >300 nm). The sunlight puts 0.2–0.3 mol photons of m^{-2} h^{-1} in the 300–400 nm range with a typical UV flux of 20–30 W m^{-2}. In addition, photons can be generated by artificial irradiation although it is the most important source of costs during the treatment of wastewater [65]. Different available TiO_2 catalysts (with different surface areas, crystal sizes and compositions) such as Degussa P25, Hombikat UV100, PC500, PC10, PC50, Rhodia and others have been tested for the photolytic degradation of pesticides in aqueous environments [39]. From them, Degussa P25 has been the most used because it has good properties (i.e. typically a 70:30 At:Rl composition, non-porous, Brunauer, Emmett and Teller (BET) around 55 m^2 g^{-1} and average particle size 30 nm) and a substantially higher photocatalytic activity than other commercial TiO_2 [65]. The higher photocatalytic activity of P25 has been attributed to its crystalline composition of Rl and At. It is known that the smaller band gap of rutile absorbs the photons and generates e^-/h^+ pairs. Then, the electron transfer takes place from the rutile to electron traps in the At phase. Thus, the recombination is inhibited and allows the hole to move to the surface of the particle to react [71].

2.2.2. Operating conditions

The e^-/h^+ formation in the photochemical reaction is strongly dependent on the light intensity at a given wavelength [72]. Therefore, the dependency of pollutant degradation rate on the light intensity has been studied in numerous investigations of various organic pollutants. According to Herrmann [73], the reaction rate is proportional to the radiant flux (ϕ) < 25 mW cm^{-2}, while above this value, the rate varies as $\phi^{1/2}$, which indicates a too high value of the flux increasing the e^-/h^+ recombination rate. When the intensity is high, the reaction rate does not depend on light intensity because at low intensity, reactions involving e^-/h^+ formation are predominant, while e^-/h^+ recombination is not significant [74].

Several authors have indicated that when the level of the target pesticide increases, a large number of molecules of the compound are adsorbed on the photocatalyst surface and, consequently, the reactive species ($^{\bullet}OH$ and $O_2^{\bullet-}$) required for pesticide degradation also increase. However, the formation of $^{\bullet}OH$ and $O_2^{\bullet-}$ on the catalyst surface remains constant for a given catalyst amount, light intensity and irradiation time. Hence, at higher concentrations the available $^{\bullet}OH$ is inadequate for pollutant degradation. Therefore, as the concentration increases, the pollutant degradation rate decreases [75]. In addition, an increase in pesticide concentration leads to the generation of intermediates (metabolites), which may be adsorbed on the surface of the catalyst.

Because TiO_2 is often used as suspension, the photocatalytic degradation rate initially increases with catalyst loading and then decreases at high concentrations due to light scattering and screening effects. Although the number of active sites in solution will increase with catalyst loading, light penetration is compromised because of excessive particle concentration. The interaction between particles (agglomeration) increases at high concentration, and, consequently, the surface area available for light adsorption is reduced and photocatalytic activity decreases. The optimum catalyst loading has to be found in order to avoid excess catalyst and ensure the maximum absorption of photons. Although the results in the literature are very different, it may be deduced that the incident radiation and path length inside the photoreactor are of special interest in determining the optimum catalyst mass.

Waste water, surface water, groundwater and drinking water pHs vary significantly and play an important role in the photodegradation of pesticides since it determines the size of aggregates it forms and the surface charge of the photocatalyst. The surface charge of the photocatalyst and the ionisation or speciation (pK_a) of a pesticide can be seriously affected by the solution pH. Electrostatic interaction between the surface of the semiconductor, substrate, solvent molecules and radicals formed during photocatalytic oxidation strongly depends on the solution pH. At pH below its pK_a value, an organic compound exists as neutral state. Above this pK_a value, organic compounds attain a negative charge, which can significantly influence their photocatalytic degradation. The point of zero charge (PZC, the pH at which the surface has a neutral net electrical charge) of TiO_2 is not very sensitive to the crystallographic structure (At *vs* Rl) and the experimental method used. The common value is pH 5.9 for both phases. Although the PZC of TiO_2 depends on the production method, the most frequent value for TiO_2 P25 is 6.3 [76]. Below or above this value, the charge of the catalyst surface is positive or negative, respectively, according to the following reactions:

$$\mathrm{pH} < \mathrm{pzc}\ \mathrm{TiOH} + \mathrm{H}^+ \rightarrow \mathrm{TiOH}_2^+, \quad \mathrm{pH} > \mathrm{pzc}\ \mathrm{TiOH} + \mathrm{OH}^- \rightarrow \mathrm{TiO}^- + \mathrm{H_2O} \tag{1}$$

At low pH, the positive holes are considered as the major oxidation step, whereas at neutral or high pHs, $^\bullet OH$ is the predominant species [77]. It is expected that the generation of $^\bullet OH$ will be higher due to the presence of more available OH^- on the TiO_2 surface. Thus, the degradation efficiency of the process will be enhanced at high pH. A very important feature of the photocatalytic process is that in many cases, a great number of different metabolites are produced, which may behave in a different way depending on the pH of the solution. As regards temperature, photocatalytic systems do not require heating and operate at room temperature because of photonic activation [65].

The e^-/h^+ recombination is one of the main drawbacks in the application of semiconductor photocatalysis as it causes waste of energy. In the absence of suitable electron acceptor, recombination step is predominant, and thus, it limits the quantum yield [39]. In HP reactions, O_2 is generally used as electron acceptor. Addition of exogenous oxidant/electron acceptors into a semiconductor suspension has been shown to improve the photocatalytic degradation of many pesticides because they can eliminate the e^-/h^+ recombination by accepting the conduction band electron, increase the $^\bullet OH$ concentration and oxidation rate of intermediate compound and produce more radicals and oxidising species to accelerate the degradation efficiency of intermediate compounds. Because •OH plays an important role in photodegradation, several researchers have investigated the effect of addition of different electron acceptors (i.e. H_2O_2, $KBrO_3$ or $Na_2S_2O_8$) on the photocatalytic degradation of many pesticides [75, 78].

The ability of peroxydisulfate is not only attributed to the promotion of charge separation but also to the production of sulphate radicals ($SO_4^{\bullet -}$), which are very strong oxidising agents (E_0 =2.6 V), and the appearance of more hydroxyl radicals (•OH) according to the following reactions:

$$S_2O_8^{=} + e^- \rightarrow SO_4^{=} + SO_4^{\bullet -} \quad (2)$$

$$SO_4^{\bullet -} + e^- \rightarrow SO_4^{=} \quad (3)$$

$$SO_4^{\bullet -} + H_2O \rightarrow SO_4^{=} + {}^{\bullet}OH + H^+ \quad (4)$$

$$SO_4^{\bullet -} + RH \rightarrow \text{Intermediates} \rightarrow SO_4^{=} + CO_2 \quad (5)$$

Besides, the addition of H_2O_2 enhances the degradation due to the increase in the •OH concentration as follows:

$$H_2O_2 + e^- \rightarrow {}^{\bullet}OH + OH^- \quad (6)$$

$$H_2O_2 + O_2^{\bullet -} \rightarrow {}^{\bullet}OH + OH^- + O_2 \quad (7)$$

$$H_2O_2 + hv \rightarrow 2\,{}^{\bullet}OH \quad (8)$$

The quantum yield of $S_2O_8^{2-}$ (1.8 mol Einstein^{-1}) is much larger than that of H_2O_2 (1 mol Einstein^{-1}), which can be related to the rate of recombination of •OH (5.3×10^9 M^{-1} s^{-1}) and $SO_4^{\bullet -}$ (8.1×10^8 M^{-1} s^{-1}) [79].

As reviewed by Ahmed et al. [39], many studies have demonstrated that water components like Ca^{2+}, Mg^{2+}, Fe^{2+}, Zn^{2+}, Cu^{2+}, HCO_3^-, PO_4^{3-}, NO_3^-, SO_4^{2-} and Cl^- and dissolved organic matter (DOM) can affect the photodegradation rate of organic pollutants since they can be adsorbed onto the surface of TiO_2 [80, 81]. These dissolved components can compete with the pesticide for the active sites depending on the solution pH, reducing the formation of •OH. Anions result in corresponding anion radicals scavenged by •OH. However, they have lower oxidation potential. In addition, DOM, ubiquitously present in storm and wastewater effluent, also plays an important role regarding pesticide degradation. The observed slowdowns are related to the inhibition (surface deactivation), competition and light attenuation effects. Moreover, the presence of humic acids in the reaction solution has been reported to significantly reduce light transmittance and consequently the photooxidation rate.

As usual, two types of photoreactors are used for photocatalytic wastewater treatment processes: (i) slurry photoreactors and (ii) fixed-bed photoreactors [82]. The slurry photoreactors utilise suspended photocatalyst particles, while the other type utilises immobilised photocatalyst particles on a surface. Currently, the catalysts are applied in the form of a slurry in most cases. However, the separation of catalyst after the reaction in the slurry systems is an expensive and tedious stage, which adds to the overall running costs of the plant. Therefore, immobilised catalyst systems are preferred in order to avoid an increase in cost and time. However, the immobilised catalyst systems have low interfacial surface areas and consequently a very low activity, being also difficult to scale [83]. It is noted that configuration of photoreactor has an important role in the efficiency of the photocatalytic wastewater treatment processes.

2.2.3. Bare, doped and immobilised application of TiO_2

Generally, the use of TiO_2 slurries has been demonstrated to have higher photocatalytic activity as compared to the same immobilised catalyst. This is due to changes on the surface of the catalyst by blocking pores and the appearance of by-products causing the loss of active sites on its surface. Usually, TiO_2 is prepared in the form of nanopowders, crystals, thin films, nanotubes and nanorods. As a rule, the immobilisation of TiO_2 onto supporting material has been carried out via one of two major routes: (i) physical (thermal treatment) and (ii) chemical (sol-gel, electrodeposition, etc.). The evolution of different supports and the benefits and drawbacks of various immobilisation techniques to obtain a high-surface-area TiO_2 can be seen in the reviews by Shan et al. [84] and Dahl et al. [85]. Nonetheless, the interest for the development of TiO_2 supported on different materials is growing because the use of the bare TiO_2 phases presents some drawbacks as (i) necessity of irradiation with UV light due to the small amount of photons absorbed in the Vis region, (ii) high recombination rate for the photoproduced e^-/h^+ pairs, (iii) difficulty to improve the performance by doping with some materials that often act as recombination centres, (iv) deactivation in the absence of H_2O vapour when aromatic molecules must be abated and (v) difficulty to support powdered TiO_2 on some materials [45]. Consequently, the research line in HP has been driven to modify some electronic and morphological properties of TiO_2 to enhance its photoefficacy. In this context, nanosized particles and films on glasses or other supports and powdered samples with high specific surface areas have been obtained to increase the possibility for the involved species to avoid the separation step [86]. Doping, loading and sensitisation of TiO_2 are methods aimed to shift the light absorption towards visible light and/or to increase the lifetime of the photoproduced e^-/h^+ pairs.

A number of approaches have been suggested in recent years to enhance photocatalytic activity of TiO_2 in the visible light region for its use in water detoxification [87]. Metal ion doping and co-doping with non-metals can improve trapping of the photoexcited conduction band electrons at the surface, thereby minimising charge carrier recombination. Several dopants used (e.g. Sn, Ag, Pd, Re, Bi, V, Mo, Th or Pt among others), have been shown to enhance photocatalytic activity for the systems examined. However, the photoactivity of the metal-doped TiO_2 photocatalyst significantly depends on the dopant ion nature and concentration, preparation method and operating conditions [88]. The deposition of metal ions on TiO_2 can modify the photoconductive properties by increasing the charge separation efficiency between electrons and holes, which will enhance the formation of both free hydroxyl radicals and active oxygen species [89]. Recent research indicates that the desired narrowing on the band gap of TiO_2 can be achieved using non-metal elements such as N, F, S and C. Thus, modified TiO_2 showed a significant improvement on the absorption in the Vis light region due to band-gap narrowing and enhancing the degradation of pesticides under Vis light irradiation, mainly under sunlight [90, 91]. Also, it is possible to produce coupled colloidal structures using TiO_2 such as TiO_2-SnO_2, TiO_2-CdS, TiO_2-Bi_2S_3, TiO_2-WO_3 or TiO_2-Fe_2O_3, in which illumination of one semiconductor produces a response in the other at the interface between them by increasing the charge separation and extending the energy range of photoexcitation. On the other hand, the coating of one semiconductor or metal nanomaterial on the surface of another semiconductor or metal nanoparticle core is called capping. Semiconductor nanoparticles of TiO_2 can be coated with another semiconductor (i.e. SnO_2) with a different band gap to enhance its emissive properties [92].

2.3. Heterogeneous photocatalytic degradation of pesticide residues in water over titanium dioxide

A desirable feature in the photodegradation of pesticides in water is the transformation of the parent compounds in order to avoid their toxicity. However, the main objective is the mineralisation of the pesticides. As previously commented, $^{\bullet}OH$ is the main species involved for organic substrate oxidation, but the free radical $HO_2^{\bullet}$ and its conjugate $O_2^{\bullet-}$ also play an important role although those radicals are much less reactive than $^{\bullet}OH$. All these free radicals react with pesticides by hydrogen abstraction or electrophilic addition to double bonds. Further, the radicals react with O_2 to give organic peroxyl radicals (ROO^-) initiating different oxidative reactions that may lead to the complete mineralisation of the pesticides. Since $^{\bullet}OH$ is non-selective, numerous and different transformation products (intermediates) can be formed at low concentrations being in certain cases more persistent and toxic than the parent compounds.

$$\text{Mineralization:} \rightarrow \text{Organic pesticide} \rightarrow \text{Intermediates} \rightarrow CO_2 + H_2O + Cl^- + NO_3^- + SO_4^{2-} + PO_4^{3-} \text{ etc.} \quad (9)$$

The presence of pesticide residues in water is usually monitored using chromatographic techniques such as gas chromatography (GC) and liquid chromatography (LC) coupled to mass spectrometry (MS^n) and time-of-flight (TOF) detection systems. However, since identification of all the transformation products generated during the photooxidation is not possible, the measure of total organic carbon (TOC) and more specifically dissolved organic carbon (DOC) is crucial in the process because determination of CO_2 must be stoichiometric with the organic carbon in the parent pesticide. This determination can be carried out in a simple and rapid way to know the mass balance and the remaining amount of metabolites. As example, **Figure 5** shows the photodegradation pathways proposed for chlorantraniliprole [93] and tebuconazole [94] dissolved in water when illuminated in the presence of TiO_2.

In the case of chlorantraniliprole, a new class of anthranilic diamide insecticide, several transformation products are generated during irradiation as a result of different reactions such as hydroxylation, deamination, hydrolysis of the amide bridge and rearrangement followed by cyclisation, methyl amine transfer and fragmentation. For tebuconazole, a common triazole fungicide with numerous agricultural and urban uses, the transformation pathway was found to proceed through tert-butyl chain cleavage, hydroxylation, oxidation and dechlorination and showed that its degradation mechanism was mainly driven by $^{\bullet}OH$ and h^+.

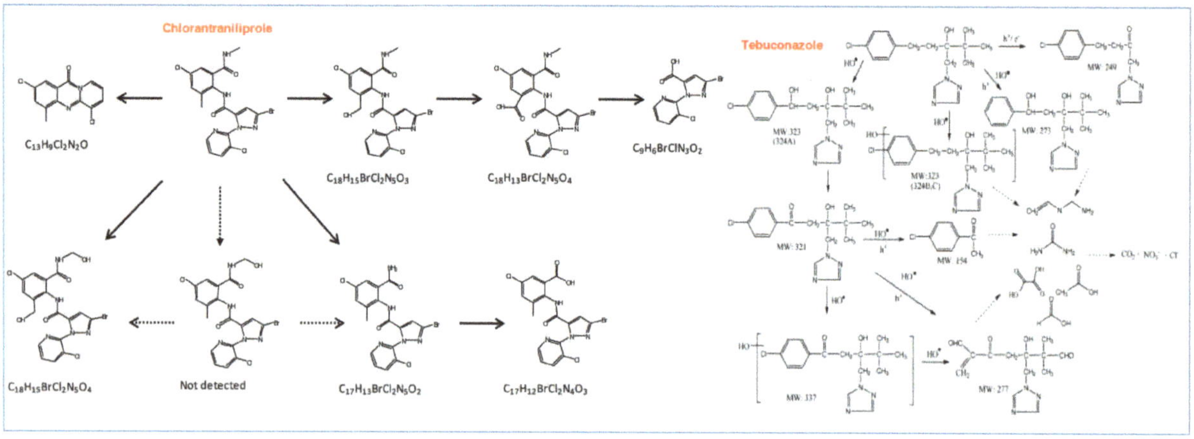

Figure 5. Photometabolic pathways proposed for chlorantraniliprole (insecticide) and tebuconazole (fungicide) in water slurries when illuminated in the presence of TiO_2.

The use of HP has showed a powerful growth in recent years. Today, pesticides constitute an important group concerning pollutant treatment. The large number of papers published in the last years proves this interest. A review to the literature extracted from the Web of Science™ (formerly ISI Web of Knowledge, www.isiknowledge.com) managed by Thomson Reuters (Philadelphia, USA) using the following keywords, *TiO_2, pesticides* and *water*, shows 463 papers only in the period 2005–2016.

Table 1 shows some of the most popular journal publishing on the topic "photocatalysis and pesticides" according to the following criteria: impact factor > 2 and Eigenfactor score > 0.01 (Journal Citation Reports (JCR) Science Edition 2015).

The *Journal Impact Factor* is defined as all citations to the journal in the current JCR year to items published in the previous 2 years, divided by the total number of scholarly items (these comprise articles, reviews and proceeding papers) published in the journal in the previous 2 years. The *Eigenfactor* score calculation is based on the number of times articles from the journal published in the past 5 years have been cited in the JCR year, but it also considers which journals have contributed these citations so that highly cited journals will influence the network more than lesser cited journals. References from one article in a journal to another article from the same journal are removed, so that *Eigenfactor* scores are not influenced by journal self-citation.

Following the criteria above, some of the most representative publications are presented in this summary (**Table 2**). Results show that sunlight photoalteration (photolysis) processes are well now to play an important role in the degradation of pesticides and other contaminants in the aquatic environment. These technologies allow the removal of pesticides by mineralisation. When the exciting energy used comes from the Sun, the process is called *solar photocatalysis* [95]. Photocatalytic oxidation by semiconductor oxides is an area of environmental interest for the treatment of polluted water, particularly relevant for Mediterranean agricultural areas, where solar irradiation is highly available making this process quite attractive. An ideal photocatalyst is characterised by photostability, biologically and chemically inert

Journal title (ISO)	Editorial/country	ISSN [a]	IF [b]	5Y-IF [c]	ES [d]
Chemical Reviews	American Chemical Society (USA)	0009-2665	37.369	51.560	0.24503
Journal of Photochemistry and Photobiology C: Photochemistry Reviews	Elsevier Science BV (Japan)	1389-5567	12.162	15.268	0.00442
Applied Catalysis B: Environmental	Elsevier Science BV (The Netherlands)	0926-3373	8.328	8.142	0.05714
Journal of Catalysis	Academic Press INC Elsevier Science (USA)	0021-9517	7.354	7.482	0.03438
Water Research	Pergamon-Elsevier Science Ltd. (England)	0043-1354	5.991	6.769	0.08394
Environmental Science and Technology	American Chemical Society (USA)	0013-936X	5.393	6.396	0.20232
Chemical Engineering Journal	Elsevier Science SA (Switzerland)	1385-8947	5.360	5.439	0.09715
Catalysis Science and Technology	Royal Society of Chemistry (England)	2044-4753	5.287	5.547	0.02367
Journal of Hazardous Materials	Elsevier Science BV (The Netherlands)	0304-3894	4.836	5.641	0.09715
Solar Energy Materials and Solar Cells	Elsevier Science BV (The Netherlands)	0927-0248	4.732	5.016	0.04275
Catalysis Today	Elsevier Science BV (The Netherlands)	0920-5861	4.312	4.105	0.03376
Applied Catalysis A: General	Elsevier Science BV (The Netherlands)	0926-860X	4.012	4.403	0.03476
Science of the Total Environment	Elsevier Science BV (The Netherlands)	0048-9697	3.976	4.317	0.08394
Journal of Molecular Catalysis A: Chemical	Elsevier Science BV (The Netherlands)	1381-1169	3.958	4.045	0.01727
Chemosphere	Pergamon-Elsevier Science Ltd. (England)	0045-6535	3.698	4.068	0.06483
Solar Energy	Pergamon-Elsevier Science Ltd. (USA)	0038-092X	3.685	4.414	0.02277
Journal of Environmental Management	Academic Press-Elsevier Science Ltd. (England)	0301-4797	3.131	4.049	0.03326
Catalysts	MDPI AG (Switzerland)	2073-4344	2.964	3.194	0.16915

Journal title (ISO)	Editorial/country	ISSN [a]	IF [b]	5Y-IF [c]	ES [d]
Journal of Agricultural and Food Chemistry	American Chemical Society (USA)	0021-8561	2.857	3.308	0.08776
Environmental Science and Pollution Research	Springer Heidelberg (Germany)	0944-1344	2.760	2.876	0.02617
Journal of Chemical Technology and Biotechnology	Wiley-Blackwell (England)	0268-2575	2.738	2.744	0.01051
Journal of Photochemistry and Photobiology A: Chemistry	Elsevier Science SA (Switzerland)	1010-6030	2.477	2.573	0.01025
Photochemical and Photobiological Science	Royal Society of Chemistry (England)	1474-905X	2.235	2.673	0.01040
Journal of Environmental Science	Science Press (Mainland China)	1001-0742	2.208	2.699	0.01282

[a]International Standard Serial Number.

[b]Impact factor.

[c]5-Year impact factor.

[d]Eigenfactor score.

Table 1. Some of the main and influential journals of interest to readers where the authors usually publish their works about 'photocatalytic degradation of pesticides in water' (source: JCR, 2015).

Pesticides	Photocatalysts	Light source	Main findings	References
Pyrimethanil	TiO_2 P25 and home made	Solar	Similar photoefficiences in mineralisation	[81]
Acephate, omethoate, methyl parathion	Fe_3O_4@SiO_2@mTiO_2 nanomicrospheres	UV	Disappearance after 45–80 min	[96]
Imidacloprid	TiO_2 and TiO_2-based hybrid	Solar	TiO_2 + fly ash is 2–3 times less active than sol-gel TiO_2	[97]
Spirotetramat, spirodiclofen, spiromesifen	ZnO, TiO_2 P25, TiO_2 Kronos vlp 7000, Zn_2TiO_4 and $ZnTiO_3$	UV	ZnO > TiO_2 P25 > TiO_2 Kronos vlp 7000 > Zn_2TiO_4 > $ZnTiO_3$	[98]
Flubendiamide	ZnO/$Na_2S_2O_8$ and TiO_2 P25/$Na_2S_2O_8$	UV	ZnO and TiO_2 oxides strongly enhance the degradation rate	[99]
Thiamethoxam, imidacloprid, acetamiprid	ZnO/$Na_2S_2O_8$ and TiO_2 P25/$Na_2S_2O_8$	UV/solar	Solar irradiation was more efficient compared to artificial light for the removal of these insecticides ($t_{½}$ = 0.3–2 min)	[100]
Chlorantraniliprole	ZnO, TiO_2 P25, TiO_2 Kronos vlp 7000, Zn_2TiO_4	UV	Half-lives of 53 and 71 min for ZnO/$Na_2S_2O_8$ and TiO_2/$Na_2S_2O_8$ systems, respectively	[93]

Pesticides	Photocatalysts	Light source	Main findings	References
Metamitron, metribuzin	Different TiO_2 nanopowders	UV	$t_{½}$ lower than 8 min using TiO_2 P25	[101]
Dichlorvos	TiO_2 supported on zeolites	UV	Complete degradation in 360–540 min	[102]
Atrazine	N-Doped TiO_2 supported by phosphors	UV	Enhanced performance compared to either pure N-TiO_2 nanoparticles or bare phosphors microparticles	[103]
Spinosad, indoxacarb	$ZnO/Na_2S_2O_8$ and TiO_2 P25/$Na_2S_2O_8$	UV	95% degradation rate was observed for both spinosyns after 2 min when using $Na_2S_2O_8$	[104]
Methabenzthiazuron	ZnO, TiO_2 P25	UV	Half-lives of 2 and 7 min for ZnO and TiO_2, respectively	[105]
Monocrotophos, endosulfan, chlorpyriphos	TiO_2 coated on polymeric beads	Solar	Rapid photodegradation using immobilised bead photoreactor	[106]
Isoproturon	TiO_2 coated on inert cement beads	UV	85% degradation rate after 6 h	[107]
Diazinon	TiO_2 doped with N, NS and FeFNS	UV LED	The FeFNS-doped TiO_2 was found to be an efficient catalyst (96% degradation after 100 min)	[108]
Methyl oxidemeton, methidathion, carbaryl, dimethoathe	TiO_2 P25, Hombikat UV 100, Millennium PC (50, 100, 105), Kronos 7101	UV	TiO_2 P25 was the most effective for pesticide degradation. Removal of pesticides in less than 300 min	[109]
30 sulfonylurea herbicides	ZnO and TiO_2 P25 in tandem with $Na_2S_2O_8$	UV	Average time required for 90% degradation about 3 and 30 min for $ZnO/Na_2S_2O_8$ and $TiO_2/Na_2S_2O_8$ systems, respectively	[110]
Chlorotoluron, diuron, fluometuron, isoproturon, linuron	ZnO, TiO_2, WO_3, SnO_2 and ZnS	Solar	The time required for 90% degradation ranged from 23 to 47 min for isoproturon and linuron, respectively, when using the tandem $ZnO/Na_2S_2O_8$	[111]
Carbofuran	ZnO, TiO_2 P25 Degussa	Solar	Half-lives ranging from 6 to 385 min	[112]
Malathion	WO_3/TiO_2		2% WO_3/TiO_2 exhibited the best photocatalytic activity achieving abatement of 76% TOC after 300 min	[113]
Ethoprophos, isoxaben, metalaxyl, metribuzin, pencycuron, pendimethalin, propanil, tolclofos-methyl	ZnO, TiO_2, WO_3, SnO_2, ZnS	Solar	Half-lives for ZnO ranged from 9 to 38 min (t_{30W} 0.3–1.7 min), while in the presence of TiO_2, ranged from 41 to 260 min (t_{30W} 1.9–16.3 min)	[114]
Simazine, prometryn, terbutryn, atrazine, terbuthylazine, propachlor, S-metolachlor, alachlor	ZnO, TiO_2	Solar	Degradation rate >70% after 240 min in the $ZnO/Na_2S_2O_8$ system	[115]

Pesticides	Photocatalysts	Light source	Main findings	References
Fenamiphos	ZnO, TiO_2, WO_3, SnO_2	Solar	Half-life <3 min for ZnO, TiO_2 in tandem with $Na_2S_2O_8$	[116]
Cyprodinil, fludioxonil	ZnO, TiO_2	Solar	DT_{75}, referred to the normalised illumination time (t_{30W}), was lower than 40 and 550 min (t_{30W} = 2 and 40 min) for both fungicides using ZnO and TiO_2, respectively	[117]
Methyl parathion, dichlorvos	N-doped and P25 TiO_2	UV/solar	N-doped TiO_2 showed higher photocatalytic activity under solar radiation compared to UV and visible light	[118]
MCPA, clopyralid, mecoprop	Fe- and N-doped TiO_2	UV	Lowering of the band gap of titanium dioxide by doping is not always favourable for increasing photocatalytic efficiency of degradation	[119]
2,4-D, Diuron, ametryne	TiO_2 slurry	Solar	Results showed that the overall model fitted the experimental data of herbicides mineralisation in the solar CPC reactor satisfactorily for both cloudy and sunny days	[120]
Carbendazim	TiO_2 slurry	UV	More than 90% of fungicide was degraded after 75 min	[121]
Glyphosate	TiO_2 slurry	UV	Addition of Fe^{3+}, Cu^{2+}, H_2O_2, $K_2S_2O_8$ or $KBrO_3$ enhances the photodegradation	[122]
Alachlor, atrazine, chlorfenvinphos, diuron, isoproturon, pentachlorophenol	TiO_2 slurry and Fe^{2+}	Solar	Photo-Fenton treatment was found to be shorter than TiO_2 and more appropriate for these compounds	[123]
Triclopyr, dantinozid	Different types of TiO_2 slurries in tandem with electron acceptors	Solar	The photocatalyst Degussa P25 was found to be more efficient as compared with other photocatalysts	[124]
Cymoxanil, methomyl, oxamyl, dimethoate, pyrimethanil, telone	TiO_2 slurry and Fe^{2+}	Solar	Total disappearance of the parent compounds and nearly complete mineralisation were attained with all pesticides tested	[125]
Dimethoate	ZnO, TiO_2	UV	Both catalysts were unable to mineralise dimethoate, but the addition of oxidants improved the efficiency of the processes	[126]
Dichlorvos, monocrotophos, parathion, phorate	$TiO_2 \cdot SiO_2$ beads	Solar	After 420 min illumination by sunlight, 0.65×10^{-4} mol dm^{-3} of four organophosphorus pesticides can be completely photocatalytically degraded into PO_4^{3-}.	[127]

Table 2. A brief summary of recent research studies in which TiO_2 was used for treating pesticide-polluted waters in the period 2005–2016.

nature, low cost and availability and capability to adsorb reactants under efficient photonic activation. Due to these characteristics, titanium dioxide (TiO_2) has been demonstrated to be an excellent catalyst, and its behaviour is very well documented for the photodegradation of pesticide residues in water.

2.4. Environmental impact and treatment cost

As previously stated, there is a very extensive literature (at laboratory and pilot plant scale) on the photocatalytic degradation of organic pollutants in water. However, there are not many works devoted to the study of the impact of the process from an environmental and economic point of view. In this context, the well-known life cycle impact assessment (LCIA) tool has been successfully used to assess the environmental impact of chemical processes. This tool finds the potential impacts associated with the entire life cycle of a product or a process. This methodology has been devised to study and compare processes at the industrial level but can be perfectly used at the laboratory level. On the other hand, there are several methods of estimating the costs of implementing each of the different AOPs. In general, the following items are proposed: (i) facility cost, (ii) project contingency, (iii) engineering project and (iv) replacement costs. The sum of these four concepts is the total installed cost, based on which the yearly economic impact can be evaluated. Then, operating costs have to be calculated. These costs are normally yearly and consist of the following items: (i) personnel, (ii) maintenance, (iii) electricity and (iv) materials and services. These costs added to the annual facility costs are the total annual costs [128].

3. Conclusion

Currently, population growth, technology development, inadequate agricultural practices and land use have created unprecedented water pollution problems. Agricultural wastewater is characterised by high organic matter content and traces of organic pollutants, mainly pesticides. In addition, effluents from the agro-food and other industries have a potentially environmental risk that requires appropriate and comprehensive treatment. The most often used methods for the treatment of the industrial wastewaters, including membrane filtration, chemical coagulation/flocculation, ion exchange, precipitation, adsorption, biological degradation and ozonation, are not efficient enough and have important limitations to remove bio-recalcitrant compounds as many pesticides. In recent years, advanced oxidation processes (AOPs) have been proposed as a most promising way for degradation of various pollutants. Among them, heterogeneous photocatalysis technology is an interesting route among AOPs, which can be conveniently used for the complete degradation of different hazardous compounds including pesticides. Results show that sunlight photoalteration processes are well now to play an important role in the degradation of pesticides and other contaminants in water. These technologies allow pesticides to be removed by mineralisation. Photocatalytic oxidation by semiconductor oxides is an area of environmental interest for the treatment of polluted water, particularly relevant for Mediterranean agricultural areas, where solar irradiation is highly available (more than 2800 h of sunshine per year on average) making this process quite attractive. Due to its specific characteristics, titanium dioxide (TiO_2) has been demonstrated to be an excellent catalyst, and its behaviour is very well documented for the photodegradation of pesticide residues in water. Recently, many authors have also developed combined AOP and biological systems to implement in wastewater treatment plants. In addition to the experimental and modelling work, the aspect most lacking of this combination systems for the treatment of bio-recalcitrant specific industrial wastewater is the performance of complete economic studies, which could present this innovative technology as a cost-competitive one.

Acknowledgements

The authors acknowledge the financial support received from the European Commission through the LIFE+ program (LIFE 13 ENV/ES/000488) and the EU's funding instrument for the environment and climate action.

Abbreviations

POPs	Persistent organic pollutants
EDCs	Endocrine-disrupting chemicals
OC	Organochlorine
EWFD	European Water Framework Directive
AOPs	Advanced oxidation processes
HP	Heterogeneous photocatalysis
E_g	Band-gap energy
D	Donor
A	Acceptor
cb	Conduction band
e^-	Electron
h^+	Hole
vb	Valence band
SC	Semiconductor
At	Anatase
Bk	Brookite
Rl	Rutile
SEM	Scanning electron microscopy
XRD	X-ray diffraction
UV	Ultraviolet

BET	Brunauer, Emmett and Teller
PZC	Point of zero charge
TOC	Total organic matter
DOM	Dissolved organic matter
GC	Gas chromatography
LC	Liquid chromatography
MS	Mass spectrometry
TOF	Time of flight
JCR	Journal Citation Reports
LCIA	Life cycle impact assessment
EU	European Union
ISSN	International Standard Serial Number
IF	Impact factor
5Y-IF	5-year impact factor
ES	Eigenfactor score

Author details

Nuria Vela[1], Gabriel Pérez-Lucas[2], José Fenoll[3] and Simón Navarro[2*]

*Address all correspondence to: snavarro@um.es

1 Applied Technology Group to Environmental Health, Faculty of Health Science, Catholic University of Murcia, Murcia, Spain

2 Department of Agricultural Chemistry, Geology and Pedology, Faculty of Chemistry, University of Murcia, Murcia, Spain

3 Sustainability and Quality Group of Fruit and Vegetable Products, Murcia Institute of Agri-Food Research and Development, Murcia, Spain

References

[1] Food and Agriculture Organization (FAO) of the United Nations. International Code of Conduct on the Distribution and Use of Pesticides. [Internet]. 2003. Available from: ftp://ftp.fao.org/docrep/fao/005/y4544e/y4544e00.pdf [Accessed: 24-01-2017]

[2] Briggs, J. Green revolution. In: Kitchin R, Thrift N, editors. International Encyclopedia of Human Geography. Amsterdam: Elsevier; 2009. pp. 634-638

[3] Carson, R. Silent Spring. Anniversary edition (2002 and 2012). Boston: Houghton Mifflin Harcourt; 2002. 400 p

[4] Food and Agriculture Organization (FAO) of the United Nations. Pesticides Registration Legislation. [Internet]. 1995. Available from: http://www.fao.org/docrep/012/T0553E/T0553E.pdf [Accessed: 24 January 2017]

[5] Shen L, Wania F, Lei YD, Teixeira C, Muir DCG, Bidleman TF. Atmospheric distribution and long-range transport behavior of organochlorine pesticides in North America. Environmental Science & Technology. 2005;**39**:409-420

[6] Dietz R, Riget FF, Sonne C, Letcher R, Born EW, Muir, DCG. Seasonal and temporal trends in polychlorinated biphenyls and organochlorine pesticides in East Greenland polar bears (*Ursus maritimus*), 1990-2001. Environmental Science & Technology. 2004;**331**: 107-124

[7] Mehemetli E, Koumanova B. The Fate of Persistent Organic Pollutants in the Environment. Dordrecht: Springer; 2008. 471 p

[8] European Community. Council Decision (2006/507/EC) of the 14 October 2004 concerning the conclusion of the Stockholm convention on persistent organic pollutants. Official Journal of the European Union. 2004;**L 209**:1-2

[9] Mnif W, Hassine AIH, Bouaziz A, Bartegi A, Thomas O, Roig B. Effect of endocrine disruptor pesticides: A review. International Journal of Environmental Research and Public Health. 2011;**8**:2265-2303

[10] Roig B, Mnif W, Hassine AI, Zidi I, Bayle S, Bartegi A, Thomas O. Endocrine disrupting chemicals and human risk assessment: A critical review. Critical Reviews in Environmental Science and Technology. 2013;**21**:2297-2351

[11] Giulivo M, López de Alda M, Capri E, Barceló D. Human exposure to endocrine disrupting compounds: Their role in reproductive systems, metabolic syndrome and breast cancer. A review. Environmental Research. 2016;**151**:251-264

[12] Food and Agriculture Organization (FAO) of the United Nations. FAOSTAT. Statistic Division. [Internet]. 2016. Available from: http://www.fao.org/faostat [Accessed: 24 January 2017]

[13] Pimentel D. Pesticides and pest control. In: Rajinder P, Dhawan A, editors. Integrated Pest Management: Innovation-Development Process. Vol. **1**. Dordrecht: Springer; 2009. pp. 83-87

[14] Liu CJ, Men WJ, Liu YJ, Hao Z. The pollution of pesticides in soils and its bioremediation. System Sciences and Comprehensive Studies in Agriculture. 2002;**18**:291-297

[15] Cai DW. Understand the role of chemical pesticides and prevent misuses of pesticides. Bulletin of Agricultural Science and Technology. 2008;**1**:36-38

[16] Aktar W, Sengupta D, Chowdhury A. Impact of pesticides use in agriculture: Their benefits and hazards. Interdisciplinary Toxicology. 2009;**2**:1-12

[17] The European Parliament and the Council of the European Union. Directive 2000/60/EC of the European Parliament and of the Council of 23 October 2000 establishing a framework for community action in the field of water policy, Official Journal of the European Community. 2000;**L 327**:1-69

[18] The European Parliament and the Council of the European Union. Directive 2008/105/EC of the European Parliament and of the Council of 16 December 2008 on environmental quality standards in the field of water policy, amending and subsequently repealing Council Directives 82/176/EEC, 83/513/EEC, 84/156/EEC, 84/491/EEC, 86/280/EEC and amending Directive 2000/60/EC of the European Parliament and of the Council. Official Journal of the European Community. 2008;**L 348**:84-97

[19] The European Parliament and the Council of the European Union. Directive 2013/39/EU of the European Parliament and of the Council of 12 August 2013 amending Directives 2000/60/EC and 2008/105/EC as regards priority substances in the field of water policy. Official Journal of the European Community. 2013;**L 226**;1-17

[20] The European Parliament and the Council of the European Union. Directive 2006/118/EC of the European Parliament and of the Council of 12 December 2006 on the protection of groundwater against pollution and deterioration. Official Journal of the European Community. 2006;**L 372**:19-31

[21] Cooper B. Central and North West Regions Water Quality Program 1995/1996: Report on Pesticide Monitoring. TS96.048. Sydney: NSW Department of Land & Water Conservation; 1996

[22] Cerejeira MJ, Viana P, Batista S, Pereira T, Silva E, Valério MJ, Silva A, Ferreira M, Silva-Fernandes AM. Pesticides in Portuguese surface and ground waters. Water Research. 2003;**37**:1055-1063

[23] Konstantinou IK, Hela DG, Albanis TA. The status of pesticide pollution in surface waters (rivers and lakes) of Greece. Part I. Review on occurrence and levels. Environmental Pollution. 2006;**141**:555-570

[24] Gilliom RJ. Pesticides in U.S. streams and groundwater. Environmental Science & Technology. 2007;**41**:3408-3414

[25] Arias-Estévez M, López-Periago E, Martínez-Carballo E, Simal-Gándara J, Mejuto JC, García-Río L. The mobility and degradation of pesticides in soils and the pollution of groundwater resources. Agriculture, Ecosystems & Environment. 2008;**123**:247-260

[26] Woudneh MB, Ou Z, Sekela M, Tuominen T, Gledhil, M. Pesticide multiresidues in waters of the Lower Fraser Valley, British Columbia, Canada. Part I. Surface water. Journal of Environmental Quality. 2009;**38**:940-947

[27] Añasco N, Un, S, Koyama J, Matsuoka T, Kuwahara N. Assessment of pesticide residues in freshwater areas affected by rice paddy effluents in Southern Japan. Environmental Monitoring and Assessment. 2010;**160**:371-383

[28] Jurado A, Vázquez-Suñe E, Carrera J, López de Alda M, Pujades E, Barceló D. Emerging organic contaminants in groundwater in Spain: A review of sources, recent occurrence and fate in the European context. Science of the Total Environment. 2012;**440**:82-94

[29] Lapworth DJ, Baran N, Stuart ME, Ward RS. Emerging organic contaminants in groundwater. A review of sources, fate and occurrence. Environmental Pollution. 2012;**163**: 287-303

[30] Bottoni P, Grenni P, Lucentini L, Caracciolo A. Terbuthylazine and other triazines in Italian water resources. Microchemical Journal. 2013;**107**:136-142

[31] Moreno-Gónzález R, Campillo JA, León VM. Influence of an intensive agricultural drainage basin on the seasonal distribution of organic pollutants in seawater from a Mediterranean coastal lagoon (Mar Menor, SE Spain). Marine Pollution Bulletin. 2013;**77**:400-411

[32] Grung M, Lin Y, Zhang H, Orderdalen Steen A, Huang J, Zhang G, Larssen T. Pesticide levels and environmental risk in aquatic environments in China—A review. Environment International. 2015;**81**:87-97

[33] Yadav IC, Devi NL, Syed JH, Cheng Z, Li J, Zhang G, Jones KC. Current status of persistent organic pesticides residues in air, water, and soil, and their possible effect on neighboring countries: A comprehensive review of India. Science of the Total Environment. 2015;**511**:123-137

[34] Ccanccapa A, Masia A, Navarro-Ortega A, Pico Y, Barceló D. Pesticides in the Ebro river basin: Occurrence and risk assessment. Environmental Pollution. 2016;**211**:414-424

[35] Dolan T, Howsam P, Parsons DJ. Diffuse pesticide pollution of drinking water sources: Impact of legislation and UK responses. Water Policy. 2012;**14**:680-693

[36] Kaushik CP, Sharma HR, Kaushik A. Organochlorine pesticide residues in drinking water in the rural areas of Haryana, India. Environmental Monitoring and Assessment. 2012;**184**:103-112

[37] Barbosa AMC, Solano M, Umbuzeiro G. Pesticides in drinking water—The Brazilian monitoring program. Frontiers in Public Health. 2015;**3**:Article 246

[38] Mekonen S, Argaw R, Simanesew A, Houbraken M, Senaeve D, Ambelu A, Spanoghe P. Pesticide residues in drinking water and associated risk to consumers in Ethiopia. Chemosphere. 2016;**162**:252-260

[39] Ahmed S, Rasul MG, Brown R, Hashib MA. Influence of parameters on the heterogeneous photocatalytic degradation of pesticides and phenolic contaminants in wastewater: A short review. Journal of Environmental Management. 2011;**92**:311-330

[40] Stanisakis AS. Use of selected advanced oxidation processes (AOPs) for wastewater treatment—A mini review. Global NEST Journal. 2008;**10**:376-378

[41] Ribeiro AR, Nunes OC, Pereira MFR, Silva AMT. An overview on the advanced oxidation processes applied for the treatment of water pollutants defined in the recently launched Directive 2013/39/EU. Environmental International. 2015;**75**:33-51

[42] Hoigné J. Inter-calibration of OH radical sources and water quality parameters. Water Science and Technology. 1997;**35**:1-8

[43] Pera-Titus M, García-Molina V, Baños MA, Giménez J, Espluga, S. Degradation of chlorophenols by means of advanced oxidation processes: A general review. Applied Catalysis B. 2004;**47**:219-256

[44] Di Paola A, García-López E, Marcì G, Palmisano, L. A survey of photocatalytic materials for environmental remediation. Journal of Hazardous Materials. 2012;**211**:3-29

[45] Babuponnusami A, Muthukumar K. A review on Fenton and improvements to the Fenton process for wastewater treatment. Journal of Environmental Chemical Engineering. 2014;**2**:557-572

[46] Andreozzi R, Caprio V, Insola A, Marotta R. Advanced oxidation processes (AOP) for water purification and recovery. Catalysis Today. 1999;**53**:51-59

[47] Petrovic M, Radjenovic J, Barceló D. Advanced oxidation processes (AOPs) applied for wastewater and drinking water treatment. Elimination of pharmaceuticals. The Holistic Approach to Environment. 2011;**1**:63-74

[48] Reddy PV, Kim K. A review of photochemical approaches for the treatment of a wide range of pesticides. Journal of Hazardous Materials. 2015;**285**:325-335

[49] Bhatkande DS, Pangarkar VG, Beenackers A. Photocatalytic degradation for environmental applications: A review. Journal of Chemical Technology and Biotechnology. 2002;**77**:102-116

[50] Fechete I. Wang, Y. Védrine C. The past, present and future of heterogeneous catalysis. Catalysis Today. 2012;**189**:2-27

[51] Qu Y, Duan X. Progress, challenge and perspective of heterogeneous photocatalysis. Chemical Society Reviews. 2013;**42**:2568-2580.

[52] Litter MI. Heterogeneous photocatalysis—Transition metal ions in photocatalytic systems. Applied Catalysis B: Environmental. 1999;**23**:89-114

[53] Serpone N, Sauve G, Koch R, Tahiri H, Pichat P, Piccinini P, Pelizzetti E, Hidaka H. Standardization protocol of process efficiencies and activation parameters in heterogeneous photocatalysis: Relative photonic efficiencies. Journal of Photochemistry and Photobiology A: Chemistry. 1996;**94**:191-203

[54] Lazar MA, Varghese S, Nair SS. Photocatalytic water treatment by titanium dioxide: Recent updates. Catalysts. 2012;**2**:572-601

[55] Konstantinou IK, Albanis TA. Photocatalytic transformation of pesticides in aqueous titanium dioxide suspensions using artificial and solar light: Intermediates and degradation pathways. Applied Catalysis B: Environmental. 2003;**42**:319-335

[56] Herrmann JM. Heterogeneous photocatalysis: State of the art and present applications. Topics in Catalysis. 2005;**34**:49-65

[57] Devipriya S, Yesodharan S. Photocatalytic degradation of pesticide contaminants in water. Solar Energy Materials Solar Cells. 2005;**86**:309-348

[58] Gaya UI, Abdullah AH. Heterogeneous photocatalytic degradation of organic contaminants over titanium dioxide: A review of fundamentals, progress and problems. Journal of Photochemistry and Photobiology C: Photochemistry Reviews. 2008;**9**:1-12

[59] Plantard G, Janin, T, Goetz V, Brosillon S. Solar photocatalysis treatment of phytosanitary refuses: Efficiency of industrial photocatalysts. Applied Catalysis B: Environmental. 2012;**115**:38-44

[60] Reddy PAK, Reddy PVL, Kwon E, Kim KH, Akter T, Kalagara S. Recent advances in photocatalytic treatment of pollutants in aqueous media. Environment International. 2016;**91**:94-103

[61] Fujishima A, Zhang X, Tryk DA. TiO_2 photocatalysis and related surface phenomena. Surface Science Reports. 2008;**63**:515-582

[62] Henderson MA. A surface science perspective on TiO_2 photocatalysis. Surface Science Reports. 2011;**66**:185-297

[63] Yu JG, Jimmy CY, Leung MKP, Zhao XJ, Ho WK, Zhao JC. Effects of acidic and basic hydrolysis catalysts on the photocatalytic activity and microstructures of bimodal mesoporous titania. Journal of Catalysis. 2003;**217**:69-78

[64] Kabra, K, Chaudhary R, Sawhney RL. Treatment of hazardous organic and inorganic compounds through aqueous-phase photocatalysis: A review. Industrial & Engineering Chemistry Research. 2004;**43**:7683-7696

[65] Malato S, Fernández-Ibáñez P, Maldonado MI, Blanco J, Gernjak W. Decontamination and disinfection of water by solar photocatalysis: Recent overview and trends. Catalysis Today. 2009;**147**:1-59

[66] Chong MN, Jin B, Chow CWK, Saint C. Recent developments in photocatalytic water treatment. Water Research. 2010;**44**:2997-3027

[67] Hashimoto K, Irie H, Fujishima A. TiO_2 photocatalysis: A historical overview and future prospects. Japanese Journal of Applied Physics. 2005;**44**:8269-8285

[68] Hanaor DAH, Sorrell CC. Review of the anatase to rutile phase transformation. Journal of Materials Science. 2011;**46**:855-874

[69] Zhang Z, Wang C, Zakaria R, Ying JY. Role of particle size in nanocrystalline TiO_2-based photocatalysts. Journal of Physical Chemistry B. 1998;**102**:10871-10878

[70] Zhang J, Zhou P, Liu J, Yu J. New understanding of the difference of photocatalytic activity among anatase, rutile and brookite TiO_2. Physical Chemistry Chemical Physics. 2014;**16**:20382-20386.

[71] Hurun DC, Agrios AG, Gray KA, Rajh T, Thurnaur MC. Explaining the enhanced photocatalytic activity of Degussa P 25 mixed-phase TiO_2 using EPR. Journal of Physical Chemistry B. 2003;**107**:4545-4549

[72] Cassano AE, Alfano OM. Reaction engineering of suspended solid heterogeneous photocatalytic reactors. Catalysis Today. 2000;**58**:167-197

[73] Herrmann JM. Heterogeneous photocatalysis: Fundamentals and applications to the removal of various types of aqueous pollutants. Catalysis Today. 1999;**53**:115-129

[74] Ollis DF, Pelizzetti E, Serpone N. Photocatalyzed destruction of wáter contaminants. Environmental Science and Technology. 1991;**25**:1522-1529

[75] Bahnemann W, Muneer M, Haque MM. Titanium dioxide-mediated photocatalysed degradation of few selected organic pollutants in aqueous suspensions. Catalysis Today. 2007;**124**:133-148

[76] Kosmulski M. Isoelectric points and points of zero charge of metal (hydr)oxides: 50 years after parks' review. Advances in Colloid and Interface Science. 2016;**238**:1-61

[77] Shifu C, Ghengyu C. Photocatalytic degradation of organophosphorus pesticides using floating photocatalyst TiO_2/SiO_2/beads by sunlight. Solar Energy. 2005;**79**:1-9

[78] Wei L, Shifu C, Wei Z, Sujuan Z. Titanium dioxide mediated photocatalytic degradation of methamidophos in aqueous phase. Journal of Hazardous Materials. 2009;**164**:154-160

[79] Kusic H, Peternel I, Ukic S, Koprivanac N, Bolanca T, Papic S, Loncaric Bozic A. Modeling of iron activated persulfate oxidation treating reactive azo dye in water matrix. Chemical Engineering Journal. 2011;**172**:109-121.

[80] Carbajo J, García-Muñoz P, Tolosana-Moranchel A, Faraldos M, Bahamonde A. Effect of water composition on the photocatalytic removal of pesticides with different TiO_2 catalysts. Environmental Science and Pollution Research. 2014;**21**:12233-12240.

[81] Carbajo J, Jiménez M, Miralles S, Malato S, Faraldos M, Bahamonde A. Study of application of titania catalysts on solar photocatalysis: Influence of type of pollutants and water matrices. Chemical Engineering Journal. 2016;**291**:64-73

[82] Kowalska E, Rau S. Photoreactors for wastewater treatment: A review. Recent Patents on Engineering. 2010;**4**:242-266

[83] Charles G, Roques–Carmes T, Becheikh N, Falk L, Commenge JM, Corbel S. Determination of kinetic constants of a photocatalytic reaction in micro–channel reactors in the presence of mass–transfer limitation and axial dispersion. Journal of Photochemistry and Photobiology A: Chemistry. 2011;**223**:202-211

[84] Shan AY, Ghazi TI, Rashid SA. Immobilisation of titanium dioxide onto supporting materials in heterogeneous photocatalysis: A review. Applied Catalysis A: General. 2010;**389**:1-8

[85] Dahl M, Liu Y, Yin Y. Composite titanium dioxide nanomaterials. Chemical Review. 2014;**114**:9853-9889

[86] Bourikas K, Kordulis C, Lycourghiotis A. Titanium dioxide (anatase and rutile): Surface chemistry, liquid–solid interface chemistry, and scientific synthesis of supported catalysts. Chemical Review. 2014;**114**:9754-9823

[87] Kumar SG, Devi LG. Review on modified TiO_2 photocatalysis under UV/visible light: Selected results and related mechanisms on interfacial charge carrier transfer dynamics. The Journal of Physical Chemistry A. 2011;**115**:13211-13241

[88] Dvoranova D, Brezova V, Mazur M, Malati MA. Investigations of metaldoped titanium dioxide titanium dioxide photocatalysts. Applied Catalysis B: Environmental. 2002;**37**:91-105

[89] Kato S, Hirano Y, Iwata M, Sano T, Takeuchi K, Matsuzawa S. Photocatalytic degradation of gaseous sulphur compounds by silver-deposited titanium dioxide. Applied Catalysis B: Environmental. 2005;**57**:109-115

[90] Peláez M, Nolan N, Pillai S, Seery M, Falaras P, Kontos AG, Dunlop PSM, Hamiltone JWJ, Byrne JA, O'Shea K, Entezari MH, Dionysiou DD. A review on the visible light active titanium dioxide photocatalysts for environmental applications. Applied Catalysis B: Environmental. 2012;**125**:331-349

[91] Chatterjee D, Dasgupta S. Visible light induced photocatalytic degradation of organic pollutants. Journal of Photochemistry and Photobiology C: Photochemistry Reviews. 2005;**6**:186-205

[92] Gupta SM, Tripathi M. A review of TiO_2 nanoparticles. Chinese Science Bulletin. 2011; **56**:1639-1657

[93] Fenoll J, Garrido I, Cava J, Hellín P, Flores P, Navarro S. Photometabolic pathways of chlorantraniliprole in aqueous slurries containing binary and ternary oxides of Zn and Ti. Chemical Engineering Journal. 2015;**264**:720-727

[94] Stamatis N, Antonopoulou M, Konstantinou I. Photocatalytic degradation kinetics and mechanisms of fungicide tebuconazole in aqueous TiO_2 suspensions. Catalysis Today. 2015;**252**:93-99

[95] Spasiano D, Marotta R, Malato S, Fernández-Ibáñez P, Di Somma I. Solar photocatalysis: Materials, reactors, some commercial and pre-industrialized applications. A comprehensive approach. Applied Catalysis B: Environmental. 2015;**170-171**:90-123

[96] Zheng LL, Pi FW, Wang YF, Xu H, Zhang YZ, Sun XL. Photocatalytic degradation of acephate, omethoate, and methyl parathion by Fe_3O_4@SiO_2@mTiO_2 nanomicrospheres. Journal of Hazardous Materials. 2016;**315**:11-22

[97] Andronic L, Isac L, Miralles-Cuevas S, Visa M, Oller I, Duta A, Malato S. Pilot-plant evaluation of TiO_2 and TiO_2-based hybrid photocatalysts for solar treatment of polluted water. Journal of Hazardous Materials. 2016;**320**:469-478

[98] Fenoll J, Garrido I, Hellín P, Vela N, Flores P, Navarro S. Photooxidation of three spirocyclic acid derivative insecticides in aqueous suspensions as catalyzed by titanium and zinc oxides. Journal of Photochemistry and Photobiology A: Chemistry. 2016;**328**:189-197

[99] Fenoll J, Vela N, Garrido I, Navarro G. Pérez-Lucas G, Navarro S. Reclamation of water polluted with flubendiamide residues by photocatalytic treatment with semiconductor oxides. Photochemistry & Photobiology. 2015;**91**:1088-1094

[100] Fenoll J, Garrido I, Hellín P, Flores P, Navarro S. Photodegradation of neonicotinoid insecticides in water by semiconductor oxides. Environmental Science and Pollution Research. 2015;**22**:15055-15066

[101] Vela N, Fenoll J, Garrido I, Navarro G, Gambín M, Navarro S. Photocatalytic mitigation of triazinone herbicide residues using titanium dioxide in slurry photoreactor. Catalysis Today. 2015;**252**:70-77

[102] Gómez S, Marchena CL, Renzini MS, Pizzio L, Pierella L. In situ generated TiO_2 over zeolitic supports as reusable photocatalysts for the degradation of dichlorvos. Applied Catalysis B: Environmental. 2015;**162**:167-173

[103] Sacco O, Vaiano V, Han C, Sannino D, Dionysiou DD. Photocatalytic removal of atrazine using N-doped TiO_2 supported on phosphors. Applied Catalysis B: Environmental. 2015;**164**:462-474

[104] Fenoll J, Vela N, Garrido I, Pérez-Lucas G, Navarro S. Abatement of spinosad and indoxacarb residues in pure water by photocatalytic treatment using binary and ternary oxides of Zn and Ti. Environmental Science & Pollution Research. 2014;**21**:12143-12153

[105] Fenoll J, Flores P, Hellín P, Hernández J, Navarro S. Minimization of methabenzthiazuron residues in leaching water using amended soils and photocatalytic treatment with TiO_2 and ZnO. Journal of Environmental Sciences. 2014;**26**:757-764

[106] Sivagami K, Krishna RR, Swaminathan T. Photo catalytic degradation of pesticides in immobilized bead photo reactor under solar irradiation. Solar Energy. 2014;**103**:488-493

[107] Verma A, Prakash NT, Toor AP. Photocatalytic degradation of herbicide isoproturon in TiO_2 aqueous suspensions: Study of reaction intermediates and degradation pathways. Environmental Progress & Sustainable Energy. 2014;**33**:402-409

[108] Hossaini H, Moussavi G, Farrokhi M. The investigation of the LED-activated FeFNS-TiO_2 nanocatalyst for photocatalytic degradation and mineralization of organophosphate pesticides in water. Water Research. 2014;**59**:130-144

[109] Vicente R, Soler J, Arques A, Amat AM, Frontistis Z, Xekoukoulotakis N, Mantzavinos D. Comparison of different TiO_2 samples as photocatalyst for the degradation of a mixture of four commercial pesticides. Journal of Chemical Technology and Biotechnology. 2014;**89**:1259-1264

[110] Fenoll J, Sabater P, Navarro G, Vela N, Pérez-Lucas G, Navarro S. Abatement kinetics of 30 sulfonylurea herbicide residues in water by photocatalytic treatment with semiconductor materials. Journal of Environmental Management. 2013;**130**:361-368

[111] Fenoll J, Martínez-Menchón M, Navarro G, Vela N, Navarro S. Photocatalytic degradation of substituted phenylurea herbicides in aqueous semiconductor suspensions exposed to solar energy. Chemosphere. 2013;**91**:571-578

[112] Fenoll J, Hellín P, Flores P, Martínez CM, Navarro S. Degradation intermediates and reaction pathway of carbofuran in leaching water using TiO_2 and ZnO as photocatalyst under natural sunlight. Journal of Photochemistry and Photobiology A: Chemistry. 2013;**251**:33-40

[113] Ramos-Delgado NA, Gracia-Pinilla MA, Maya-Trevino L, Hinojosa-Reyes L, Guzman-Mar JL, Hernandez-Ramirez A. Solar photocatalytic activity of TiO_2 modified with WO_3 on the degradation of an organophosphorus pesticide. Journal of Hazardous Materials. 2013;**263**:36-44

[114] Fenoll J, Flores P, Hellín P, Martínez CM, Navarro S. Photodegradation of eight miscellaneous pesticides in drinking water after treatment with semiconductor materials under sunlight at pilot plant scale. Chemical Engineering Journal. 2012;**204-206**:54-64

[115] Fenoll J, Hellín P, Martínez CM, Flores P, Navarro S. Semiconductor-sensitized photodegradation of s-triazine and chloroacetanilide herbicides in leaching water using TiO_2 and ZnO as catalyst under natural sunlight. Journal of Photochemistry and Photobiology A: Chemistry. 2012;**238**:81-87

[116] Fenoll J, Hellín P, Martínez CM, Flores P, Navarro S. Semiconductor oxides-sensitized photodegradation of fenamiphos in leaching water under natural sunlight. Applied Catalysis B: Environmental. 2012;**115-116**:31-37

[117] Fenoll J, Ruiz E, Hellín P, Flores P, Navarro S. Heterogeneous photocatalytic oxidation of cyprodinil and fludioxonil on leaching water under solar irradiation. Chemosphere. 2011;**85**:1262-1268

[118] Senthilnathan J, Philip L. Photodegradation of methyl parathion and dichlorvos from drinking water with N-doped TiO_2 under solar radiation. Chemical Engineering Journal 2011;**172**:678-688

[119] Sojic DV, Despotovic VN, Abazovic ND, Comor MI, Abramovic BF. Photocatalytic degradation of selected herbicides in aqueous suspensions of doped titania under visible light irradiation. Journal of Hazardous Materials. 2010;**179**:49-56

[120] Colina-Márquez J, Machuca-Martínez F, Puma GL. Photocatalytic mineralization of commercial herbicides in a pilot-scale solar CPC reactor: Photoreactor modeling and reaction kinetics constants independent of radiation field. Environmental Science & Technology. 2009;**43**:8953-8960

[121] Saien J, Khezrianjoo S. Degradation of the fungicide carbendazim in aqueous solutions with UV/TiO_2 process: Optimization, kinetics and toxicity studies. Journal of Hazardous Materials. 2008;**57**:269-276

[122] Chen SF, Liu YZ. Study on the photocatalytic degradation of glyphosate by TiO_2 photocatalyst. Chemosphere. 2007;**67**:1010-1017

[123] Maldonado MI, Passarinho PC, Oller I, Gernjak W, Fernández P, Blanco J, Malato S. Photocatalytic degradation of EU priority substances: A comparison between TiO_2 and Fenton plus photo-Fenton in a solar pilot plant. Journal of Photochemistry and Photobiology A: Chemistry. 2007;**185**:354-363

[124] Qamar M, Muneer M, Bahnemann D. Heterogeneous photocatalysed degradation of two selected pesticide derivatives, triclopyr and daminozid in aqueous suspensions of titanium dioxide. Journal of Environmental Management. 2007;**80**:99-106

[125] Oller I, Gernjak W, Maldonado MI, Pérez-Estrada LA, Sánchez-Pérez JA, Malato S. Solar photocatalytic degradation of some hazardous water-soluble pesticides at pilot-plant scale. Journal of Hazardous Materials 2006;**138**:507-517

[126] Evgenidou E, Fytianos K, Poulios I. Photocatalytic oxidation of dimethoate in aqueous solutions. Journal of Photochemistry and Photobiology A: Chemistry. 2006;**175**:29-38

[127] Chen SF, Cao GY. Photocatalytic degradation of organophosphorus pesticides using floating photocatalyst TiO_2 center dot SiO_2/beads by sunlight. Solar Energy. 2005;**79**:1-9

[128] Giménez J, Bayarri B, González O, Malato S, Peral J, Esplugas S. Advanced oxidation processes at laboratory scale: Environmental and economic impacts. ACS Sustainable Chemistry & Engineering. 2015;**3**:3188-3196

5

Heat Transfer Applications of TiO_2 Nanofluids

Hafiz Muhammad Ali, Muhammad Usman Sajid and Adeel Arshad

Abstract

To achieve acme heat transfer is our main disquiet in many heat transfer applications such as radiators, heat sinks and heat exchangers. Due to furtherance in technology, requirement for efficient systems have increased. Usually cooling medium used in these applications is liquid which carries away heat from system. Liquids have poor thermal conductivity as compared to solids. In order to improve the efficiency of system, cooling medium with high thermal conductivity should be used. Quest to improve thermal conductivity leads to usage of different methods, and one of them is addition of nanoparticles to base liquid. Application of nanofluids (a mixture of nanoparticles and base fluid) showed enhancement in heat transfer rate, which is not possible to achieve by using simple liquids. Different researchers used TiO_2 nanoparticles in different heat transfer applications to observe the effects. Addition of titanium oxide nanoparticles into base fluid showed improvement in the thermal conductivity of fluid. This chapter will give an overview of usage of titanium oxide nanoparticles in numerous heat transfer applications.

Keywords: titanium oxide nanoparticles, nanofluids, heat transfer enhancement, cooling medium, heat exchangers

1. Introduction

Nanoparticles, which were named as ultra-fine particles during the 1970s and 1980s, have size usually less than 100 nm. When a bulk material is considered then its physical properties remain nearly constant, but in case of nanoparticles it is not true. Nanoparticles are being used in many consumer goods such as paints, cosmetics and textiles. Nanoparticles are mixed with base fluid such as water, ethylene glycol and oil, to improve its properties. This mixture of nanoparticles and base fluid which is known as nanofluid can be used in different heat transfer applications.

In the following chapter effects of using titanium oxide nanofluids in heat transfer applications have been presented. Titanium oxide nanofluids have better thermal properties as compared to simple liquids. Due to their better heat transfer characteristics they can be used as an alternate for simple liquids in many heat transferring systems such as radiator of cars, heat exchangers and heat sinks. The disadvantage associated with these nanoparticles is their potential toxicity.

2. Thermal performance enhancement by using TiO_2 nanoparticles

Liquids have poor thermal properties, which is a barrier in development of energy efficient systems. Hence, some kind of technique should be adopted to overcome this problem. Addition of nanoparticles to base fluid showed better thermo-physical properties as compared to simple fluid. The reasons attributed to this enhancement in heat transfer performance could include Brownian motion and reduction in thermal boundary layer. This behaviour of nanofluids attracts many researchers to do research work in this field.

Hamid et al. [1] experimentally found performance factor $\eta = \frac{h_{nf}}{h_{bf}} = h'$ and friction factor of TiO_2 nanofluids. Reynolds number, temperature, concentration and thermal properties have significant effect on the performance factor. At 30°C and volume concentration less than 1.2%, the performance factor was less than base fluid. However, at 50 and 70°C for all concentrations, nanofluid showed better improvement as compared to base fluid. Enhancement in Reynolds number, temperature and concentration enhanced the performance factor, while pressure drop increased as Reynolds number and concentration increased. We can convert weight concentration to volume concentration by using Eq. (1)

$$\varnothing = \frac{\omega \rho_{bf}}{[(1 - 0.01\omega)\rho_p + 0.01\omega \rho_{bf}]} \tag{1}$$

where ω is weight concentration, ρ_{bf} is density of base fluid and ρ_p is density of nanoparticles.

To find volume concentration if mass is given, Eq. (2) can be used

$$\varnothing = \frac{m_p/\rho_p}{m_p/\rho_p + m_{bf}/\rho_{bf}} \times 100 \tag{2}$$

Density of nanofluid ρ_{nf} can be calculated from Eq. (3)

$$\rho_{nf} = \varnothing \rho_p + (1 - \varnothing)\rho_{bf} \tag{3}$$

Vakili et al. [2] measured enhancement in convective heat transfer coefficient of TiO_2 nanofluid flowing through a vertical pipe. According to experimental findings, thermal conductivity has nonlinear dependence on concentration. Increment in values of Reynolds number, nanoparticle concentration and heat flux also improved convective heat transfer coefficient. TiO_2 nanofluid with water/ethylene glycol as base fluid showed more enhancement in convective heat transfer coefficient as compared to TiO_2 nanofluid with distilled water as base fluid. Azmi et al. [3] used

TiO_2 and Al_2O_3 nanofluids in his experimentation to find and compare heat transfer coefficient and friction factor. At 30°C, Al_2O_3 nanofluid showed higher enhancement in viscosity and its viscosity varies with temperature whereas TiO_2 nanofluid has viscosity independent of temperature. Lower heat transfer coefficient than water and ethylene glycol is obtained for TiO_2 nanofluid for all concentrations (0–1%) at 30°C. Similar trend in heat transfer coefficient for both nanofluid is achieved at higher temperature. Friction factor augmentation for both nanofluids with volume concentration is not considerable.

Wang et al. [4] added TiO_2 nanoparticles in paraffin wax (a phase changing material) to improve its thermal properties. Thermal properties varied with the concentration of nanoparticles. A drop in phase change temperature is observed when loading of nanoparticle was less than 1 wt% while a drop in latent heat capacity is observed when nanoparticle loading was greater than 2 wt %. Thermal conductivity of composite decreased as the temperature is increased and it is lower in liquid state than in solid state. Azmi et al. [5] experimentally investigated the effects of working fluid temperature and concentration on thermal conductivity, viscosity and heat transfer coefficient. These thermo-physical properties were greatly influenced by temperature and concentration. Thermal conductivity has direct relation with temperature at low concentration. Range of variation in viscosity is 4.6–33.3% depending on temperature and concentration.

Sajadi et al. [6] study the turbulent heat transfer behaviour of TiO_2/water base nanofluid. The basic aim was to study the effects of volume concentration on heat transfer coefficient and on pressure drop. Dispersion of nanoparticle is improved by mixing an ultrasonic cleaner. Increasing concentration of nanoparticle has no significant effect on heat transfer but pressure drop increased. When the Reynolds number is increased then the ratio of heat transfer coefficient for nanofluid to base fluid is decreased while the Nusselt number is increased for both base and nanofluids. Wei et al. [7] did experimentation to find thermal conductivity and stability of TiO_2/diathermic oil nanofluid. Effect of temperature and concentration on thermal conductivity had been examined. Thermal conductivity was having a linear correlation with concentration. Thermal conductivity of nanofluid increased with an increase in temperature. Zeta potential values of different samples indicated good stability of nanofluids. To calculate thermal conductivity, classical models are available such as Hamilton-Crosser (H-C) [8] is presented in Eq. (4)

$$\frac{k_{nf}}{k_f} = \frac{k_p + (n-1)k_f + (n-1)\varnothing(k_p - k_f)}{k_p + (n-1)k_f - \varnothing(k_p - k_f)} \tag{4}$$

where k_{nf} is thermal conductivity of nanofluid, k_f is thermal conductivity of base fluid, k_p is thermal conductivity of nanoparticles, $\varnothing$ is volume fraction and $n = \frac{3}{\psi}$. Ψ is the sphericity.

Yu et al. [9] also gave a model to find thermal conductivity of nanofluids and it is presented in Eq. (5) as follows:

$$\frac{k_{nf}}{k_f} = \frac{k_p + 2(k_p - k_f)(1+\beta)^3\varnothing + 2k_f}{k_p - 2(k_p - k_f)(1+\beta)^3\varnothing + 2k_f} \tag{5}$$

where β is the ratio of the nano-layer thickness to the original particle radius.

Researcher	Nanoparticle and base fluid	Size	Reynolds number range	Temperature range	Flow type	Nanoparticle concentration	Flow rates	Experimental setup	Results
Hamid et al. [1]	TiO_2-water and ethylene glycol (60:40)	50 nm	3000–24,000	30, 50 and 70°C	–	0.5, 0.7, 1.0, 1.3 and 1.5% by volume	2–20 LPM	A tube with length of 1.25 m, inner diameter of 16 mm and outer diameter of 19 mm	Maximum performance factor obtained is 1.29 times that of base fluid
Vakili et al. [2]	TiO_2-Ethylene glycol and water (60:40)	25 nm	2030, 2960 and 3960	–	Laminar and Turbulent	0.5, 1.0 and 1.5 % by volume	0.5–5.0 LPM	A vertical copper tube with a length of 120 cm, inner diameter of 6 mm and outer diameter of 8 mm	A maximum of more than 44% enhancement in convective heat transfer coefficient is achieved.
Azmi et al. [3]	TiO_2-water and ethylene glycol (60:40) Al_2O_3-water and ethylene glycol (60:40)	50 nm 13 nm	–	30, 50 and 70°C	Turbulent	0.5–1.0% by volume	2–20 LPM	A tube with inner diameter of 16 mm and outer diameter of 19 mm	Maximum enhancement in heat transfer coefficient is 24.2% for TiO_2 nanofluids.
Wang et al. [4]	TiO_2 (anatase)-paraffin wax	20 nm	–	15–65°C	–	0–7% by weight	–	–	Considerable increase in latent heat is achieved by addition of nanoparticles at around 0.7% by weight
Azmi et al. [5]	TiO_2-water and ethylene glycol (60:40)	50 nm	–	30–80°C	Turbulent	0.5–1.5% by volume	3.9–21.15 LPM	Copper tube with length of 1.5 m, inner diameter of 16 mm and outer diameter of 19 mm	Maximum enhancement in Nusselt number is about 28.9%
Sajadi and Kazemi [6]	TiO_2-water	30 nm	5000–30,000	–	Turbulent	0.05, 0.1, 0.15, 0.20, and 0.25%	–	Copper tube with inner diameter of 5 mm, length of 1800 mm and thickness of 0.675 mm	Enhancement in heat transfer coefficient was about 22%
Wei et al. [7]	TiO_2 (anatase)-diathermic oil	10 nm	–	20–50°C	–	0.1–1% by volume	–	Test tube	Thermal conductivity achieved at 50°C and 1% volume concentration is 0.136 W/m·k

Table 1. TiO_2 nanofluids used to enhance thermal performance.

Timofeeva [10] gave a model, which is given in Eq. (6)

$$\frac{k_{\text{eff}}}{k_{\text{f}}} = 1 + 3\varnothing \tag{6}$$

Basic information such as nanoparticle size, concentration of nanoparticles in base fluid and results, related to these research works can be obtained from **Table 1**.

3. TiO_2 nanofluids as coolant for radiator and electronic devices

Radiator (usually a cross flow heat exchanger) is an important component of automobile. It cools down the liquid which is carrying heat from engine block and protecting it from damage. For high heat transfer rate, if the size of radiator is increased then it will increase both the volume and weight, which is undesirable. Researchers are interested to increase the effectiveness and compactness of radiators by using coolants with additives such as nanoparticles to the base fluids.

Hussein et al. [11] used SiO_2 and TiO_2 nanofluids in automotive cooling system to check the effects of volumetric flow rate, inlet temperature and volumetric concentration on Nusselt number. Statistical models have been obtained by statistical softwares using multiple linear regression methods and factorial methodology. Nusselt number increases as the volume flow rate, inlet temperature and volume concentration is increased. Wadd et al. [12] performed experimentation on automobile radiator to check the performance of metal (copper/water) and non-metal (titania TiO_2/water) nanofluids. Sodium lauryl sulphate was used as dispersant. Copper-based nanofluids showed more thermal conductivity than TiO_2. The stability of metal nanoparticle was found to be less than non-metal nanoparticles. Friction factor and pressure drop was found to be nearly same for both.

Figure 1 shows the flow of nanofluid through radiator.

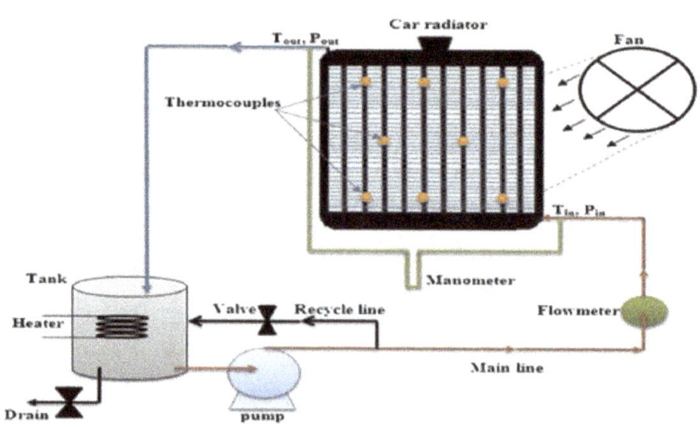

Figure 1. Flow diagram showing flow of nanofluid through radiator [11].

Sandhya et al. [13] used TiO_2 water/ethylene glycol base nanofluid in car radiator to check the improvement in cooling performance. Nusselt number showed enhancement by increasing volume flow rate, volume concentration and Reynolds number. By increasing the volumetric flow rate, outlet temperature of the nanofluid also increased. The inlet temperature of nanofluid has slight effect on Nusselt number. Bhimani et al. [14] used TiO_2/water nanofluid as a coolant in automobile radiator to study heat transfer enhancement. Chemical treatment is done to avoid agglomeration and sedimentation because of hydrophobic nature to TiO_2. Heat transfer coefficient enhanced as the flow rate and volume concentration increased.

According to Newton's law of cooling, heat transfer can be calculated as given by Eq. (7)

$$Q = hA_s\Delta T = hA_s(T_b - T_s) \tag{7}$$

where h is heat transfer coefficient, A_s is surface area of tube, T_b is bulk temperature and T_s is tube wall temperature.

Heat transfer rate can be calculated as given by Eq. (8)

$$Q = mC\Delta T \tag{8}$$

and heat transfer coefficient can be calculated as given by Eq. (9)

$$h_{\text{exp}} = \frac{mC(T_{\text{in}} - T_{\text{out}})}{nA_s(T_b - T_s)} \tag{9}$$

where n is number of tubes.

While Nusselt number can be calculated as given by Eq. (10)

$$Nu = \frac{h_{\text{exp}} \times D_h}{k} \tag{10}$$

Chen and Jia [15] experimentally checked the enhancement in thermal conductivity and convective heat transfer coefficient by using TiO_2 nanofluid in automobile radiator. Pump damage due to application of nanofluid is studied by using cavitation corrosion test. Nanofluid showed good corrosion impediment capability under circulation. Hamid et al. [16] did experimental work to find pressure drop by application of TiO_2 nanofluid. Increase in pressure drop will lead to higher pump power requirement, which is not desired at all. Experimental findings showed no significant increase in pressure drop. Friction factor decreased at high Reynolds number.

Darcy equation to calculate pressure drop is given by Eq. (11), and to calculate friction factor Eq. (12) can be used as follows:

$$\Delta P = \frac{f\rho v^2 L}{2D} \tag{11}$$

$$f = \frac{0.3164}{Re^{0.25}} \tag{12}$$

Researcher	Nanoparticle and base fluid	Size	Reynolds number range	Temperature range	Flow type	Nanoparticle concentration	Flow rates	Design	Results
Hussein et al. [11]	TiO_2-water SiO_2-water	–	–	60–80°C	Laminar	1–2% by volume	2–8 LPM	Radiator with louvered fins and 32 flat vertical copper tubes with flat cross sectional area	Maximum enhancement of nusselt number for TiO_2 obtained is 11%. Maximum enhancement of nusselt number for SiO_2 obtained is 22.5%.
Wadd et al. [12]	Cu-water TiO_2-water	20 nm	3000–20,000	65–70°C	Turbulent	0.1% by weight	1–5 LPM	Louvered fin and tube type, 37 vertical tubes with ellipse-shaped cross section	For copper nanoparticles maximum enhancement in Nusselt number is about 15%. For TiO_2 nanoparticles maximum enhancement in Nusselt number is about 13%.
Sandhya et al. [13]	TiO_2-water and ethylene glycol (60% :40%)	21 nm	4000–15,000	35, 40 and 45°C	Turbulent	0.1, 0.3 and 0.5% by volume	2–5 LPM	Radiator consists of three rows of 104 tubes with a diameter of 5 mm and length of 0.3 m.	The enhancement in heat transfer at concentration of 0.5% is about 35% as compared to the base fluid.
Bhimani et al. [14]	TiO_2-Water	15 nm	Nearly between 5000 and 10,000	80°C	Turbulent	0.1, 0.3, 0.5, 0.7, and 0.1% by volume	90–120 LPM	Louvered fin and tube type, 34 vertical tubes with stadium-shaped cross section	Enhancement in heat transfer coefficient is 40–45% as compared to base fluid at concentration of 1% by volume.
Chen et al [15]	TiO_2—85% wt. % ethylene glycol and 15% wt.% water	18 nm	–	–	–	1% by weight	15–30 LPM	Straight copper tube with 2 m length and 12 mm diameter	Enhancement in thermal conductivity is 3% and in convective heat transfer is about 10%.
Hamid et al. [16]	TiO_2-water and ethylene glycol (60:40)	50 nm	Less than 30,000	50 and 70°C	Turbulent	0.5, 1 and 1.5% by volume	Not specified	Copper tube with length of 1.5 m having inner diameter of 16 mm	No significant increase in pressure drop for TiO_2 nanofluid.
Ijam et al. [18]	TiO_2-water SiC-water	–	–	–	Turbulent	0.8, 1.6, 2.4, 3.2 and 4%	2, 4 and 6 m/s	Heat sink with bottom of 20 cm × 20 cm	Enhancement in thermal conductivity by using SiC/water and TiO_2/water nanofluid at 4% concentration is about 12.44. and 9.99%, respectively.

Researcher	Nanoparticle and base fluid	Size	Reynolds number range	Temperature range	Flow type	Nanoparticle concentration	Flow rates	Design	Results
									Enhancement in heat flux by using SiC-water and TiO_2-water nanofluids with inlet velocity of 6 m/s was 12.43 and 12.77%, respectively.
Rafati et al. [17]	Silica(SiO_2) Alumina (Al_2O_3) Titania (TiO_2)/ water (75% vol.) and ethylene glycol (25%	SiO_2 particle size is 14 nm Al_2O_3 particle size is 40 nm TiO_2 particle size is 21 nm	–	–	Laminar	0.5, 1.0 and 1.5% for silica 0.1, 0.25 and 0.5% for titania 0.5, 0.75 and 1.0% for alumina	0.5, 0.75 and 1 LPM	Quadcore processor (Phenom II X4 965)	Twice enhancement in convective heat transfer coefficient is found for alumina with 0.5% volumetric concentration at flow rate of 1 L/min.

Table 2. TiO_2 nanofluids used in radiator and electronic devices.

where ΔP is pressure drop, f is friction factor, ρ is density, v is velocity, L is length and D is diameter.

3.1. Cooling of electronic devices

Nowadays cooling of electronic devices is a challenging task because of compactness and high heat dissipation. Different approaches are being used to increase the thermal performance of electronic systems. One of such way is to enhance the thermal performance of coolant being used in the system. Nanofluids have showed better thermal performance than base fluid.

Rafati et al. [17] used three different types of nanofluids as coolant for cooling of microchips. A high conductive thermal paste is used between block and processor's integrated heat spreader. For computer cooling, the selection of nanofluid is based on factors such as better thermal performance, economic aspect and having no chemical and corrosion impact. The highest decrease in temperature was observed for alumina nanofluid, which was about 5.5°C.

Ijam et al. [18] used SiC and TiO_2 nanofluids as coolant in electronic devices. Enhancement in thermal conductivity and heat flux is achieved by increasing volume concentration. Pressure drop increases as flow rate is increased, which results in increase in pumping power. Pressure drop for SiC and TiO_2 nanofluid increases from 2159.26 and 2170 Pa, respectively, at 0.8% volume fraction to 2319.58 and 2375.07 Pa, respectively, at 4% volume fraction with 2 m/s inlet velocity. Pumping power increased from 0.28 to 5.49 W for SiC and from 0.26 to 5.64 W and for TiO_2 with 4% volume fraction when inlet velocity increased from 2 to 6 m/s. **Table 2** provides information about different parameters and result obtained in cooling of radiators and electronic devices.

4. Application of TiO_2 nanofluids in heat exchangers

The basic function of heat exchanger is to transfer heat energy from one fluid (which is at high temperature) to other fluid (which is at low temperature). Two fluids in heat exchanger did not come into direct contact or mix with each other. To achieve high heat transfer rate in heat exchanger is our main concern. Use of nanofluid is one of simple way to attain this purpose.

Duangthongsuk et al. [19] used a double tube counter flow heat exchanger to check heat transfer and flow characteristics of TiO_2 nanofluid. Nanofluid flows in inner copper tube while hot water flows in annular. Heat transfer coefficient increased as mass flow rate of hot water and Reynolds number increased while temperature of nanofluid is decreased. Hot water temperature has no significant effect on heat transfer coefficient. When compared with base fluid (water) pressure drop and friction factor are nearly the same. To calculate Nusselt number, different equations are available in literature such as Gnielinski equation [20] is defined as in Eq. (13)

$$Nu = \frac{\left(\frac{f}{8}\right)(Re - 1000)Pr}{1 + 12.7\left(\frac{f}{8}\right)^{0.5}\left(Pr^{\frac{2}{3}} - 1\right)} \tag{13}$$

where Nu is Nusselt number, Re is Reynolds number, Pr is Prandtl number and *f* is friction factor in above equation.

To predict Nusselt number Pak and Cho [21] correlation is given in Eq. (14) as follows:

$$Nu_{\mathrm{nf}} = 0.21 Re_{\mathrm{nf}}^{0.8} Pr_{\mathrm{nf}}^{0.5} \tag{14}$$

Another correlation in Eq. (15) to predict Nusselt number is given by Xuan and Li [22]

$$Nu_{\mathrm{nf}} = 0.0059\left(1 + 7.6286 Ø^{0.6886} Pe_{\mathrm{d}}^{0.001}\right) Re_{\mathrm{nf}}^{0.9238} Pr_{\mathrm{nf}}^{0.4} \tag{15}$$

where Pe is Peclet number of nanofluid in above relation.

Singh et al. [23] did experimental studies on double pipe heat exchanger by using CuO/TiO_2 nanofluids with different flow rates and volume concentrations. Application of CuO/TiO_2 nanofluids enhanced heat transfer rate as concentration and flow rate is increased. CuO nanofluid showed better results than TiO_2 nanofluids because of high thermo-physical properties. Reddy et al. [24] did experimentation to check heat transfer coefficient and friction factor in double pipe heat exchanger with and without helical coil inserts by using TiO_2 nanofluid. Nanofluid flows in the inner tube while hot fluid flows in the outer tube. Enhancement in heat transfer coefficient and friction factor (in terms of pressure drop) is measured. New correlations for Nusselt number and friction factor developed are given in Eqs. (16) and (17)

$$Nu_{\mathrm{Reg}} = 0.007523 Re^{0.8} Pr^{0.5} (1 + Ø)^{7.6} (1 + P/d)^{0.037} \tag{16}$$

$$f_{\mathrm{Reg}} = 0.3250 Re^{-0.2377} (1 + Ø)^{2.723} (1 + P/d)^{0.041} \tag{17}$$

Khedkar et al. [25] study TiO_2/water nanofluid heat transfer characteristics in concentric heat exchanger. Nanofluid with the highest concentration has the highest overall heat transfer coefficient. Flow diagram of apparatus used in experimentation [25] is shown in **Figure 2**.

Duangthongsuk et al. [26] found that enhancement in heat transfer coefficient and pressure drop is related to nanoparticle concentration. If nanoparticle concentration is increased beyond the

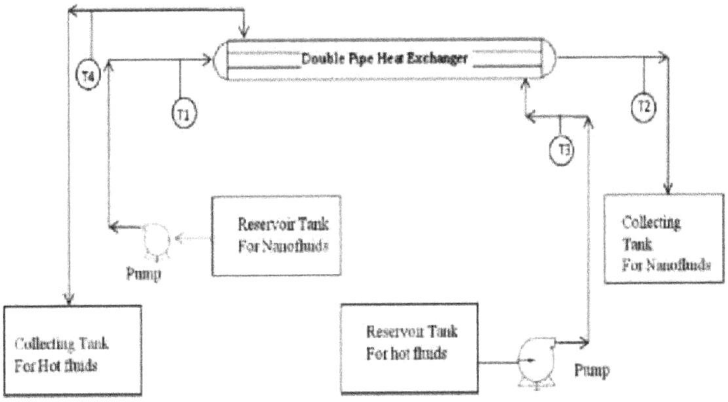

Figure 2. Schematic representation of nanofluid flow through concentric tube heat exchanger.

limit then a decrease in the heat transfer coefficient is observed. This is attributed to increase in viscosity. In this experiment, value of heat transfer coefficient increases as volume concentration is increased up to 1% and after that decrease in heat transfer coefficient is observed. Proposed correlations to predict Nusselt number and friction factor are mentioned in Eqs. (18) and (19), respectively.

$$Nu = 0.07Re^{0.707}Pr^{0.385}\emptyset^{0.074} \tag{18}$$

$$f = 0.961\emptyset^{0.052}Re^{-0.375} \tag{19}$$

Barzegarian et al. [27] used brazed plate heat exchanger to check enhancement in overall heat transfer coefficient and pressure drop by using TiO_2 nanofluid. For a specified Reynolds number, increment in weight concentration of nanoparticle in nanofluid increased the overall heat transfer coefficient and pressure drop also. Increase in Reynolds number also enhanced the overall heat transfer coefficient and pressure drop. Up to 20% enhancement in pressure drop has been observed during experimentation. Tiwari et al. [28] used four different types of nanofluids (including CeO_2, Al_2O_3, TiO_2 and SiO_2) with different concentrations and volume flow rates. Better heat transfer behaviour of TiO_2 and CeO_2 nanofluid is observed at low concentrations while that of Al_2O_3 and SiO_2 at high concentrations. $CeO_2 > Al_2O_3 > TiO_2 > SiO_2$ is the order of nanofluids for which maximum heat transfer coefficient is obtained.

Overall heat transfer coefficient is calculated in Ref. [29] as by using Eqs. (20)–(22)

$$U = \frac{Q_{avg}}{AF\Delta T_{LMTD}} \tag{20}$$

$$A = N_t HW \tag{21}$$

$$\Delta T_{LMTD} = \frac{(T_{h,o} - T_{c,i}) - (T_{h,i} - T_{c,o})}{Ln\frac{(T_{h,o}-T_{c,i})}{(T_{h,i}-T_{c,o})}} \tag{22}$$

where U is over all heat transfer coefficient, A is total surface area, F is temperature correction factor, N_t is total number of plates, and H and W are height and width of plates.

Taghizadeh-Tabari et al. [29] performed experimentation on plate heat exchanger of milk pasteurization industry by using TiO_2/water nanofluid. Peclet number is used in experiment to compare performance of nanofluid with different concentrations. Nusselt number and pressure drop increased as the Peclet number or concentration or both are increased. Experimental results showed dramatic increase in heat transfer coefficient while theoretical calculated results did not. Reasons behind this could include increase of nanoparticle Brownian motions, particle migration and reduction of boundary layer thickness. The performance index $\eta = (h_{nf}/h_{bf})/(\Delta P_{nf}/\Delta P_{bf})$ is greater than 1 for all type of nanofluid concentrations used in the experimentation. Benefit of using nanofluid in milk pasteurization industry is to reduce energy consumption. Javadi et al. [30] did study work by using three different nanofluids in plate heat exchanger to compare thermo-physical and heat transfer characteristics with base fluid. The results confirmed that overall heat transfer coefficient, thermal conductivity, pressure drop,

Researcher	Nanoparticle and base fluid	Size	Reynolds number range	Temperature range	Flow type	Nanoparticle concentration	Flow rates	Design	Results
Duangthongsuk and Wongwises [19]	TiO_2-water	21 nm	4000–18,000	Nanofluid temperature range 15–25°C Hot water temperature range is 35–50°C	Turbulent	0.2% by volume	Hot water flow rate is 3 LPM and 4.5 LPM	Double tube with counter flow For inner tube, outer diameter is 9.53 mm and inner diameter is 8.13 mm. For outer tube, inner diameter is 27.8 mm and outer diameter is 33.9 mm.	Enhancement in convective heat transfer coefficient is about 6–11% as compared to base fluid.
Singh et al. [23]	TiO_2-water CuO-water	10–20 nm 30–50 nm	3500–13,500	Nanofluid temperature range 30 ± 3°C	Laminar and turbulent	0.1–0.3% by volume	Nanofluid flow rate range is 1–4 LPM	Double pipe For inner tube, outer diameter is 18 mm and inner diameter is 15 mm. For outer tube, inner diameter is 32 mm and outer diameter is 36 mm.	At volumetric concentration of 0.3% and flow rate of 4 LPM, a maximum enhancement in coefficient of heat transfer is 5% for CuO nanofluids.
Reddy et al. [24]	TiO_2	21 nm	4000–15,000	–	–	0.004–0.02% by volume	–	Double tube with and without helical coil inserts. For inner tube, outer diameter is 9.53 mm and inner diameter is 8.13 mm. For outer tube, inner diameter is 27.8 mm and outer diameter is 33.9 mm.	Enhancement in Nusselt number and friction factor for 0.02% concentration and 15,000 Reynolds number without helical coils are 10.73 and 8.73%, respectively. However, by using helical coils this enhances up to 17.71 and 16.58%, respectively.

Researcher	Nanoparticle and base fluid	Size	Reynolds number range	Temperature range	Flow type	Nanoparticle concentration	Flow rates	Design	Results
Khedkar et al. [25]	TiO_2-water	20 nm	300–4000	Temperature of hot fluid is 55, 65 and 75°C	Laminar	2 and 3% by volume	0.1–1 LPM for nanofluids. 1–3 LPM for hot fluid.	Concentric tube 8 mm diameter of inner pipe and16 mm diameter of outer pipe	Maximum enhancement in overall heat transfer coefficient is 47.37% with 3% volume concentration and 3992 Reynolds number.
Duangthongsuk et al. [26]	TiO_2-water	21 nm	3000–18,000	15, 20 and 25°C	Turbulent	0.2, 0.6, 1.0, 1.5 and 2.0% by volume	3 and 4.5 LPM	Double tube with counter flow. For inner tube, outer diameter is 9.53 mm and inner diameter is 8.13 mm. For outer tube, inner diameter is 27.8 mm and outer diameter is 33.9 mm.	Maximum enhancement in heat transfer coefficient is about 26% at 1% volume concentration.
Barzegarian et al. [27]	TiO_2-water	20 nm	Nearly 159–529	–	Turbulent	0.3, 0.8 and 1.5% by weight	2.5–7.5 LPM	Brazed plate heat exchanger Thickness, length and width of BPHE is 0.042, 0.194 and 0.08 m, respectively.	Maximum enhancement in convective heat transfer coefficient of nanofluid at weight concentration of 0.3, 0.8 and 1.5% is 6.6, 13.5 and 23.7%, respectively. And for over all heat transfer coefficient this enhancement is 2.2, 4.6 and 8.5%, respectively.

Researcher	Nanoparticle and base fluid	Size	Reynolds number range	Temperature range	Flow type	Nanoparticle concentration	Flow rates	Design	Results
Tiwari et al. [28]	CeO_2-water Al_2O_3-water TiO_2-water SiO_2-water	30 nm 45 nm 10 nm 10 nm	–	25–30°C	–	0.5, 0.75, 1.0, 1.25, 1.5, 2.0 and 3% by volume	1–4 LPM	Plate heat exchanger 10 plates and heat exchanger are of 0.3 m^2.	For CeO_2/water, Al_2O_3/water, TiO_2/water and SiO_2/water nanofluids, the maximum enhancement in heat transfer coefficient at optimum volume concentration are about 35.9, 26.3, 24.1 and 13.9%, respectively.
Taghizadeh-Tabari et al. [29]	TiO_2-water	10–15 nm	Nearly (70–220) Critical Reynolds number is 100	–	Laminar and turbulent	0.25, 0.35 and 0.8% by weight	–	Plate heat exchanger. 11 plates with the length of 0.2 m each and width of 0.11 m Spacing between plates is 0.0025 m.	Maximum ratio of Nusselt number of nanofluid to distilled water obtained is about 1.17. $Nu_{nf}/Nu_{water}= 1.17$ Maximum pressure drop is about 8% as compared to distilled water.
Javadi et al. [30]	SiO_2-liquid nitrogen TiO_2-liquid nitrogen Al_2O_3-liquid nitrogen	–	–	Inlet and outlet temperatures are 310 and 124.26 K for hot fluid while 99.719 and 301.54 K for cold fluid, respectively	–	0.2–2% by volume	Mass flow rate of SiO_2 at 0.2% concentration is 0.278 kg/s and at 2% is 2.079 kg/s. Mass flow rate of TiO_2 and Al_2O_3 at 0.2% concentration is 0.456 and 0.439 kg/s while for 2% concentration is	Plate fin heat exchanger.	Al_2O_3 has the highest overall heat transfer coefficient which is 308.69 W/m^2k in 2% concentration.

Researcher	Nanoparticle and base fluid	Size	Reynolds number range	Temperature range	Flow type	Nanoparticle concentration	Flow rates	Design	Results
							3.861 and 3.691 kg/s, respectively.		
Ashrafi and Bashirzadeh [31]	TiO_2-water	20 nm	82–497	–	Laminar	0.1, 0.5 and 1% by weight	–	Shell and tube heat exchanger, 39 tubes with 29 cm length of each, 11 mm outside diameter and 10 mm inside diameter of tube. Shell with 0.253 m^2 exchanging area and 6.35 cm inside diameter.	The average augmentation of convective heat transfer coefficient for 1% by weight concentration is about 31.6%
Kumar et al. [32]	CuO-water Monoethylene glycol TiO_2-water Monoethylene glycol	26 nm 9 nm	1000–5000	27°C	Laminar and turbulent	0.02, 0.04 and 0.06% by volume.	0.5–3 LPM	Shell and tube heat exchanger.	Maximum enhancement in heat transfer coefficient is about 29% at 0.06% concentration and 80°C with 3 LPM.

Table 3. Application of TiO_2 nanofluids in different types of heat exchangers.

heat transfer rate and entropy generation increased with increase in volume concentration of nanoparticles. While Prandtl number decreased as concentration of nanoparticles in base fluid is increased, pressure drop is lowest for SiO_2. Overall heat transfer coefficient is highest for Al_2O_3. Prandtl number is highest for lowest concentration of nanoparticles. Entropy generation is lowest for SiO_2 as 25, while for TiO_2 and Al_2O_3 this value increases to 40 and 38.7, respectively.

Ashrafi et al. [31] used nanofluid as coolant in heat exchangers of swimming pool. In shell and tube heat exchanger, nanofluid flows through tubes while cold water in shell. Results show that when weight concentration of nanoparticle and Peclet number is increased, the convective heat transfer coefficient is also increased. Kumar et al. [32] used shell and tube heat exchanger to check heat transfer characteristics of TiO_2/water, CuO/water, TiO_2/ethylene glycol and CuO/ethylene glycol nanofluids with different concentrations. Hot water flows through shell and nanofluid flows through tubes. CuO/water nanofluids showed highest enhancement among all nanofluids used in the experimentation. Convective heat transfer coefficient is improved by increasing Reynolds number, volume concentration, volume flow rate and temperature.

Different enhancements achieved by researchers are given in **Table 3**.

5. Application of TiO_2 nanofluids in heat sinks

Heat sink (which is a form of heat exchanger) is used to absorb excessive heat from a system to maintain its temperature at an optimum value and to avoid from overheating. These are made of conductive metals. Air or liquid is used to remove the heat from heat sink. Design of heat sink is such as to maximize surface area contact with air or cooling liquid. Due to limitations, we cannot increase area beyond limits rather we can use cooling liquid with higher thermal conductivity. This enhancement in thermal conductivity will result in higher heat transfer and can be achieved by addition of nanoparticles to base fluid.

Ali et al. [33] experimentally compared performance of staggered and inline pin fin heat sinks under laminar flow of TiO_2 (rutile) and TiO_2 (anatase). TiO_2 (rutile) with staggered pin fin heat sinks showed best performance. The arrangement of staggered pin fin allows more liquid to interact with pin fins, which makes its performance better than inline pin fin. By using TiO_2 (rutile) with staggered pin fin heat sinks minimum temperature of base obtained is 29.4°C. The lowest thermal resistance is obtained for TiO_2 (rutile) with staggered pin fin heat sinks at Reynolds number of 587 which was 0.012°C/W. Schematic diagram of nanofluid flow is shown in **Figure 3**.

Mohammed et al. [34] used six different nanofluids in the experimentation to find enhancement in the heat transfer coefficient, wall shear stresses, friction factor and pressure drop in triangular micro-channel heat sink. Order of achieved enhancement in heat transfer coefficient is Diamond> SiO_2 > CuO > TiO_2 > Ag > Al_2O_3. CuO/water and TiO_2/water showed same performance in terms of heat transfer coefficient. Order of pressure drop occurred along the length of the channel in experiment is SiO_2 > Diamond > Al_2O_3 > TiO_2 > CuO > pure water > Ag.

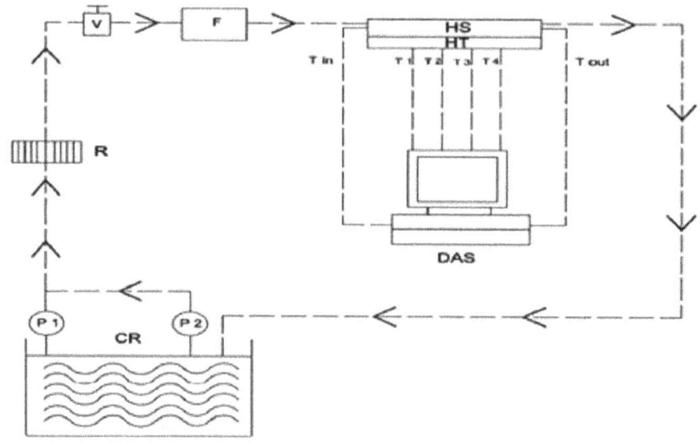

Figure 3. Schematic diagram of nanofluid flow through heat sink. CR is coolant reservoir, DAS is data acquisition system, F is flow meter, HS is heat sink, HT is heater, P is pump, R is radiator, T is thermocouple and V is valve in flow diagram.

Ag also has lowest wall shear stress. Diamond/water nanofluid has lowest thermal resistance among six nanofluids.

Naphon and Nakharintr [35] performed experiments on mini- rectangular fin heat sinks with different widths by using TiO_2/de-ionized water nanofluid to check heat transfer enhancement. Average outlet temperature and plate temperature decreased as Reynolds number is increased. Average heat transfer rate is increased with mass flow rate of nanofluids. Nusselt number has direct relation with Reynolds number. Increase in Reynolds number decreased the thermal resistance while slight increase in pressure drop is observed. Average heat transfer rate of heat sink with largest width is higher than the sinks with smaller width.

To calculate thermal resistance in Ref. [35], Eq. (23) can be used as

$$R_{th} = \frac{1}{\overline{h} \times A_s} = \frac{1}{\frac{Q_{avg}}{\Delta T_{LMTD}}} \tag{23}$$

Parameters used in above equation include R_{th} as thermal resistance, $\overline{h}$ as average heat transfer coefficient and A_s as total heat transfer surface area of heat sink.

Sohel et al. [36] used three different nanofluids to check thermo-physical properties and heat transfer performance of nanofluids in a circular copper micro-channel. CuO/water nanofluid showed best thermo-physical properties and heat transfer performance among the three nanofluids. Reduction in friction factor for CuO/water nanofluid is 9.38%, for Al_2O_3/water is 1.13% and for TiO_2/water is 1.79%. Reduction in thermal resistance for CuO/water nanofluid is 11.62%, Al_2O_3/water is 6.37% and TiO_2/water is 5.84%. Khaleduzzaman et al. [37] performed experimental work to find out the effect of nanoparticles volumetric concentration on flow rates, heat transfer coefficient and thermal resistance for water block heat sink. The interface temperature reduced as the volume flow rate increased. When the volume fraction and flow rate increased, thermal resistance decreased. Augmentation in heat transfer coefficient occurred when volume fraction and flow rate increased. Ijam et al. [38] performed cooling of copper mini-channel heat sink using two different types of nanofluid. Effects of nanofluid

Researcher	Nanoparticle and base fluid	Size	Reynolds number range	Temperature range	Flow type	Nanoparticle concentration	Flow rates	Design	Results
Ali and Arshad [33]	TiO_2 (anatase)-water TiO_2 (rutile)-water	5–30 nm 5–30 nm	Nearly between 260 and 600	–	Laminar	4.31% for TiO_2 (anatase) and 3.99% TiO_2 (rutile) by volume	–	Staggered and inline pin fin heat sinks	A maximum enhancement of 37.78% in Nusselt number is obtained for TiO_2 (rutile)/H_2O nanofluid with staggered pin fin heat sinks.
Mohammed et al. [34]	Al_2O_3-water Ag-water CuO-water Diamond-water SiO_2-water TiO_2-water	–	100–1000	–	Laminar	2% by volume	10–20 mL/min	Triangular micro channels 20 micro channels with 10 mm length	Highest and lowest heat transfer coefficient is obtained for diamond/water and Al_2O_3/water nanofluid, respectively. Highest pressure drop is obtained for SiO_2/water nanofluid while Ag/water nanofluids have lowest pressure drop and wall shear stresses.
Naphon and Nakharintr [35]	TiO_2-water	21 nm	80–200	20, 22 and 24°C	Laminar	0.4% by volume	4.5–8 g/s	Three mini-rectangular fin heat sinks with three different height of fins 1, 1.5 and 2 mm	An increase of about 42.3% in average heat transfer rate is obtained for heat sink with $w = 2$ mm.
Sohel et al. [36]	Al_2O_3-water TiO_2-water CuO-water	–	Maximum up to 1000	–	Laminar	0.5–4% by volume	Nanofluid inlet velocity = 1.5 m/s	Circular micro-channel heat sinks with diameter of each is 0.4 mm	Maximum improvement as compared to pure water in mass flow rate, Reynolds number, heat transfer coefficient, thermal conductivity and heat flux at 4% concentration for TiO_2/water nanofluid, we have 12.76, 1.84, 7.74 9.97 and 6.20%, respectively.

Researcher	Nanoparticle and base fluid	Size	Reynolds number range	Temperature range	Flow type	Nanoparticle concentration	Flow rates	Design	Results
Khaleduzzaman et al. [37]	TiO_2-water	21 nm	–	–	–	0.1% by volume	1, 1.25 and 1.5 LPM	Dimension of copper block is (94 mm × 94 mm × 20 mm).	Heat transfer coefficient improvement was about 18.91% at flow rate of 1 LPM.
Ijam et al. [38]	Al_2O_3-water TiO_2-water	–	–	–	Laminar	0.8, 1.6, 2.4, 3.2 and 4% by volume	Nanofluid inlet velocity is 0.1 and 1.5 m/s	Mini-channel heat sink	Improvement in thermal conductivity of Al_2O_3/water nanofluids with 4% volume concentration is 11.98%, while for TiO_2/water nanofluid it is 9.97%. With 4% volume fraction and 1.5 m/s inlet fluid velocity, maximum pressure drop and pumping power required is 1105.540 Pa and 0.12437 W, respectively.
Xia et al. [39]	TiO_2-water Al_2O_3-water	5 nm 5 nm	200–500	Temperature of nanofluid is 25°C	Laminar	0.1, 0.5, and 1.0% by volume	150 mL/min	Fan-shaped micro-channel Number of micro-channels are 30 with whole width of 6 mm.	Improvement in thermal conductivity for TiO_2 nanofluid is 6.55% at 1% volume concentration.

Table 4. TiO_2 nanofluids used for heat sink cooling.

volume fraction and inlet velocity on thermal conductivity, heat transfer coefficient, pumping power and pressure drop had been investigated. Fluid at low velocity absorbs more heat than the fluid at higher velocity. Mass flow rate and particle volume fraction has direct relation with heat transfer coefficient. By increasing mass flow rate and inlet velocity of fluid, pressure drop also increases. When volume fraction is increased then thermal resistance is decreased. Improvement in heat flux with volume fraction of 0.8% and nanofluid inlet velocity of 0.1 m/s for Al_2O_3 is 17.3% while for TiO_2 is 16.53%.

Xia et al. [39] compared heat transfer performance of fan-shaped micro-channel heat sink with rectangular micro-channel heat sink by using TiO_2/water and Al_2O_3/water nanofluids. Pressure drop and convection of heat transfer is higher in fan-shaped micro-channel heat sink. Enhancement in heat transfer is greater for Al_2O_3/water nanofluids when compared with TiO_2/water nanofluids, while thermal conductivity behaviour of TiO_2/water nanofluid is better.

To calculate friction resistance coefficient in Ref. [39] Eq. (24) is used

$$f = \frac{2\Delta P D_h}{\rho L u_{\mathrm{m}}^2} \tag{24}$$

where ΔP is pressure drop between inlet and outlet, D_{h} is hydrodynamic diameter, ρ is density, L is length of micro-channel and u_{m} is mean velocity. Different enhancements obtained by researchers are given in **Table 4**.

6. Application of TiO_2 nanofluid in nucleate pool boiling

Nucleate pool boiling is a boiling type which takes place when temperature of surface is about 5°C greater than the saturation temperature of liquid. This boiling region is the most desirable as we can obtain high heat transfer rates with a small value of ΔT_{excess}. Usually two methods are used to increase heat transfer rate in this region. One way is to increase the nucleation sites by doing surface treatment and other way is use of nanofluids.

Ali et al. [40] experimentally found boiling heat transfer coefficient enhancement by using TiO_2 (Rutile)/water nanofluid. Two different concentrations of 12 and 15% by weight are used in experimentation. Experimental setup accuracy is checked by using Pioro [41] correlation, which is given by Eq. (25) as follows:

$$\frac{hl}{k} = 0.075 C_{sf} \left(\frac{q}{h_{\mathrm{fg}} \rho_{\mathrm{g}}^{0.5} [\sigma g (\rho - \rho_{\mathrm{g}})]^{0.25}} \right)^{0.66} Pr^n \tag{25}$$

In above equation C_{sf} and n are constants, which are dependent on fluid and heating surface.

By increasing wall super heat a decrease in heat flux enhancement is observed. Average heat flux and boiling heat transfer coefficient enhancement obtained at 15% concentration is 2.22 and 1.38 while for 12% concentration is 1.89 and 1.24, respectively.

Trisaksri et al. [42] experimentally investigated nucleate pool boiling heat transfer at different concentration and pressure of a refrigerant-based nanofluid on cylindrical copper tube. TiO_2-R141 nanofluid with three different concentrations had been used. When concentration of nanoparticle is increased, a decline in boiling heat transfer for R141 is observed. Effect of pressure is dominant at low concentrations. Rohsenow [43] correlation used in experimentation to predict nucleate boiling heat transfer is given by Eq. (26) as follows:

$$\frac{C_{p,l}(T_s - T_{sat})}{h_{fg} Pr_l^m} = C_{sf}\left(\frac{q}{\mu_l h_{fg}}\sqrt{\frac{\sigma}{g(\rho_l - \rho_v)}}\right)^{0.33} \tag{26}$$

Suriyawong and Wongwises [44] performed experimentation on two different circular plates made of copper and aluminium with different roughness (0.2–4 μm) to check nucleate boiling heat transfer characteristics. When concentration was greater than 0.0001% by volume, a decrease in heat transfer coefficient had been observed. The reason for this deterioration is sedimentation of nanoparticles on heating surface and decrease in nucleation sites. Rough surfaces provide more heat transfer coefficient as compared to smooth surfaces because more nucleation sites are presented on such surfaces. Aluminium plate showed high heat transfer coefficient than copper plate.

Das et al. [45] checked the effects of surface modification on nucleate boiling heat transfer. In experimentation, Cu surface is coated with crystalline TiO_2 nanostructure. Increase in surface roughness, surface wet ability or surface coating thickness provide enhancement in boiling heat transfer coefficient.

7. Conclusion

This chapter gives an overview of titanium oxide nanofluids application in different heat transfer systems. Because of high thermal conductivity of these fluids as compared to simple water or other fluids, heat transfer systems using titanium oxide nanofluids performed more efficiently. Pressure drop due to the presence of nanoparticles was not significant. Therefore, no extra pumping power was required for circulation of nanofluids.

Nomenclature

Weight concentration	ω
Density of base fluid	ρ_{bf}
Density of nanoparticles	ρ_p
Density of nanofluid	ρ_{nf}
Thermal conductivity of nanofluid	k_{nf}
Thermal conductivity of base fluid	k_f

Thermal conductivity of nanoparticles	k_p
Volume fraction	Ø
Heat transfer coefficient	h
Surface area of tube	A_s
Bulk temperature	T_b
Tube wall temperature	T_s
Heat transfer rate	Q
Mass	m
Specific heat capacity	C
Nusselt number	Nu
Pressure drop	ΔP
Friction factor	f
Density	ρ
Velocity	v
Reynolds number	Re
Prandtl number	Pr
Peclet number	Pe
Over all heat transfer coefficient	U

Author details

Hafiz Muhammad Ali[1,2]*, Muhammad Usman Sajid[1] and Adeel Arshad[3,4]

*Address all correspondence to: h.m.ali@uettaxila.edu.pk

1 Mechanical Engineering Department, University of Engineering and Technology, Taxila, Pakistan

2 Center of Research Excellence in Renewable Energy (CoRE-RE), King Fahd University of Petroleum & Minerals (KFUPM), Dhahran, Saudi Arabia

3 Department of Mechanical Engineering, HITEC University, Taxila, Pakistan

4 Fluids & Thermal Engineering Research Group, Faulty of Engineering, University of Nottingham, Nottingham, UK

References

[1] Abdul Hamid K, Azmi WH, Mamat R, Sharma KV. Experimental investigation on heat transfer performance of TiO_2 nanofluids in water–ethylene glycol mixture. International Communications in Heat and Mass Transfer. 2016;**73**:16–24. DOI: 10.1016/j.icheatmasstransfer.2016.02.009

[2] Vakili M, Mohebbi A, Hashemipour H. Experimental study on convective heat transfer of TiO_2 nanofluids. Heat Mass Transfer. 2013;**49**(8):1159–1165. DOI: 10.1007/s00231-013-1158-3

[3] Azmi WH, Abdul Hamid K, Usri NA, Mamat R, Mohamad MS. Heat transfer and friction factor of water and ethylene glycol mixture based TiO_2 and Al_2O_3 nanofluids under turbulent flow. International Communications in Heat and Mass Transfer. **2016**;76:24–32. DOI: 10.1016/j.icheatmasstransfer.2016.05.010

[4] Wang J, Xie H, Guo Z, Guan L, Li Y. Improved thermal properties of paraffin wax by the addition of TiO_2 nanoparticles. Applied Thermal Engineering. 2014;**73**:1541–1547. DOI: 10.1016/j.applthermaleng.2014.05.078

[5] Azmi WH, Abdul Hamid K, Mamat R, Sharma KV, Mohamad MS. Effects of working temperature on thermo-physical properties and forced convection heat transfer of TiO_2 nanofluids in water-ethylene glycol mixture. Applied Thermal Engineering. 2016;**106**: 1190–1199. DOI: 10.1016/j.applthermaleng.2016.06.106

[6] Sajadi AR, Kazemi MH. Investigation of turbulent convective heat transfer and pressure drop of TiO_2/water nanofluid in circular tube. International Communications in Heat and Mass Transfer. 2011;**38**:1474–1478. DOI: 10.1016/j.icheatmasstransfer.2011.07.007

[7] Wei B, Zou C, Li X. Experimental investigation on stability and thermal conductivity of diathermic oil based TiO_2 nanofluids. International Journal of Heat and Mass Transfer. 2017;**104**:537–543. DOI: 10.1016/j.ijheatmasstransfer.2016.08.078

[8] Hamilton RL, Crosser OK. Thermal conductivity of heterogeneous two-component systems. Industrial & Engineering Chemistry Fundamentals. 1962;**1**(3):187–191. DOI: 10.1021/i160003a005

[9] Yu W, Choi SUS. The role of interfacial layers in the enhanced thermal conductivity of nanofluids: A renovated maxwell model. Journal of Nanoparticle Research. 2003;**5**(1): 167–171. DOI: 10.1023/A:1024438603801

[10] Timofeeva EV, Gavrilov AN, McCloskey JM, Tolmachev YV. Thermal conductivity and particle agglomeration in alumina nanofluids. Experiment and theory. Physical Review E. 2007;**76**(6). 061203. DOI: 10.1103/PhysRevE.76.061203

[11] Hussein AM, Bakar RA, Kadirgama K, Sharma KV. Heat transfer enhancement using nanofluids in an automotive cooling system. International Communications in Heat and Mass Transfer. 2014;**53**:195–202. DOI: 10.1016/j.icheatmasstransfer.2014.01.003

[12] Wadd MS, Warkhedkar RM, Choudhari VG. Comparing performance of nanofluids of metal and nonmetal as coolant in automobile radiator. International Journal of Advance Research in Science and Engineering, IJARSE, 2015;**4**(4):182–190, ISSN-2319-8354(E)

[13] Sandhya D, Reddy MCS, Rao VV. Improving the cooling performance of automobile radiator with ethylene glycol water based TiO_2 nanofluids. International Communications in Heat and Mass Transfer. 2016;**78**:121–126. DOI: 10.1016/j.icheatmasstransfer.2016.09.002

[14] Bhimani VL, Rathod PP, Sorathiya AS. Experimental study of heat transfer enhancement using water based nanofluids as a new coolant for car radiators. International Journal of Emerging Technology and Advanced Engineering. 2013;**3**(6):295–302

[15] Chen J, Jia J. Experimental study of TiO_2 nanofluid coolant for automobile cooling applications. Materials Research Innovations. 2016. DOI: 10.1080/14328917.2016.1198549

[16] Abdul Hamid K, Azmi WH, Mamat R, Usri NA, Najafi G. Effect of titanium oxide nanofluid concentration on pressure drop. ARPN Journal of Engineering and Applied Sciences. 2015;**10**(17)

[17] Rafati M, Hamidi AA, Shariati Niaser M. Application of nanofluids in computer cooling systems (heat transfer performance of nanofluids). Applied Thermal Engineering. 2012;**45-46**:9–14. DOI: 10.1016/j.applthermaleng.2012.03.028

[18] Ijam A, Saidur R. Nanofluid as a coolant for electronic devices (cooling of electronic devices). Applied Thermal Engineering. 2012;**32**:76–82. DOI: 10.1016/j.applthermaleng.2011.08.032

[19] Duangthongsuk W, Wongwises S. Heat transfer enhancement and pressure drop characteristics of TiO_2–water nanofluid in a double-tube counter flow heat exchanger. International Journal of Heat and Mass Transfer. 2009;**52**:2059–2067. DOI: 10.1016/j.ijheatmasstransfer.2008.10.023

[20] Gnielinski V. New equations for heat and mass transfer in turbulent pipe and channel flow. International Journal of Chemical Engineering. 1976;**16**(2):359–368

[21] Pak BC, Cho YI. Hydrodynamic and heat transfer study of dispersed fluids with submicron metallic oxide particles. Experimental Heat Transfer. 1998;**11**(2):151–170. DOI: 10.1080/08916159808946559

[22] Xuan Y, Li Q. Investigation on convective heat transfer and flow features of nanofluids. Journal of Heat Transfer. 2003;**125**(1):151–155. DOI: 10.1115/1.1532008

[23] Singh RN, Rajat P, Lav I, Pandey PK. Experimental studies of nanofluid TiO2/CuO in a heat exchanger (Double Pipe). Indian Journal of Science and Technology. 2016;**9**(31). DOI: 10.17485/ijst/2016/v9i31/93623

[24] Reddy MCS, Rao VV. Experimental investigation of heat transfer coefficient and friction factor of ethylene glycol water based TiO2 nanofluid in double pipe heat exchanger with and without helical coil inserts. International Communications in Heat and Mass Transfer. 2014;**50**:68–76. DOI: 10.1016/j.icheatmasstransfer.2013.11.002

[25] Khedkar RS, Sonawane SS, Wasewar KL. Heat transfer study on concentric tube heat exchanger using TiO2–water based nanofluid. International Communications in Heat and Mass Transfer. 2014;**57**:163–169. DOI: 10.1016/j.icheatmasstransfer.2014.07.011

[26] Duangthongsuk W, Wongwises S. An experimental study on the heat transfer performance and pressure drop of TiO2-water nanofluids flowing under a turbulent flow regime. International Journal of Heat and Mass Transfer. 2010;**53**:334–344. DOI: 10.1016/j.ijheatmasstransfer.2009.09.024

[27] Barzegarian R, Moraveji MK, Aloueyan A. Experimental investigation on heat transfer characteristics and pressure drop of BPHE (brazed plate heat exchanger) using TiO2-water nanofluid. Experimental Thermal and Fluid Science. 2016;**74**:11–18. DOI: 10.1016/j.expthermflusci.2015.11.018

[28] Tiwari AK, Ghosh P, Sarkar J. Performance comparison of the plate heat exchanger using different nanofluids. Experimental Thermal and Fluid Science. 2013;**49**:141–151. DOI: 10.1016/j.expthermflusci.2013.04.012

[29] Taghizadeh-Tabari Z, Heris SZ, Moradi M, Kahani M. The study on application of TiO2/water nanofluid in plate heat exchanger of milk pasteurization industries. Renewable and Sustainable Energy Reviews. 2016;**58**:1318–1326. DOI: 10.1016/j.rser.2015.12.292

[30] Javadi FS, Sadeghipour S, Saidur R, BoroumandJazi G, Rahmati B, Elias MM, Sohel MR. The effects of nanofluid on thermophysical properties and heat transfer characteristics of a plate heat exchanger. International Communications in Heat and Mass Transfer. 2013;**44**:58–63. DOI: 10.1016/j.icheatmasstransfer.2013.03.017

[31] Ashrafi N, Bashirzadeh Y. Heat transfer of Nano-Fluids as working fluids of swimming pool heat exchangers. International Journal of Advanced Design and Manufacturing Technology. 2014;**7**(1):75–81

[32] Kumar N, Sonawane SS. Experimental study of thermal conductivity and convective heat transfer enhancement using CuO and TiO2 nanoparticles. International Communications in Heat and Mass Transfer. 2016;**76**:98–107. DOI: 10.1016/j.icheatmasstransfer.2016.04.028

[33] Ali HM, Arshad W. Thermal performance investigation of staggered and inline pin fin heat sinks using water based rutile and anatase TiO2 nanofluids. Energy Conversion and Management. 2015;**106**:793–803. DOI: 10.1016/j.enconman.2015.10.015

[34] Mohammed HA, Gunnasegaran P, Shuaib NH. The impact of various nanofluid types on triangular microchannels heat sink cooling performance. International Communications in Heat and Mass Transfer. 2011;**38**:767–773. DOI: 10.1016/j.icheatmasstransfer.2011.03.024

[35] Naphon P, Nakharintr L. Heat transfer of nanofluids in the mini-rectangular fin heat sinks. International Communications in Heat and Mass Transfer. 2013;**40**:25–31. DOI: 10.1016/j.icheatmasstransfer.2012.10.012

[36] Sohel MR, Saidur R, Sabri MFM, Kamalisarvestani M, Elias MM, Ijam A. Investigating the heat transfer performance and thermophysical properties of nanofluids in a circular micro-channel. International Communications in Heat and Mass Transfer. 2013;**42**:75–81. DOI: 10.1016/j.icheatmasstransfer.2012.12.014

[37] Khaleduzzaman SS, Sohel MR, Saidur R, Selvaraj J. Convective performance of 0.1% volume fraction of TiO2/water nanofluid in an electronic heat sink. Procedia Engineering. 2015;**105**:412–417. DOI: 10.1016/j.proeng.2015.05.027

[38] Ijam A, Saidur R, Ganesan P. Cooling of minichannel heat sink using nanofluids. International Communications in Heat and Mass Transfer. 2012;**39**:1188–1194. DOI: 10.1016/j.icheatmasstransfer.2012.06.022

[39] Xia GD, Liu R, Wang J, Du M. The characteristics of convective heat transfer in microchannel heat sinks using Al_2O_3 and TiO_2 nanofluids. International Communications in Heat and Mass Transfer. 2016;**76**:256–264. DOI: 10.1016/j.icheatmasstransfer.2016.05.034

[40] Ali HM; Generous MM, Ahmad F, Irfan M. Experimental investigation of nucleate pool boiling heat transfer enhancement of TiO_2-water based nanofluids. Applied Thermal Engineering. 2017;**113**:1146–1151. DOI: 10.1016/j.applthermaleng.2016.11.127

[41] Pioro I. Boiling heat transfer characteristics of thin liquid layers in a horizontally flat two-phase thermosyphon. In: Preprints of the 10th International Heat Pipe Conference; September 1997; Stuttgart, Germany. Paper H1–5

[42] Trisaksri V, Wongwises S. Nucleate pool boiling heat transfer of TiO_2–R141b nanofluids. International Journal of Heat and Mass Transfer. 2009;**52**:1582–1588. DOI: 10.1016/j.ijheatmasstransfer.2008.07.041

[43] Rohsenow WM. A method of correlating heat transfer data for surface boiling of liquids. Transactions of ASME. 1952;**74**:969

[44] Suriyawong A, Wongwises S. Nucleate pool boiling heat transfer characteristics of TiO_2 water nanofluids at very low concentrations. Experimental Thermal and Fluid Science. 2010;**34**:992–999. DOI: 10.1016/j.expthermflusci.2010.03.002

[45] Das S, Saha B, Bhaumik S. Experimental study of nucleate pool boiling heat transfer of water by surface functionalization with crystalline TiO_2 nanostructure. Applied Thermal Engineering. 2017;**113**:1345–1357. DOI: 10.1016/j.applthermaleng.2016.11.135

6

Titanium Dioxide as Food Additive

Marie-Hélène Ropers, Hélène Terrisse,
Muriel Mercier-Bonin and Bernard Humbert

Abstract

Titanium dioxide is a white metal oxide used in many food categories as food additives to provide a whitening effect. If its use complies with the five specifications including synthesis pathway, crystallographic phase, purity, amount and innocuousness, all other parameters are not defined and were hardly documented. However, in the last 3 years, two studies have deeply characterized food-grade TiO_2 and converged to the fact that the size distribution of food-grade TiO_2 spans over the nanoparticle range (<100 nm) and the surface is not pure TiO_2 but covered by phosphate and eventually silicon species or aluminium species, which modify the surface chemistry of these particles. Until now, this material was considered as safe. However, the toxicological studies later to the last re-evaluation by the European Food Safety Agency reveal some concerns due to the ability of TiO_2 particles to alter the intestinal barrier. This reinforces the idea to go on reinforcing the risk assessment about food-grade TiO_2.

Keywords: TiO_2, anatase, food, E171, food additive, food safety, intestinal barrier

1. Introduction

Titanium dioxide (TiO_2) is a transition metal oxide with two main applications as either pigment or photocatalyst, in many sectors including buildings (self-cleaning windows, cements, paints and anti-fouling paints), paper industry, cosmetics (sunscreens and tooth paste), pharmaceutics (tablets), food (colouring agent) and others (air-purification system, rubbers, inks and ceramics). Pigmentary applications are by far the most important application of TiO_2. The interest in TiO_2 lies in the scattering of visible light controlled by its high refractive index and its granulometry (size and shape). These requirements also apply for food where TiO_2 provides a whitening effect. Although this compound has been used for a very long time in food products (with a recent re-approval for a permitted use in food by the European Food Safety Agency), the use of titanium dioxide in food has risen some concerns in Western populations due to the presence of nanoparticles, that is, particles having one or more external dimensions in the size range of 1–100 nm. This review chapter targets an audience of practicing researchers, academics and PhD students, who are interested in the food applications of this compound and the reasons of controversy.

2. Use of TiO_2 in foods: function, properties and safety

Titanium dioxide is a food additive without any nutritive value and added in processed foods to provide a whitening effect. It was first approved for use in food by the United States Food and Drug Administration (FDA) in 1966, then by the European Union in 1969, on the basis of the Codex Alimentarius of the Food and Agriculture Organization/World Health Organization (FAO/WHO). When used as a food colouring, it is labelled as E171 in Europe or INS171 in USA. In other fields, it is also called titanium white, Pigment White 6 or CI 77891. Time to time, it was re-evaluated for minor revisions of specifications in 2006, 2009, 2010 and 2012. In particular, the European Union decided in 2006 to allow the crystalline structure rutile in food in addition to the former authorized form anatase (COMMISSION DIRECTIVE 2006/33/ EC of 20 March 2006). Then, it was subjected to an in-depth evaluation in 2016 (EFSA 2016).

2.1. Food categories with permitted use of TiO_2

The food colours of Group II, including titanium dioxide, are authorized in most food categories,[1,2] such as (i) dairy products and analogues (flavoured fermented milk products and some creams), (ii) cheese and cheese products such as unripened cheese (Mozzarella, Codex Stan 262-2006 or fresh cheese, Codex Stan 221-2001), edible cheese rind, whey cheese, processed cheese, cheese products and dairy analogues including beverage whiteners, (iii) edible ices, (iv) confectionary (chewing gum, decorations, coatings and non-fruit-based fillings), (v) surimi and similar products and salmon substitutes, (vi) seasonings and condiments, mustard, soups and broths and sauces, and (vii) food supplements (Official Journal of the European Union, No 1129/2011). This list, despite its length, is in fact not exhaustive and the whole list with some restrictions of use is available on specialized websites.

Titanium dioxide was actually identified in chewing gums [1–3], confectionary [4, 5], sauces and dressings [5], non-dairy creamers [2, 5] and in dietary supplements [6]. According to a database collecting the details of new products (278,705) introduced on the market in 62 of the world's major economies, the use of TiO_2 increased constantly until 2014, representing a labelling on more than 3500 foods or drinks (Mintel GNPD database cited by the European Food Safety Agency EFSA [7]). If TiO_2 is found in only 1.3% of new products, it is nevertheless found in 51% of gums, 25% of stick, liquid and sprays, 21% of mixed assortments, 10% of pastilles, gums, jellies and chews and 10% of lollipops [7]. Chewing gums and confectionary, including pastilles, gums, jellies and chews, are the most widely concerned food categories, both in number of products labelled with TiO_2 per category and in number of new products available on the market. Cakes and pastries represent a second category of importance. This scenario has to be regularly refined as the composition of food products may evolve [8, 9].

2.2. Levels of consumption

The amount of TiO_2 consumed in the USA on a daily basis was estimated around 0.2–0.7 mg of TiO_2 per kg of body weight per day (mg/kg bw/d), while the UK and German populations consume around 1 mg TiO_2/kg bw/day [4, 10]. These data were refined for all food categories, subpopulations and exposure scenarios in Netherlands [11, 12], in Germany [10] and in Europe [7]. For example, the estimate of the median long-term exposure to titanium dioxide (E 171) ranges from 0.5 (upper limit 1.1 mg/kg bw/d) for elderly adults to 1.4 mg/kg bw/d (upper limit 3.2 mg/kg bw/d) for children in Netherlands [12], close to the estimate in Germany [10].

Whatever the scenario of exposure and methodological choices, the biggest consumers of TiO_2 are children (3–9 years) and teenagers (10–17 years) [4, 7, 10–12]. In the scenario exposure of EFSA, the contribution of chewing gums is weak in comparison to other confectionary including breath-refreshing microsweets, or sauces, salads and savoury-based sandwich spreads [7]. In the study based on the Dutch National Food Consumption Survey, the products most contributing to TiO_2 intake for young children (2–6-year-olds) are confectionary (sweets, chocolate products and chewing gums) and fine bakery wares (biscuits). For 7–69-year-olds and elderly (70+), the same food items are identified but in a different decreasing order: chewing gums, coffee creamers, sauces, then fine bakery wares. As 10 food items most contributing to TiO_2 intake represent 55%, we must keep in mind that TiO_2 intake is spread over many products, chewing gums contributing by only a few percentage points more than other food categories [11]. In a similar study performed in Germany, the food products that contribute the most to the total titanium intake by adults are savoury sauces, dressings, soft drinks and cheese (more than 75%) [10]. In addition to food products, tablets such as medicine and food supplements contain TiO_2 up to 3.6 mg/g [13], resulting in a higher total daily intake of TiO_2.

2.3. Specifications of TiO_2 for food applications

In addition to the respect of the permitted use in the above-mentioned food categories, the powder introduced in these food products must respect five criteria, namely synthesis pathways, structure, purity, amounts and, certainly, absence of toxicity (Commission Regulation (EU) No 231/2012 and Joint FAO/WHO Expert Committee on Food Additives (JECFA) [14]). Firstly, these criteria are described according to the recommended specifications, then they are commented and discussed with literature data.

2.3.1. Synthesis: sulphate and chloride processes

Depending on the desired crystalline phase, titanium dioxide is produced by either the sulphate or the chloride process. The anatase phase of titanium dioxide can only be made by the sulphate process, while the rutile phase of titanium dioxide can be obtained from both processes but the chloride process is more sustainable and provides crystals with a narrower particle size distribution than the sulphate process [15].

Briefly, in the sulphate process, sulphuric acid is used to digest the ilmenite ore ($FeTiO_3$ or FeO/TiO_2) into iron(II) sulphate and titanium salt ($Ti(SO_4)_2$). Iron(II) sulphate is removed from the liquor after dilution and crystallization/filtration to yield only the titanium salt ($Ti(SO_4)_2$) in the digestion solution. Then, some microcrystals of anatase are introduced into the liquor which is then hydrolysed under carefully controlled conditions to produce crystals of anatase. These are subsequently filtered, washed, calcined and micronized [13, 15, 16]. The chloride process, which generates rutile crystals, consists of a chlorination of the ore into titanium and iron chlorides which are then separated by distillation. Titanium chloride is then treated to remove impurities and oxidized in a controlled flame reactor to yield TiO_2 rutile crystals with the desired size [15, 16]. In addition, titanium dioxide may be coated with small amounts of alumina and/or silica to improve the technological properties of the product, which are described as blocker for photocatalytic activity [14].

Certain rutile grades of titanium dioxide as platelet form are produced using mica as a template. The specific properties of this pigment (interference colour) are controlled by the thickness of the coated titanium dioxide layer and by the coating process [13].

2.3.2. *Crystallographic structure*

Currently, E171 forms consist essentially of pure anatase and/or rutile. Until 2006, only the anatase form was authorized for food applications. Rutile has been authorized to replace anatase in food products especially in film coatings for food supplement tablets and foodstuffs [13]. In both anatase and rutile structures, the basic building block consists of a titanium atom surrounded by six oxygen atoms (**Figure 1**). The structures differ by the distortion and assembly of the octahedra [17]. In rutile, these octahedra are connected via their corners and edges (**Figure 1**) and the unit cell dimensions are $a = b = 4.587$ and $c = 2.953$ Å. For anatase, the octahedra are linked via edges and planes forming a unit cell with $a = b = 3.782$ and $c = 9.502$ Å (**Figure 1**). In each structure, the two bonds between the titanium and the apical oxygen atoms are slightly longer than the others (1.983 and 1.946 Å in the rutile structure, 1.966 and 1.937 Å in the anatase structure). Moreover, a sizeable deviation from a 90° bond angle was observed in anatase (92.6 and 102.3 Å, **Figure 1**).

Although both forms are authorized in foods, the characterization of samples in American and European laboratories shows that anatase is the predominant crystalline structure found in food applications [1, 3, 4, 18–20]. For example, five out of six chewing gums contained TiO_2 as anatase and only one contained a mixture of anatase and rutile [1]. Thus, the sulphate process seems to be predominant for obtaining pigmentary TiO_2 for food applications.

In the bulk structure, the titanium cations have a coordination number of 6 meaning the oxygen anions have a coordination number of 3 resulting from the trigonal planar coordination (**Figure 1**). But at the surface, anions and cations are said to be 'coordinatively unsaturated'. The lowly coordinated cations (Ti_{5c}) thus act as Lewis acids (electron pair acceptor) and are able to interact with electron donors like H_2O. Similarly, twofold-coordinated O atoms

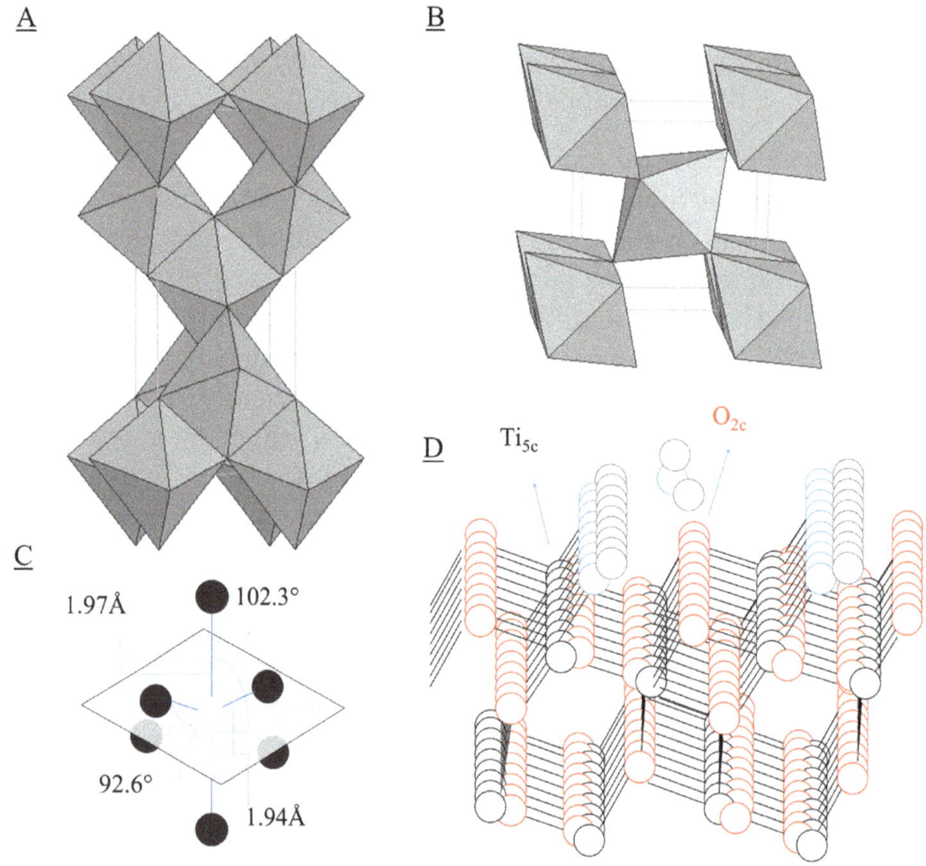

Figure 1. Bulk structures of (A) anatase, (B) rutile with (C) bond lengths and angles of the octahedrally coordinated Ti atoms in anatase, arranged from Diebold [17] and (D) arrangement of atoms on the (101) surface of anatase after adsorption and dissociation of water with Ti (grey filled balls), O from TiO_2 structure (empty balls), O from water (big hatched balls) and H from water (small hatched balls).

(O_{2c}) and named bridging oxygen atoms are Lewis base sites and are able to interact with electron acceptors like H^+. Thus, once the oxide surface is exposed to moisture present in the atmosphere, it becomes fully covered with adsorbed water and hydroxyl groups. Molecularly adsorbed water in vacancies partly dissociates to form two kinds of hydroxyl groups: (1) terminal hydroxyls which are adsorbed onto Ti_{5c} sites (TiOH) and (2) bridging hydroxyls which result from protonation of O_{2c} atoms (Ti_2OH) [21]. Surface hydroxyl groups are able to behave as Brønsted acid or base sites when TiO_2 particles are dispersed in water.

2.3.3. *Purity*

In Europe as well as in the USA, the content in titanium dioxide must be no less than 99.0% on an aluminium oxide and silicon dioxide-free basis (Commission Regulation (EU) No 231/2012) and the amount of alumina and/or silica must not exceed 2%. The investigated samples complied with these specifications [18–20, 22]. Additionally, the Commission specifies that the loss on drying must be lower than 0.5% (105°C, 3 h) and the loss on ignition must represent less than 1.0% (800°C) on the dried basis. The acid-soluble substances must represent less than 0.5% (less than 1.5% for products containing alumina or silica) and the water-soluble matter must represent less than 0.5%. For impurities soluble in 0.5 N hydrochloric acid, their amount must be lower than 1 mg/kg for arsenic, cadmium and mercury, lower than 2 mg/kg for antimony and lower than 10 mg/kg for lead. These specifications are very similar to those given by JECFA [14].

2.3.4. *Amounts*

In Europe, titanium dioxide is authorized *at quantum satis*, whereas it is used in the USA in the limit of 1% by weight of food. Although no maximum use level is specified for this additive in Europe, it shall be used in accordance with the good manufacturing practices (GMPs), that is, at a level not higher than is necessary to achieve the intended technical effect. This decision was motivated by the fact that TiO_2 was considered as an inactive ingredient in human food, and that neither significant absorption nor tissue storage following the ingestion of TiO_2 was possible. In its last report, the Panel of EFSA concluded that definitive and reliable data on the reproductive toxicity of E 171 are not yet available to enable the Panel to establish an acceptable daily intake (ADI) [7].

The quantification of TiO_2 in commercial products indicates that chewing gums are the food products richest in titanium dioxide [2, 4]. They contain between 0.7 and 5.4 mg Ti/g of food. The next category is sweets with 0–2.5 mg Ti/g food, followed by pastry with 0–0.5 mg Ti/g food [2]. In the report of EFSA, including more numerous food categories and data provided by industry, the highest maximum level in TiO_2 is in decorations, coatings and fillings [7] with 20 mg TiO_2/g food which corresponds to 12 mg Ti/g food, a little bit above the maximum level reported for chewing gums (16 mg TiO_2/g, i.e., 9.6 mg Ti/g food). Considering the mean use level, it is a little bit higher in processed nuts (3.8 mg Ti/g food) than in chewing gums (3.4 and 2.8 mg Ti/g food, depending on manufacturers), food supplements (2.8 mg Ti/g food) and salads and savoury-based sandwich spreads (2.5 mg Ti/g food).

2.3.5. *Innocuousness of TiO_2*

Since the early 1960s, TiO_2 is considered as safe for use in food. Since this time, some authors called this fact into question [23]. In the recent re-evaluation of titanium dioxide (E171) as food additive [7], the EFSA Panel estimated that the absorption of orally administered TiO_2 particles, including micro- and nano-sized (less than 3.2% by mass) fractions, was negligible, reaching at most 0.02–0.1% of the administered dose. They also indicated that no adverse effect resulting from the eventual accumulation of the absorbed particles was expected, based on the results of long-term studies which did not highlight any toxicity up to the highest administered dose. The lowest value found in the literature for the no-observed adverse effect levels (NOAEL) was 2250 mg TiO_2/kg bw/d.

2.4. Other physicochemical properties of food-grade TiO_2

Titanium dioxide is insoluble in water, hydrochloric acid, dilute sulphuric acid and organic solvents. It dissolves slowly in hydrofluoric acid and hot concentrated sulphuric acid. It is almost insoluble in aqueous alkaline media (COMMISSION DIRECTIVE 2008/128/EC).

The physicochemical characteristics of particles, including morphology (spherical and cylindrical), size (smaller or larger than <100 nm), surface charge (negative, neutral or positive), structure (crystallinity), agglomeration (aggregates, agglomerates and primary particles) and surface composition, are assumed or demonstrated to play a role in nanoparticle uptake through the gut [24]. Therefore, the five criteria detailed before have to be completed by a deeper characterization of food-grade TiO_2, which unfortunately received much less attention.

2.4.1. Content in nanoparticles and size distribution

Considering the food use of TiO_2 as whitening agent, the size distribution is expected to be centred on a mean pigment size of 250 nm to obtain an optimal effect [25]. However, the mean size of food-grade TiO_2 is actually rather comprised between 106 and 145 nm and the size distribution spans between 30 and 300 nm [4, 18, 19, 26, 27] or 60 and 300 nm [2]. For example, several size distribution spans and mean sizes are reported in **Figure 2**. Overall, they span between 30 and 300 nm. In these batches, the fraction of nanoparticles (<100 nm) ranged from 17 to 36%. In the whole set of samples investigated in the literature, the nanoparticle size distribution expressed in number was always smaller than 50%. In chewing gums, this fraction mounts to 43.7% [1].

To determine the exposure scenario, the equivalent mass of NPs is more interesting. According to several studies, the mass (wt%) of nanoparticles present in E171 ranges between 0.31 and 12.5% [7, 10, 11, 18]. This explains some discrepancies in the different exposure to TiO_2 nanoparticles in the literature and, for example, the factor of 10 in the estimate of NP

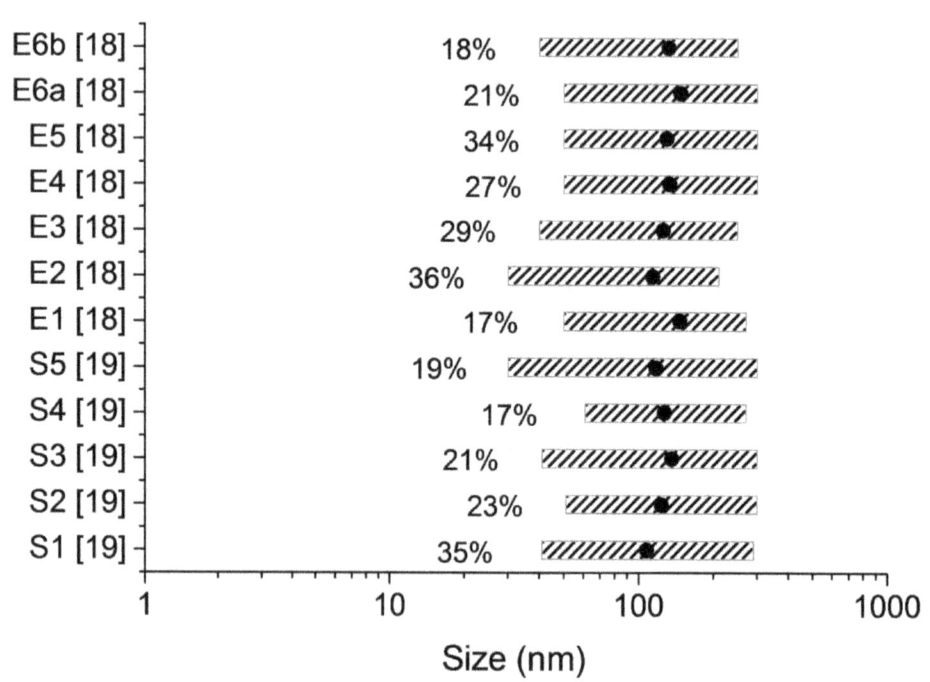

Figure 2. Size distribution (dashed rectangles), mean size (black dots) and percentage of nanoparticles in number (%) of food-grade TiO_2 particles characterized by (E) Dudefoi et al. [18] and (S) Yang et al. [19]. The mean sizes of the distribution vary between 106 and 145 nm.

consumptions between the study by Rompelberg et al. [11] who considered 0.31% of NPs and the evaluation of EFSA [7] who used a weight ratio of 3.2%.

2.4.2. Specific surface area

The specific surface area (SSA) of a material is defined as the total surface area of the material per unit of mass. It is reversely proportional to the size of the particles: the smaller the size of a material, the higher its specific surface area and its reactivity with the environment. The SSA is usually determined from the volumetric adsorption isotherms at 77 K of nitrogen gas followed by the Brunauer-Emmett-Teller (BET) adsorption treatment (the so-called N_2-BET isotherm) assuming a multilayer of adsorbates. The specific surface area of food-grade TiO_2 ranges between 8.6 and 10.7 m^2/g [18, 20] with an average of 9.3 m^2/g. These values are quite low in comparison to anti-caking agents, for example, which are around 200 m^2/g. This hints that TiO_2 offers a low contact surface with its environment.

2.4.3. Surface chemical composition

The surfaces of food-grade TiO_2 were found to be mainly covered by hydroxyl groups [18], phosphate groups [18, 19] and potassium ions [18]. Some phosphate groups may not be tightly bound to the surface and be released after washing [19]. In a few cases, TiO_2 was covered by silica [18] and alumina [19], thus modifying the surface chemistry.

2.4.4. Surface potential

As mentioned previously, surface hydroxyl groups, which behave as Brønsted acid or base sites, confer a charge to the particle surface. When TiO_2 particles are dispersed in an aqueous medium, this charge is mainly determined by two phenomena: protonation/deprotonation of surface hydroxyls controlled by pH and adsorption of electrolyte ions [28]. An electrostatic potential, exponentially decaying away from the surface, is associated to the overall charge distribution in the interfacial region. The experimental determination of this potential, called zeta potential, is generally performed by electrophoretic mobility measurements. All models converting electrophoretic mobility into zeta potential consider ideal spherical particles, which is a delicate assumption in the case of TiO_2 due to the formation of agglomerates with non-spherical particles (subsequent section). An improved model exists to convert electrophoretic mobility measurements to zeta potential values taking into account the effect of the agglomerate size and surface conductance of TiO_2 [29]. Zeta potential values depend not only on the parameters controlling the surface charges, namely, the nature of the medium where TiO_2 particles are dispersed (pH, ionic strength and adsorbed species [20]) but also on the primary particle size [29, 30] and the crystallographic face [31]. The point where the zeta potential is zero defines the isoelectric point (IEP).

The isoelectric point of food-grade TiO_2 samples measured by electrophoretic mobility measurements was found between 3 and 4 for most samples (**Table 1**), far below the classical value for anatase. Such a difference is interpreted by the presence of phosphate groups on the surface of TiO_2 particles [18, 19] or by silica coating [18], which decrease the isoelectric point

Reference	[18]	[20]	[19]
Experimental conditions	Ultrapure water, without fixing ionic strength	Ultrapure water, without fixing ionic strength	KNO_3 10^{-2} mol.L^{-1}
IEP	$4.0 \leq pH \leq 4.2$	pH = 5.1	$3.2 \leq pH \leq 4.0$
ζ at pH 7	−42 to −50 mV	−35 mV	−42 to −50 mV

Table 1. Isoelectric point (IEP) and zeta potential at pH 7 of various food-grade TiO_2 (E171) dispersed in water, without any protein.

towards lower pH values. It is interesting to note that the isoelectric point of a food-grade sample measured through electroacoustic measurements gave a value of 5.1 [20], close to the classical data for anatase. For all these samples, the zeta potential of their suspensions varies between −35 and −45 mV at a physiological pH value. Faust et al. compared the zeta potential of a food-grade TiO_2 and an extract of chewing gum, and observed that the gum extract presented a largely more negative potential (−45 mV at pH 7) than food-grade TiO_2 (−20 mV at pH 7), which may be due to coating of TiO_2 in chewing-gum formulation [26].

2.4.5. *Agglomeration*

The dispersion state of particles in aqueous solution is governed by the surface chemistry of the oxide and depends on the composition of the dispersion medium (pH, ionic strength, nature of electrolyte and presence of proteins). Traditionally, zeta potential measurements are used to assess the stability of colloidal dispersions: the higher the zeta potential absolute value, the more stable the dispersion. Around the IEP or when ionic strength is high in solution, the system is unstable and agglomeration of particles occurs, leading to settling of the suspension. It is thus important to consider agglomeration in the experimental medium, as this may alter the size of the particles which will be 'seen' later by the organism after ingestion.

In usual conditions of pH and ionic strength, TiO_2 particles tend to form large-sized agglomerates (particles relatively loosely bound) which settle after a few hours, partially due to the large density of TiO_2 (3.9 g/cm^3 for anatase as powder). For neutral pH values (around 6–7) and in the absence of any salt, E171 particles present agglomerates with a diameter of 200–400 nm, in agreement with the largely negative-measured zeta potential. When pH becomes closer to IEP, the measured diameter is larger than 1 μm, which is the sign of agglomeration due to low electrostatic repulsions [18].

Once particles are agglomerated or aggregated, they do not fragment easily and are difficult to disperse as primary particles. Ultrasound sonication can be used to break the agglomerates prior to zeta potential and size measurements, providing ultrasounds do not alter the surface chemistry of the material [32]. The hydrodynamic diameter of E171 particles dispersed in ultrapure water (pH not mentioned) and bath sonicated (for 5–30 min) comprises between 120 and 400 nm [19, 26]. Another possibility to stabilize the suspension and avoid agglomeration of particles consists in adding a dispersant which is able to cover the particles and create steric hindrance between them [33]. Bovine serum albumin (BSA) was typically used to stabilize E171 TiO_2 particles, in combination with ultrasound sonication (30 min), leading to a mean hydrodynamic diameter of 150 nm [4]. In solutions added with salts (NaCl and $NaHCO_3$), E171 particles dispersed by sonication presented a moderate stability, with a particle size of agglomerates remaining between 360 and 390 nm for at least 2 h. The same experiment conducted with P25 sample showed rapid and extensive aggregation of the particles [4].

2.4.6. Specificities of food-grade TiO_2

Food-grade TiO_2 powders are finally characterized by a low specific surface area (around 10 m^2/g), a pure crystalline anatase phase (sometimes traces of rutile), a low isoelectric point (around 4.1 in water) related to the phosphate found at its surface, a mean size of 140 nm with a distribution spanning from 30 to 300 nm and a fraction of nanoparticles comprised between 17 and 36%. For toxicological studies, including toxicity assessment by oral exposure, another kind of TiO_2, called P25, is commonly used as it is considered as a reference material [34]. This compound is characterized by 100% NPs, a mean size of 23 nm, a specific surface area of 50 m^2/g, a mixture of anatase and rutile grains (85/15) and an isoelectric point at pH 6.5 [18, 19]. In **Figure 3**, some physical and chemical properties of E171 and P25 samples, extracted from two studies [18, 19], are reported.

The P25 samples clearly distinguish from E171 samples by all parameters taken into account. A peculiar sample of E171, rich in rutile phase, is observed as well. E171 TiO_2 being strongly different from the reference material P25, we thus concluded that P25 does not appear to be the most suitable reference material for toxicity studies by ingestion [18]. It is, moreover, not the most relevant material to represent the nanoparticle fraction of E171.

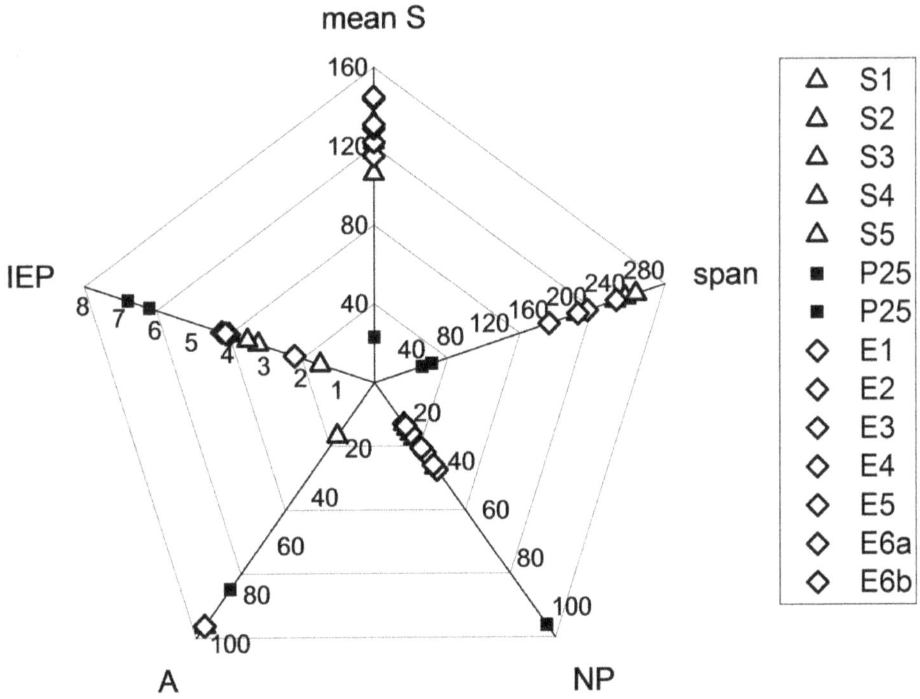

Figure 3. Physical and chemical parameters describing E71 and P25 forms of TiO_2, namely the content in NPs, isoelectric point (IEP), the mean size of the distribution deduced by transmission electron microscopy (mean S) and the span of the particle size distribution (span). Data come from references [8, 19].

2.5. Fate of TiO_2 after ingestion

Among the different routes of exposure to TiO_2, the oral uptake route remains the less documented. Once ingested, TiO_2 particles pass through the digestive tract, starting with the port of entry, the oral cavity followed by the gastrointestinal tract, comprising oesophagus, stomach, small and large intestines and rectum (**Figure 4**).

During the transit through the digestive fluids, TiO_2 particles were not metabolized and were found to be mainly agglomerated, mediated by proteins and electrolytes [35, 36], but according to some studies, a small fraction is still in the nanosized range [35, 37, 38]. The low absorption of TiO_2 and reversely the high percentage of titanium dioxide excreted from the body in faeces [39, 40] were believed to be the proof of any adverse effect. However, the recent data on the intestinal compartment call this belief into question. Indeed, the intestinal barrier, which involves epithelium, mucus and microbiota in its luminal side (**Figure 4**), provides a physical, chemical and biological line of defence for the host, probably through an orchestrated manner [41, 42]. Taken together or independently, these three partners exhibit some alterations due to the presence of TiO_2 particles, which are briefly reported from the microbiota to the epithelium.

2.5.1. *TiO_2 in interaction with the intestinal microbiota*

The effects of TiO_2 on the gut microbiota composition and metabolic activity in animal models or humans are largely unknown, whereas the intestinal microbiota contributes actively to the maintenance of host homeostasis. Indeed, it plays a key role in the gut, fulfilling protection, maturation and production functions. In particular, it acts as a barrier against pathogens, preventing their implantation, and participates in xenobiotic metabolism [43].

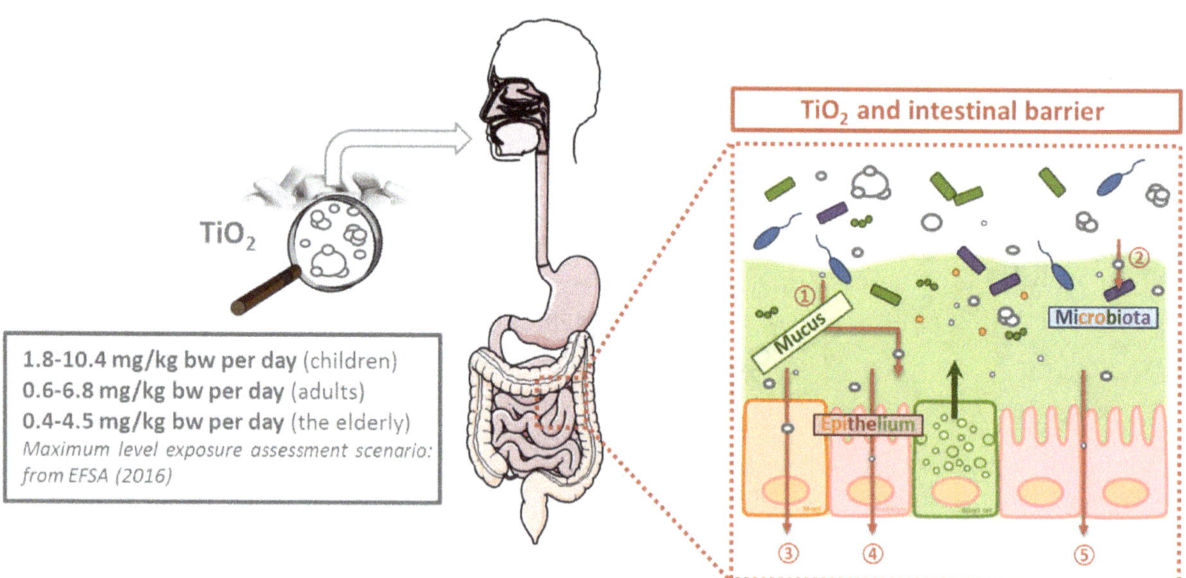

Figure 4. Schematic representation of the fate of TiO_2 within the digestive tract illustrating potential mechanisms by which ingested nanoparticles interact with the intestinal barrier; (1) mucus; (2) microbiota; (3–5) epithelium with (3) internalization and active transport to Peyer's patch lymphoid follicles by M-cells, (4) transcellular transport and (5) paracellular transport through intercellular tight junctions (intercellular space between adherent epithelial cells). The unrestricted migration through foci of damaged epithelium is not represented. For the sake of clarity, scheme is not to scale.

Studies reported to date were mainly focused on the antibacterial activity of TiO_2 nanoparticles in *in vitro* pure cultures using *Escherichia coli* as the bacterial representative [44–46]. Such an activity is generally associated with the photocatalytic effects of TiO_2, although increasing experimental evidence also demonstrated TiO_2-mediated cell alterations without UV illumination [50, 51]. Taylor et al. [47] investigated the *in vitro* exposure of a gut microbial community from a healthy donor to three different types of metal oxide nanoparticles, including TiO_2, in a model colon. Such exposure-induced changes in the phenotypic traits of the gut community, including short-chain fatty acid production (particularly for butyric acid), cell hydrophobicity, sugar content of extracellular polymers, cell size and electrophoretic mobility. In a further study, Waller et al. [48] evaluated the impact of food-grade TiO_2 (vs industrial-grade TiO_2) on the composition and phenotype of a human gut microbiota. An inhibition of the control-induced shift in microbial composition from Proteobacteria to Firmicutes phyla was observed. TiO_2 exposure also resulted in a lower value of the colonic pH (~pH 4) as compared to the control (>5). Additionally, similar trends in microbial community hydrophobicity and electrophoretic mobility were obtained between control and food-grade exposures. Interestingly, different microbial responses were observed with the industrial-grade form, underlying the significance of physical and chemical properties of TiO_2 in intestinal homeostasis.

2.5.2. *TiO_2 in interaction with the intestinal mucus*

Mucus is the viscoelastic gel that lines and protects the intestinal epithelium. It is secreted continuously along the whole intestine by specialized goblet cells in the epithelium (**Figure 4**), and is present in larger amounts in the colon than elsewhere. Mucus was long considered to act as a 'simple' physical barrier, but it is now known to have other key functions essential for the preservation of intestinal homeostasis [49–51], including (i) lubrication of the epithelium, facilitating the progress of material along the digestive tract, (ii) maintenance of a stable microenvironment at the epithelial surface, (iii) protection of the epithelium through the presence of immune system molecules and (iv) provision of an ecological niche for the intestinal microbiota.

Interactions between TiO_2 and intestinal mucus are far from being understood. Variable capacities for absorption and transport of TiO_2 nanoparticles have been described *in vitro* [52], depending on whether epithelial cells are cultured alone or in the presence of mucus-secreting goblet cells. In fact, Caco-2 cells in monoculture only displayed low levels of intracellular nano-TiO_2 accumulation after 24-h exposure, whereas the same treatment in Caco-2/HT29-MTX mucus-producing co-culture led to 50 times higher levels of accumulation [52]. In ex vivo studies on porcine buccal mucosa [36, 38], TiO_2 nanoparticles, regardless of their size and hydrophilicity/hydrophobicity, were able to permeate mucus and penetrate underlying tissues.

2.5.3. *TiO_2 in interaction with the intestinal epithelium*

Epithelium is in charge of nutrients and water absorption while restricting the access for potentially noxious substances to the internal organs. Thus, it constitutes a selective—and dynamic—barrier, mediating transport of compounds through the transcellular pathway (i.e., across the cells) and/or the paracellular pathway (i.e., between the cells). It is polarized into an apical and basolateral surface with the apical surface covered with microvilli to increase the absorptive surface area. There are at least three pathways enabling uptake/translocation of TiO_2 nanoparticles (**Figure 4**): first, they can disrupt the cell junctions (paracellular route), second they can be internalized by the cells (transcellular route, e.g., endocytosis) and finally they can exert a toxic effect on the cells or alter their function, resulting in cell death [53]. In addition, many studies underlined the involvement of the M-cell-rich layer of Peyer's patches which are epithelial cells specialized for the transcytosis of macromolecules and particles [40, 53–56] (**Figure 4**). However, this mechanism of translocation is still under debate since contradictory results were obtained for *in vitro* cells [36, 40, 53, 55–57].

In vitro studies, mainly on Caco-2 cells, converge on the possible TiO_2-mediated disruption of the epithelial barrier. Indeed, subtle or more substantial alterations were depicted, including cytotoxicity [58], alteration of the brush-border microvilli [26, 53], upregulation of nutrient transporters and efflux pumps [59], production of reactive oxygen species [59, 60], misbalance of redox repair systems [59], increase in epithelial permeability [60] and uptake/translocation of TiO_2 nanoparticles [53, 55, 60], at a different extent according to the type of TiO_2 nanoparticles (size and crystal phase) and experimental conditions used.

In line with the findings of Faust et al. [26], recent piece of evidence suggests some adverse effects of oral exposure to E171 on the intestinal mucosa barrier with a putative additional impact on intestinal diseases and colorectal cancer [61–63]. Proquin et al. [63] showed *in vitro* that E171-induced ROS formation and DNA damage through its micro-sized and/or nano-sized fractions in Caco-2 and HCT116 cells. In rodents, Bettini et al. [61] found TiO_2 particles present in Peyer's patches along the small intestine as well as in the colonic mucosa of rats orally given E171 at human relevant levels. No significant change in epithelial paracellular permeability was observed.

2.5.4. Biodistribution of TiO_2

When TiO_2 particles overcome the mucus/microbiota/epithelium-protective triad, they may enter systemic circulation [64, 65] but in an extremely limited amount [36] and infiltrate organs like liver and kidney which are the organs for exogenous chemicals metabolism and for the excretion of metabolic wastes, respectively. But they were also found in lung, spleen and brain [66, 67] and presented a poor clearance [67]. With a half-life of 12.7 days [66], TiO_2 particles may be thus regularly renewed in the organism, suggesting a bioaccumulation [23] but there is an absence of toxicological effects in the conditions of the study [66]. In the terminal ileum of children suspected of having inflammatory bowel disease, the amount of pigment in Peyer's patches became denser with increasing age [68].

3. Conclusion

With the aim to ensure a healthy food, the knowledge about TiO_2 as food additive increased in the last 5 years. Among the large set of TiO_2 samples, E171 food-grade materials have different physicochemical properties from the reference material P25. Indeed, it is characterized by a low specific surface area (around 10 m^2/g), a pure anatase crystalline phase (sometimes traces of rutile), a low isoelectric point (around 4.1 in ultrapure water) mainly related to the phosphate found at its surface, a mean size of around 140 nm with a distribution spanning from 30 to 300 nm and a fraction of nanoparticles comprised between 17 and 36%. Due to the lack of data on E171, the risk assessment of oral exposure to TiO_2 has been mainly performed with TiO_2 nanomaterials like P25 which possess a different surface chemistry. As TiO_2 has a low absorption rate, it is mostly excreted in the faeces, suggesting that it does not present any toxicity concern. Nevertheless, there is an increasing awareness of proved or suspected deleterious effects of TiO_2 during its transit in the digestive tract, by compromising intestinal homeostasis before absorption in the upper compartments and/or throughout the entire intestine by the non-absorbed fraction. Albeit increasingly recognized as key players in gut health, mucus and microbiota have often been neglected in food nanotoxicology and should now be more deeply investigated. The link with some intestinal diseases needs to be confirmed as well. For all further studies, the use of food-grade forms of TiO_2 is more relevant than that of the nanomaterial P25.

Author details

Marie-Hélène Ropers[1]*, Hélène Terrisse[2], Muriel Mercier-Bonin[3] and Bernard Humbert[2]

*Address all correspondence to: marie-helene.ropers@inra.fr

1 INRA, Biopolymers Interactions and Assemblies, Nantes, France

2 Institute of Materials Jean ROUXEL (IMN), Université de Nantes, CNRS, Nantes, France

3 INRA, Toxalim (Research Centre in Food Toxicology), University of Toulouse, INRA, ENVT, INP-Purpan, UPS, Toulouse, France

References

[1] Chen X-X, Cheng B, Yang Y-X, Cao A, Liu J-H, Du L-J, et al. Characterization and preliminary toxicity assay of nano-titanium dioxide additive in sugar-coated chewing gum. Small. 2013;**9**(9-10):1765-1774

[2] Peters RJB, van Bemmel G, Herrera-Rivera Z, Helsper HPFG, Marvin HJP, Weigel S, et al. Characterization of titanium dioxide nanoparticles in food products: Analytical methods to define nanoparticles. Journal of Agriculture & Food Chemistry. 2014;**62**(27):6285-6293

[3] Periasamy VS, Athinarayanan J, Al-Hadi AM, Juhaimi F Al, Mahmoud MH, Alshatwi AA. Identification of titanium dioxide nanoparticles in food products: Induce intracellular oxidative stress mediated by TNF and CYP1A genes in human lung fibroblast cells. Environmental Toxicology & Pharmacology. 2015;**39**(1):176-186

[4] Weir A, Westerhoff P, Fabricius L, Hristovski K, von Goetz N. Titanium dioxide nanoparticles in food and personal care products. Environmental Science & Technology. 2012;**46**(4):2242-2250

[5] Lomer MCEMCE, Thompson RPHRPH, Commisso J, Keen CL, Powell JJ. Determination of titanium dioxide in foods using inductively coupled plasma optical emission spectrometry. Analyst. 2000;**125**(12):2339-2343

[6] Lim J-H, Sisco P, Mudalige TK, Sánchez-Pomales G, Howard PC, Linder SW. Detection and characterization of SiO_2 and TiO_2 nanostructures in dietary supplements. Journal of Agriculture & Food Chemistry. 2015;**63**(12):3144-3152

[7] EFSA ANS Panel (EFSA Panel on Food Additives and Nutrient Sources added to Food). Scientific opinion on the re-evaluation of titanium dioxide (E 171) as a food additive. EFSA Journal. 2016;**14**(9):83. DOI: 10.2903/j.efsa.2016.4545

[8] William D, Ana V, Stéphane P, Céline M, Marie-Hélène R, Gontard Nathalie CB. Nanosciences and nanotechnologies for biobased materials, packaging and food applications: New opportunities and concerns. Innovative Food Science and Emerging Technologies. 2017; in revision

[9] Rydström Lundin C. Nanoparticles in Food—With a Focus on the Toxicity of Titanium Dioxide. Uppsala: Uppsala Universitet; 2012

[10] Bachler G, von Goetz N, Hungerbuhler K. Using physiologically based pharmacokinetic (PBPK) modeling for dietary risk assessment of titanium dioxide (TiO_2) nanoparticles. Nanotoxicology. 2015;**9**(3):373-380

[11] Rompelberg C, Heringa MB, van Donkersgoed G, Drijvers J, Roos A, Westenbrink S, et al. Oral intake of added titanium dioxide and its nanofraction from food products, food supplements and toothpaste by the Dutch population. Nanotoxicology. 2016;**10**(10):1404-1414

[12] Sprong C, Bakker M, Niekerk M, Vennemann M. Exposure Assessment of the Food Additive Titanium Dioxide (E 171) based on Use Levels Provided by the Industry [Internet]. 2015. Available from: http://www.rivm.nl/dsresou

[13] EFSA. Opinion of the scientific panel on food additives, flavourings, processing aids and materials in contact with food on a request from the commission related to the safety in use of rutile titanium dioxide as an alternative to the presently permitted anatas. EFSA Journal [Internet]. 2004;**163**:1-12. Available from: http://www.efsa.europa.eu/sites/default/files/scientific_output/files/main_documents/163.pdf

[14] Joint FAO/WHO Expert Committee on Food Additives. Compendium of Food Additive Specifications, 77th Meeting [Internet]. FAO JECFA Monographs. 2013. Available from: http://www.fao.org/docrep/019/i3544e/

[15] Braun JH. Titanium dioxide—A review. Journal of Coatings Technology. 1997;**69**(868): 59-72

[16] Kuznesof PM, Rao MV. Titanium Dioxide. Chemical and Technical Assessment [Internet]. 2006. Available from: ftp://ftp.fao.org/ag/agn/jecfa/cta_tio2.pdf

[17] Diebold U. The surface science of titanium dioxide. Surface Science Reports. 2003;**48**: 53-229

[18] Dudefoi W, Terrisse H, Richard-Plouet M, Gautron E, Popa F, Humbert B, et al. Criteria to define a more relevant reference sample of titanium dioxide in the context of food: A multiscale approach. Food Additives & Contaminants: Part A. 2017; in press. Available from: https://www.tandfonline.com/doi/full/10.1080/19440049.2017.1284346.

[19] Yang Y, Doudrick K, Bi X, Hristovski K, Herckes P, Westerhoff P, et al. Characterization of food-grade titanium dioxide: The presence of nanosized particles. Environmental Science & Technology. 2014;**48**(11):6391-6400

[20] Rezwan K, Studart AR, Vörös J, Gauckler LJ. Change of ζ potential of biocompatible colloidal oxide particles upon adsorption of bovine serum albumin and lysozyme. Journal of Physical Chemistry B. 2005;**109**(30):14469-14474

[21] Kung HH. Transition Metal Oxides: Surface Chemistry and Catalysis. In: Delmon B, Yates TJ, editors. Amsterdam: Elsevier Science Publishers B.V.; 1989. pp. 1-299

[22] Mutsuga M, Sato K, Hirahara Y, Kawamura Y. Analytical methods for SiO_2 and other inorganic oxides in titanium dioxide or certain silicates for food additive specifications. Food Additives & Contaminants Part A. 2011;**28**(4):423-427

[23] Jovanović B. Critical review of public health regulations of titanium dioxide, a human food additive. Integrated Environmental Assessment and Management. 2015;**11**(1):10-20

[24] Yada RY, Buck N, Canady R, DeMerlis C, Duncan T, Janer G, et al. Engineered nanoscale food ingredients: Evaluation of current knowledge on material characteristics relevant to uptake from the gastrointestinal tract. Comprehensive Reviews in Food Science and Food Safety. 2014;**13**(4):730-744

[25] Theissmann R, Kluwig M, Koch T. A reproducible number-based sizing method for pigment-grade titanium dioxide. Beilstein Journal of Nanotechnology. 2014;**5**:1815-1822

[26] Faust JJ, Doudrick K, Yang Y, Westerhoff P, Capco DG. Food grade titanium dioxide disrupts intestinal brush border microvilli in vitro independent of sedimentation. Cell Biology and Toxicology. 2014;**30**(3):169-188

[27] Athinarayanan J, Alshatwi AA, Periasamy VS, Al-Warthan AA. Identification of nanoscale ingredients in commercial food products and their induction of mitochondrially mediated cytotoxic effects on human mesenchymal stem cells. Journal of Food Science. 2015;**80**(2):N459-N464

[28] Panagiotou GD, Petsi T, Bourikas K, Garoufalis CS, Tsevis A, Spanos N, et al. Mapping the surface (hydr)oxo-groups of titanium oxide and its interface with an aqueous solution: The state of the art and a new approach. Advances in Colloid and Interface Science. 2008;**142**(1-2):20-42

[29] Leroy P, Tournassat C, Bizi M. Influence of surface conductivity on the apparent zeta potential of TiO_2 nanoparticles. Journal of Colloid and Interface Science. 2011;**356**(2): 442-453

[30] Suttiponparnit K, Jiang J, Sahu M, Suvachittanont S, Charinpanitkul T, Biswas P. Role of surface area, primary particle size, and crystal phase on titanium dioxide nanoparticle

dispersion properties. Nanoscale Research Letters. 2010;**6**(27):28271. Available from: http://www.nanoscalereslett.com/content/6/1/27

[31] Bullard JW, Cima MJ. Orientation dependence of the isoelectric point of TiO_2 (rutile) surfaces. Langmuir. 2006;**22**(24):10264-10271

[32] Taurozzi JS, Hackley VA, Wiesner MR. Ultrasonic dispersion of nanoparticles for environmental, health and safety assessment—Issues and recommendations. Nanotoxicology. 2011;**5**(4):711-729

[33] Guiot C, Spalla O. Stabilization of TiO_2 nanoparticles in complex medium through a pH adjustment protocol. Environmental Science and Technology. 2013;**47**(2):1057-1064

[34] Orts-Gil G, Natte K, Osterle W. Multi-parametric reference nanomaterials for toxicology: State of the art, future challenges and potential candidates. RSC Advances. 2013;**3**(40):18202-18215

[35] Teubl BJ, Schimpel C, Leitinger G, Bauer B, Frohlich E, Zimmer A, et al. Interactions between nano-TiO_2 and the oral cavity: Impact of nanomaterial surface hydrophilicity/hydrophobicity. Journal of Hazardous Materials. 2015;**286**:298-305

[36] Jones K, Morton J, Smith I, Jurkschat K, Harding A-H, Evans G. Human in vivo and in vitro studies on gastrointestinal absorption of titanium dioxide nanoparticles. Toxicology Letters. 2015;**233**(2):95-101

[37] Teubl BJ, Leitinger G, Schneider M, Lehr CM, Frohlich E, Zimmer A, et al. The buccal mucosa as a route for TiO_2 nanoparticle uptake. Nanotoxicology. 2014;**9**(2):253-261

[38] Tay CY, Fang WR, Setyawati MI, Chia SL, Tan KS, Hong CHL, et al. Nano-hydroxyapatite and nano-titanium dioxide exhibit different subcellular distribution and apoptotic profile in human oral epithelium. ACS Applied Materials & Interfaces. 2014;**6**(9):6248-6256

[39] Cho W-S, Kang B-C, Lee JK, Jeong J, Che J-H, Seok SH. Comparative absorption, distribution, and excretion of titanium dioxide and zinc oxide nanoparticles after repeated oral administration. Part Fibre Toxicology. 2013;**10**(1):9

[40] MacNicoll A, Kelly M, Aksoy H, Kramer E, Bouwmeester H, Chaudhry Q. A study of the uptake and biodistribution of nano-titanium dioxide using in vitro and in vivo models of oral intake. Journal of Nanoparticle Research. 2015;**17**(2):66

[41] Burger-van Paassen N, Vincent A, Puiman PJ, van der Sluis M, Bouma J, Boehm G, et al. The regulation of intestinal mucin MUC2 expression by short-chain fatty acids: implications for epithelial protection. Biochemistry Journal. 2009;**420**:211-219

[42] Tomas J, Wrzosek L, Bouznad N, Bouet S, Mayeur C, Noordine ML, et al. Primocolonization is associated with colonic epithelial maturation during conventionalization. FASEB Journal. 2013;**27**(2):645-655

[43] Gill SR, Pop M, DeBoy RT, Eckburg PB, Turnbaugh PJ, Samuel BS, et al. Metagenomic analysis of the human distal gut microbiome. Science 2006;**312**(5778):1355-1359

[44] Liu P, Duan WL, Wang QSI, Li X. The damage of outer membrane of Escherichia coli in the presence of TiO_2 combined with UV light. Colloids and Surfaces B: Biointerfaces. 2010;**78**(2):171-176

[45] Kumar A, Pandey AK, Singh SS, Shanker R, Dhawan A. Engineered ZnO and TiO_2 nanoparticles induce oxidative stress and DNA damage leading to reduced viability of Escherichia coli. Free Radical Biology & Medicine. 2011;**51**(10):1872-1881

[46] Zhukova LV, Kiwi J, Nikandrov VV. TiO_2 nanoparticles suppress Escherichia coli cell division in the absence of UV irradiation in acidic conditions. Colloids and Surfaces B: Biointerfaces. 2012;**97**:240-247

[47] Taylor A, Marcus MI, Guysi LR, Walker LS. Metal oxide nanoparticles induce minimal phenotypic changes in a model colon gut microbiota. Environmental Engineering Science. 2015;**32**(7):602-612

[48] Waller T, Chen C, Walker S. Food and Industrial Grade Titanium Dioxide Impacts Gut Microbiota. Environmental Engineering Science [Internet]. 2017; in press. Available from: http://online.liebertpub.com/doi/10.1089/ees.2016.0364

[49] Johansson MEV, Sjovall H, Hansson GC. The gastrointestinal mucus system in health and disease. Nature Reviews Gastroenterology & Hepatology. 2013;**10**(6):352-361

[50] Juge N. Microbial adhesins to gastrointestinal mucus. Trends in Microbiology. 2012;**20**(1): 30-39

[51] Ouwerkerk JP, de Vos WM, Belzer C. Glycobiome: Bacteria and mucus at the epithelial interface. Best Practice & Research: Clinical Gastroenterology. 2013;**27**(1):25-38

[52] Brun E, Barreau F, Veronesi G, Fayard B, Sorieul S, Chaneac C, et al. Titanium dioxide nanoparticle impact and translocation through ex vivo, in vivo and in vitro gut epithelia. Particle and Fibre Toxicology. 2014;**11**(1):13. Available from: http://particleandfibretoxicology.biomedcentral.com/articles/10.1186/1743-8977-11-13

[53] Koeneman BA, Zhang Y, Westerhoff P, Chen Y, Crittenden JC, Capco DG. Toxicity and cellular responses of intestinal cells exposed to titanium dioxide. Cell Biology and Toxicology. 2010;**26**(3):225-238

[54] Pietroiusti A, Magrini A, Campagnolo L. New frontiers in nanotoxicology: Gut microbiota/microbiome-mediated effects of engineered nanomaterials. Toxicology and Applied Pharmacology. 2016;**299**:90-95

[55] Gitrowski C, Al-Jubory AR, Handy RD. Uptake of different crystal structures of TiO_2 nanoparticles by Caco-2 intestinal cells. Toxicology Letters. 2014;**226**(3):264-276

[56] Janer G, Mas del Molino E, Fernández-Rosas E, Fernández A, Vázquez-Campos S. Cell uptake and oral absorption of titanium dioxide nanoparticles. Toxicology Letters. 2014;**228**(2):103-110

[57] McCracken C, Dutta PK, Waldman WJ. Critical assessment of toxicological effects of ingested nanoparticles. Environmental Science: Nano. 2016;**3**(2):256-82. Available from: http://dx.doi.org/10.1039/C5EN00242G

[58] Gerloff K, Fenoglio I, Carella E, Kolling J, Albrecht C, Boots AW, et al. Distinctive toxicity of TiO2 rutile/anatase mixed phase nanoparticles on Caco-2 cells. Chemical Research in Toxicology. 2012;**25**(3):646-655

[59] Dorier M, Brun E, Veronesi G, Barreau F, Pernet-Gallay K, Desvergne C, et al. Impact of anatase and rutile titanium dioxide nanoparticles on uptake carriers and efflux pumps in Caco-2 gut epithelial cells. Nanoscale. 2015;**7**(16):7352-7360

[60] Ruiz PA, Morón B, Becker HM, Lang S, Atrott K, Spalinger MR, et al. Titanium dioxide nanoparticles exacerbate DSS-induced colitis: Role of the NLRP3 inflammasome. Gut. 2016. Available from: http://gut.bmj.com/lookup/doi/10.1136/gutjnl-2015-310297.

[61] Bettini S, Boutet-Robinet E, Cartier C, Coméra C, Gaultier E, Dupuy J, et al. Food-grade TiO_2 impairs intestinal and systemic immune homeostasis, initiates preneoplastic lesions and promotes aberrant crypt development in the rat colon. Scientific Reports. 2017;**7**:40373

[62] Urrutia-Ortega IM, Garduno-Balderas LG, Delgado-Buenrostro NL, Freyre-Fonseca V, Flores-Flores JO, Gonzalez-Robles A, et al. Food-grade titanium dioxide exposure exacerbates tumor formation in colitis associated cancer model. Food and Chemical Toxicology. 2016;**93**:20-31

[63] Proquin H, Rodríguez-Ibarra C, Moonen CG, Urrutia Ortega IM, Briedé JJ, de Kok TM, et al. Titanium dioxide food additive (E171) induces ROS formation and genotoxicity: Contribution of micro and nano-sized fractions. Mutagenesis. 2017;**32**(1):139-149

[64] Pele LC, Thoree V, Bruggraber SFA, Koller D, Thompson RPH, Lomer MC, et al. Pharmaceutical/food grade titanium dioxide particles are absorbed into the bloodstream of human volunteers. Particle and Fibre Toxicology. 2015;**12**:26

[65] Böckmann J, Lahl H, Eckert T, Unterhalt B. Titan-Blutspiegel vor und nach Belastungs-versuchen mit Titandioxid. Pharmazie. 2000;**55**(2):140-3.

[66] Elgrabli D, Beaudouin R, Jbilou N, Floriani M, Pery A, Rogerieux F, et al. Biodistribution and clearance of tio2 nanoparticles in rats after intravenous injection. PLoS One. 2015;**10**(4):e0124490

[67] Disdier C, Devoy J, Cosnefroy A, Chalansonnet M, Herlin-Boime N, Brun E, et al. Tissue biodistribution of intravenously administrated titanium dioxide nanoparticles revealed blood-brain barrier clearance and brain inflammation in rat. Particle and Fibre Toxicology. 2015 Dec 4;**12**(1):27. Available from: http://particleandfibretoxicology.biomedcentral.com/articles/10.1186/s12989-015-0102-8

[68] Hummel TZ, Kindermann A, Stokkers PCF, Benninga MA, ten Kate FJW. Exogenous pigment in Peyer patches of children suspected of having IBD. Journal of Pediatric Gastroenterology and Nutrition. 2014;**58**(4):477-480

7

Titanium Dioxide in Sunscreen

Megha Trivedi and Jenny Murase

Abstract

Titanium dioxide has been used in various industrial and cosmetic applications due to its unique elemental properties. This substance has a refractive index higher than most other compounds (n = 2.6142, in comparison to water at 20°C n = 1.33). This allows titanium dioxide to have an iridescent and bright quality. It has been used in the cosmetics industry for "whitening" and "thickening" in multiple make-up brands. As titanium dioxide has the ability to greatly absorb UV light, it has been used as a physical sunscreen for many years. Over time, newer formulations, including "nanoparticle" formulations, have been marketed for ease of consumer use. We aim to discuss the evolution of titanium dioxide in sunscreens over time, discuss its mechanisms of actions, and comment on the efficacy and safety of these products.

Keywords: titanium dioxide in sunscreens, titanium dioxide, skin, allergic contact dermatitis

1. Introduction

A systematic search with the keywords "titanium dioxide" and "sunscreen" all in the title was conducted using the Google Scholar database. No restrictions on year of publication, language, full-text availability, human- or animal-based studies were applied to the initial search. The results were then manually filtered using a systematic approach. Articles that did not pertain to the topic of titanium dioxide and sunscreens were excluded. Articles which were published only in a language other than English or in which full text was not available were excluded. A similar search and systematic review was conducted using the PUBMED database with the same keywords. However, the filters "clinical trials," "review," "full text," and "humans" were applied to result in 133 results which were then manually filtered excluding articles in non-English languages and not pertaining to the topic.

Titanium dioxide (TiO_2) is a Food and Drug Administration (FDA)-approved sunscreen, which is considered broad spectrum (protects against UVB radiation as well as UVA2 radiation 320–340 nm). Previously, TiO_2 had a suboptimal cosmetic profile, appearing thick and white on application. Current formulations are micronized or nanoparticle formulations, which blend in with the skin tone and attain better cosmesis. Concern about dermal penetration of these smaller particulate formulations and safety has been raised. However, current data are controversial. Several studies demonstrate negligible penetration beyond the stratum corneum, whereas other studies demonstrate cytotoxicity and oxidative stress in cell models. More studies are needed to definitively comment on long-term use of TiO_2 sunscreens and health effects. TiO_2 has interestingly been implicated in allergic contact dermatitis in gold allergic patients in recent years. We discuss a newly hypothesized mechanism of TiO_2 and gold particulate matter interaction on the surface of the skin.

2. History of photoprotection and titanium dioxide

For generations, humans have sought various methods of sun protection, extending to ancient times. Ancient Egypt, India, Greece, and indigenous American populations used physical barriers such as hats, umbrellas, cloth wrapped over the head and face, and even topical materials such as tars and oils. During those times, these efforts were mostly directed at preventing darkening effects from the sunlight as in some of these cultures, fairer complexion, especially for female members of society, was desired [1]. The more important effects of ultraviolet radiation, that is, damage to the skin's cells and subsequent cancerous/mutagenic potential from prolonged sunlight exposure were not studied until 1800s [1]. The development and use of sunscreens on a population-wide scale was initiated during and after WWII as American troops experienced high intensity sunlight exposure from spending months to years in equatorial and tropical regions [2]. Multiple products and agent have been used in sunscreens to date and the FDA has currently approved 17 active ingredients, which are allowed in sunscreens bought and sold in the USA. With the variety of sunscreen agents used in cosmetic and UV protection products, Australia, Canada, and the European Union (EU) have also developed regulatory protocols on safe sunscreen product use. Unlike the USA though, Australia has approved 34 active sunscreen ingredients and the EU has approved 28 of these ingredients. This is thought to be due to the designation of sunscreen products in the USA as over-the-counter agents unlike in other countries which designate many of these ingredients as cosmetic. The former type of designation requires greater scrutiny in the approval process by a regulatory body [2, 3]. The introduction of titanium dioxide use in sunscreens came in 1952, around which time other physical blockers such as zinc oxide and its derivatives were also being introduced. Even so, the FDA only recently approved the use of titanium dioxide in sunscreens in 1999 [4]. However, titanium dioxide along with zinc oxide, although effective in sun protection, were not popular commercially when first introduced. This is because of the large particle size of these agents and thick white, opaque appearance upon application [3]. This characteristic of the products made them cosmetically unappealing and less preferred among consumers. However, in the 1990s, micronized versions of titanium dioxide sunscreens were marketed for their cosmetic superiority to older versions. These versions along with "nanoparticle" formulations of the product quickly rose to popularity.

However, the newer formulations came with their own set of troubles with questions about safety, toxicity, and efficacy arising from the consumer as well as scientific community as is discussed in the following sections.

3. Mechanism of action

Titanium dioxide is a widely used substance due to its convenient chemical properties, ease of mass production, and relatively low cost. The substance is chemically inert, which means that it rarely interacts with other chemical substances to undergo a reaction which can change its own chemical composition or the chemical milieu of the surrounding environment. This is true for biological surfaces and environments as well, thus leading to its safe use in food additives, cosmetics, and sunscreen products [4]. TiO_2 is a semiconducting material with very high refractive indices for all its crystalline forms. The high refractive index is what allows the substance to scatter visible light as well as has a significant ability to whiten substances in which it is an additive. There are three main crystalline forms in which TiO_2 can occur: rutile, anatase, and brokite. The rutile form of TiO_2 has an impressive refractive index with the average of rutile polycrystalline films being $n = 4.0$ from experimentation. This refractive index is higher than that of zinc oxide derivatives, and therefore TiO_2 has a greater whitening effect [5]. Titanium dioxide is used in sunscreens mainly due to its ability to reflect and scatter ultraviolet radiation. However, the high refractive index is not the only property that determines how well a substance blocks light. The film thickness in which the substance is suspended as well as the size of the individual particles also affects the efficacy of the sunscreen. However, the reflectant property, a numerical value often used to compare efficacies of different sunscreen agents does rely wholly on refractive index of the substance. The formula for this measure is defined by the refractive index of the substance divided by the refractive index of the surrounding medium [6]. This high refractive index of TiO_2 as discussed earlier also confers it an opaque, cosmetically unappealing appearance upon topical application. Therefore, most formulations are micronized, and the properties of these agents are more relevant to discuss. Older forms of TiO_2 products were created with particle sizes of 150–300 nm. The newer micro-formulations which are transparent upon topical application have TiO_2 particle sizes of 20–150 nm [2]. The micronized particle formula of TiO_2 does not cross the stratum corneum and confers broad spectrum UV protection against both UVB and UVA2 which includes wavelengths from 315–340 nm. However, there is no protection against UVA1 wavelength ranges [7]. The caveat when manufacturing micronized forms of inorganic particulate sunscreens is the possibility of agglomeration. This describes a phenomenon in which the TiO_2 particles have a tendency to join together and form larger clusters. This physically causes the sunscreen to appear opaque once again on application and reduces its photoprotection efficacy defeating the purpose of the micronized formula. Therefore, many manufacturers coat the TiO_2 particles with either silica or dimethicone to prevent aggregation. In addition to micronized formulas, it should be mentioned that the addition of other ingredients into TiO_2 formulations has also been tried. Iron oxide, another inorganic blocking agent, naturally has a reddish tinge which can resemble the complexion of human skin. This substance is sometimes added to TiO_2 products to reduce the white opacity of the final application. Furthermore, iron oxide, itself being an inorganic blocking agent, enhances the photoprotection of TiO_2 formulations when included [2, 3, 6].

For the sake of comprehensive discussion, the mechanism of action of organic sunscreens, also known as chemical sunscreens should also be noted. Whereas, the inorganic blockers, mainly titanium dioxide, zinc oxide, and iron oxide compounds, scatter and reflect light, the organic sunscreens actually absorb UV light which tends to be high energy in the natural environment and release it in a lower energy state that does not damage skin cells. Chemically, most of these compounds are aromatic and have a conjugated carbonyl group. These include aminobenzoates, anthralates, cinnamates, salicylates, benzophenones, dibenzoylmethane, and camphor to name a few [7].

4. Ultraviolet radiation and its effects on human skin

It is worthwhile to briefly discuss ultraviolet radiation, its definition and categorization, and its effects on human skin. The main source of ultraviolet radiation exposure for human beings is natural sunlight. There are three main categories of ultraviolet radiation including UVA, UVB, and UVC radiation. UVC radiation (100–280 nm)[1] will not be emphasized as this form of radiation does not generally reach the earth's surface to a significant degree. The earth's atmosphere blocks most UVC rays, and therefore, they are not thought to be important contributors to the biological effects on human skin [8]. UVB radiation (280–315 nm),[1] on the other hand, has been implicated in three major adverse effects in human beings: sunburn (radiation-induced erythema and inflammation of the skin), skin cancer, and immunosuppression. However, it should be noted that UVB does play a beneficial role in human health and wellness as it is important in the conversion of vitamin D in the skin to useable forms for human metabolism. Recent evidence in the literature also suggests that it may lower the risk of certain cancers including colon, prostate, and breast cancer [8]. UVA radiation (315–400 nm)* was previously regarded as having insignificant carcinogenic potential and mainly implicated as a cause of aging and wrinkling. However, recent evidence has suggested that it too has an important role in carcinogenesis and immunosuppression [8, 9]. Although, the major source of human exposure to UVA is still natural sunlight, tanning beds have had an increasing role in UVA exposure during recent years [8].

The molecular effects of UV radiation on human skin can be broadly separated into short-term (acute) effects, and long-term, continuous effects. When the skin is acutely exposed to UV radiation, DNA damage in the form of pyrimidine dimers occur. However, most humans with normal cell functioning are able to fix these lesions through DNA repair mechanisms. Furthermore, when individual cells recognize that too much damage has ensued and normal DNA repair processes will not suffice, cells engage in active, purposeful cell death to prevent the cell from multiplying and propagating mutagenic DNA which would lead clinically to cancer. On a physical level, acute UV radiation exposure can lead to two main changes: sunburn and tanning. Sunburn describes the phenomenon thickening of the outer skin layer (stratum corneum) and swelling in the layers of the skin containing blood vessels. This clinically appears as swollen, red skin that is painful to the touch. In addition to sunburn, tanning

[1]These wavelength ranges vary depending on the source of the text. However, they are all within range of ±5 nm.

is a known side effect of UV radiation exposure as the cells in the skin responsible for pigment production (melanocytes) are stimulated to produce more pigment for unclear physiological reasons. Certain theories suggest that the hormone melanocyte-stimulating hormone and its receptor may have a role depending on individual skin type [9]. Long-term effects of sunlight exposure include both photoaging and increased risk of skin cancer including both non-melanoma type as well as melanoma. Photoaging results from thickening of outer skin layers and breakdown in the support structure proteins of the skin, mainly collagen and elastin. The clinical manifestations of photoaging include freckles, dryness, hyperpigmentation, wrinkling, and dilated blood vessels [9]. Malignancy resulting from chronic UV radiation exposure is thought to result from two major processes. Chronic UV exposure can lead to accumulating genetic mutations which can affect normal DNA repair and cell self-destruction mechanisms over time. The accumulation of genetic changes can lead to cell transformation and dysplasia which can eventually lead to clinical malignancy. In addition to genetic changes, chronic UV exposure is also thought to play a role in immune suppression in the outer layers of the skin which can make it easier for cancerous cells to escape detection and subsequent cell death as the immune system plays a role in the elimination of genetically damaged cells [9].

5. Efficacy of titanium dioxide sunscreens

The efficacy of sunscreens is determined by the ability to protect against both UVB radiation and UVA radiation. As discussed in previous paragraphs, titanium dioxide sunscreens are effective broad-spectrum agents, meaning that they are capable of protection against both types of radiation. A sunscreen's efficacy in protecting against UVB radiation is assessed based on two values: the sun protection factor (otherwise known as SPF) and substantivity. The SPF is calculated as a ratio of the amount of ultraviolet radiation needed to cause sunburn when the skin is protected with a certain sunscreen to the amount of ultraviolet radiation needed to cause sunburn when the skin is left unprotected. Therefore, if a product is SPF 30, then 30 times the UV radiation is needed to cause sunburn if the patient is wearing the sunscreen with that SPF value [10]. Generally, in order to be considered broad spectrum, a sunscreen must be at least SPF 30 or greater. Substantivity is another measure that sunscreens are subjected to when being evaluated for use in the United States. This measurement evaluates the ability of a sunscreen to withstand exposure to factors which can cause the physical removal of the agent off the skin (and still remain effective afterward). These factors include sweat and water, among others. By measuring the substantivity of a product, manufacturers can have insight into the product's actual usability in the real-world setting. Interestingly, until 1933, many products were being labeled as "waterproof" based on superior product stability when tested with water immersion. However, the FDA released a statement disapproving the terminology as misleading since all products lose effectiveness after a certain amount of exposure. Now, the term "water-resistant" is more commonly seen on products [2].

Efficacy of UVA radiation protection is often measured by immediate pigment darkening (IPD), persistent pigment darkening (PPD), and protection factor (PFA). Both IPD and PPD measure photo-oxidation of melanin in the skin which contributes to skin pigmentary changes.

However, PPD is seen as the better measure in most cases, as the measurement takes place 2–24 h following exposure to radiation unlike IPD which reaches peak levels within 1 min of exposure. PPD also measures the photostability of the sunscreen being tested. Protection factor measures redness and tanning 24 h after UV radiation exposure and can also be used to test UVA protection efficacy [2].

Both titanium dioxide and zinc oxide sunscreen formulations have several advantages as inorganic blockers. These agents are not only effective against UV radiation but also protect against infrared radiation and visible light. As mentioned before, titanium dioxide is chemically inert—this is useful since in most cases, titanium dioxide is combined with other organic sunscreens and does not react with other ingredients in the compound. In the past, both titanium dioxide and zinc oxide sunscreens were less preferred due to their chalky, white appearance on application. However, micronization of particles and more recently "nanoparticle" formulations have made TiO_2 sunscreens more popular in recent years [11]. Early studies comparing micronized versions of TiO_2 versus micronized versions of zinc oxide sunscreens concluded that micronized zinc oxide sunscreens were superior as they conferred better protection against longer wave UVA and had greater cosmetic outcomes [12]. However, nanoparticle formulations of TiO_2 have virtually replaced all other forms in the sunscreen market and these sunscreens have demonstrated great effectiveness. Nanoparticle TiO_2 sunscreens not only have superior cosmetic outcomes as these product are virtually invisible upon application, but they also possibly confer greater UV protection. One study published by Tyner and colleagues in 2011 demonstrated that nanoparticle TiO_2 may have greater UV radiation attenuation. The authors also commented that overall formulation of the sunscreen is important when attaining high levels of efficacy. The best TiO_2 sunscreens occur when particles are coated, stabilized, and distributed evenly on the surface of the skin. Coated nanoparticles are very effective at non-agglomeration and therefore distribute effectively on the skin surface [13]. It is also important to note that the vehicle of the sunscreen can also affect effectiveness. The best vehicles avoid significant interaction between active and inactive ingredients. Vehicle also affects the substantivity measure of the sunscreen since the vehicle is often the main determinant of how well the sunscreen resists water, sweat, and other environmental factors. The vehicle types currently available include lotions and creams, gels, sticks, and sprays. Gels are often the most sensitive to water and sweat removal [3].

6. Safety of titanium dioxide sunscreens

The safety of sunscreens has various dimensions and is assessed most commonly by toxicity studies. However, irritation and allergic sensitization as well as dermal penetration are also commonly studied when assigning safety profiles to different products [10].

Titanium dioxide and zinc oxide have always been considered safe alternatives to their organic counterparts because of their chemically inert status. However, with the introduction of micro and nanoparticulate formulations, titanium dioxide sunscreens have fallen under greater scrutiny. The two most important concerns noted have been the potential of TiO_2 nanoparticles to generate reactive oxygen species when exposed to UV radiation which can cause DNA damage within skin cells and the dermal penetration of TiO_2 nanoparticles themselves.

Other concerns related to the size of the particles themselves include greater surface reactivity due to the fact that nanoparticles have a greater proportional surface area compared to their larger particle counterparts. Catalytic reactions with proteins and translocation into tissues, which may lead to autoimmune consequences, and evasion of regular immune surveillance mechanisms have also been discussed [14].

Since the 1990s, there have been several studies with evidence supporting the claim that TiO_2 nanoparticles can generate potentially dangerous free radicals. In 2001, Serpone and colleagues demonstrated that TiO_2 particles absorbed significant UV radiation to generate hydroxyl-radical species. These free radicals can cause DNA strand breaks and lead to increased susceptibility to genetic aberrations [15]. In 2006, Hidaka and colleagues published similar results describing specific damage to DNA induced by titanium dioxide and zinc oxide which had absorbed UV radiation. This led the subsequent production of free radicals, mainly hydroxyl species which then caused conformational changes in DNA at a rapid rate [16]. However, even though there was ample evidence for reactive oxygen species forming from irradiation of titanium oxide and zinc oxide nanoparticles, Newman and colleagues aptly noted in a recent review that the true biological consequences on humans was not truly known. In order for the reactive oxygen species to be concerning, they would have to penetrate deeper levels of skin. The question of whether nanoparticles penetrate the outer layer of skin to any significant degree is controversial, but there have been several published studies in recent years which provide evidence against such a phenomenon [14].

In 1999, Lademan and colleagues conducted a study in which penetration of coated titanium dioxide microparticles into the horny layer of the skin and orifice of the hair follicle was evaluated. They took biopsies of relevant skin areas and noted that penetration of particles had occurred in the open segments of the hair follicles, but this was less than one percent of the total sunscreen applied. They also noted that no particles were found in the viable sections of skin tissue [17]. Pflucker and colleagues published a study in 2001 which used electron and light microscopy to evaluate whether dermal penetration of micronized titanium dioxide particles occurred to any measurable degree. Their results suggested that particles could only be found in the top most layer of the stratum corneum and further penetration into deeper layers of stratum corneum, epidermis, or dermis was not noted [18]. Sadrieh and colleagues conducted a similar more recent study in 2010 using electron microscopy-energy dispersive x-ray analysis and again noted that there was no significant penetration of particles past the intact epidermis [19]. A general review of the toxicology of titanium dioxide nanoparticles by Shi and colleagues was published in 2013. They concluded that most dermal exposure studies with TiO_2 had not found significant penetration of particles into deeper layers of skin. However, the review noted that long term studies of TiO_2 inhalation were associated with lung tumors in rats. Several reviewed studies also suggested that intravenous injection of TiO_2 nanoparticles could potentially result in pathological lesions to a variety of internal organs [20].

The current opinion and recommendation on nanoparticles is controversial with the data that has been presented in recent years. The European Union's Scientific Committee on Emerging and Newly Identified Health Risks notes that TiO_2 nanoparticle use in sunscreens and cosmetics appears to be safe, but they advise caution when using such products on areas of impaired barrier function until more data can be obtained [21].

Another potential health and safety concern with the general use of sunscreens is the potential of vitamin D deficiency. Appropriate vitamin D synthesis necessitates certain levels of UVB exposure and the American Academy of Dermatology has recently revised its position on sunscreen effects on adequate vitamin levels. They note several populations are at higher risk for vitamin D deficiency including the elderly, darker skinned those who live in areas of low sunlight exposure, the obese, and the photosensitive. It is generally recommended that individuals obtain 1000 international units of vitamin D daily and some may have to supplement this through oral formulations [3].

Unlike other organic sunscreens, the inert nature of titanium dioxide sunscreens makes it one of the least likely compounds to cause irritation and sensitization of the skin. In fact, because of its chemical inertness and lack of photoreactivity, it is often the preferred sunscreen for individuals with sunscreen allergies and children [3]. However, in recent years an interesting interaction between gold allergic patients and titanium dioxide has been proposed and studied. In 2005, Nedorost and Wagman hypothesized that patients with facial and eyelid dermatitis may be experiencing reactions due to an interaction between their gold jewelry (earrings, bracelets, etc.) and the titanium dioxide in their cosmetic products. Mechanistically, titanium dioxide was thought to adsorb gold particulate matter which then caused contact dermatitis in the locations of makeup application. Their results demonstrated a subgroup of individuals who used TiO_2 containing makeup on areas then affected by contact dermatitis, benefiting from gold jewelry avoidance, lending credence to the interaction theory [22]. In 2015, Danesh and Murase published a clinical pearl in the Journal of the American Academy of Dermatology regarding a trial of gold avoidance in patients presenting with eyelid allergic contact dermatitis. They noted that the North American Contact Dermatitis Group implicated gold as the most common allergen to cause eyelid dermatitis most likely through a mechanism in which eyelid cosmetics containing titanium dioxide particles adsorbed gold particles from elsewhere on the body [23].

7. General recommendations on appropriate sunscreen use

The American Academy of Dermatology has set forth guidelines for appropriate sunscreen use and sun protection. They recommend using sunscreen with broad-spectrum properties (protection against both UVA and UVB radiation) daily. These include sunscreens with sun protection factors greater than 30. Sunscreen should be applied 15–30 min prior to sun exposure and be reapplied as needed when environmental factors (swimming and sweating) cause sunscreen removal [3].

8. Conclusions

In conclusion, titanium dioxide sunscreens are effective agents against UV radiation. They also have an advantage as they protect against multiple wavelengths of energy including UV radiation, visible light, and infrared radiation. Furthermore, unlike organic compounds, titanium dioxide, like its zinc counterpart, is a chemically inert substance, and therefore, there is less concern about reactivity in terms of toxicity profiles and allergic reactions. Although these agents were not heavily marketed in previous decades due to their thick, chalky appearance on the skin, the development of nanoparticle formulas improved their cosmetic profile. Nanoparticle formulations are considered effective sun protection agents as well as cosmetically acceptable. Concerns about safety with induction of radical oxygen species on the skin as well as absorption have been discussed. However, current evidence does not support claims of nanoparticle toxicity to humans. With this said, a greater number of studies with longitudinal data will be helpful in further informing this topic.

Author details

Megha Trivedi[1,2]* and Jenny Murase[2,3]

*Address all correspondence to: mtrivedi41@gmail.com

1 University of Michigan Medical School, Ann Arbor, MI, USA

2 Department of Dermatology, University of California, San Francisco, CA, USA

3 Department of Dermatology, Palo Alto Medical Foundation, CA, USA

References

[1] Urbach F. The historical aspects of sunscreens. Journal of Photochemistry and Photobiology B: Biology. 2001;**64**(2-3): 99-104.

[2] Palm MD, O'Donoghue MN. Update on photoprotection. Dermatologic Therapy. 2007;**5**:360-376

[3] Sambandan DR, Ratner D. Sunscreens: An overview and update. Journal of the American Academy of Dermatology. 2011;**64**:748-758

[4] Skocaj M, Filipic M, Petkovic J, Novak S. Titanium dioxide in our everyday life; is it safe? Radiology and Oncology. 2011;**45**(4):227-247

[5] Smijs TG, Pavel S. Titanium dioxide and zinc oxide nanoparticles in sunscreens: Focus on their safety and effectiveness. Nanotechnology, Science and Applications. 2011;**4**:95-112

[6] Murphy GM. Sunblocks: Mechanisms of action. Photodermatology, Photoimmunology & Photomedicine. 1999;**15**(1):34-36

[7] Rai R, Shanmuga SC, Srinivas CR. Update on photoprotection. Indian Journal of Dermatology. 2012; **57**(5):335

[8] Gallagher RP, Lee TK. Adverse effects of ultraviolet radiation: A brief review. Progress in Biophysics and Molecular Biology. 2006;**92**:119-131

[9] Matsumura Y, Ananthaswamy HN. Toxic effects of ultraviolet radiation on the skin. Toxicology and Applied Pharmacology. 2004;**195**:298-308

[10] Latha MS, Martis J, Shobha V, Shinde RS, Bangera S, Krishnankutty B, et al. Sunscreening agents: A review. Journal of Clinical and Aesthetic Dermatology. 2013;**6**(1):16-26

[11] Fathi-Azarbayjani A, Tan PL, Chan YY, Chan SY. Ascorbic acid for the safe use of a sunscreen agent: Accumulation of nano zinc oxide and titanium dioxide on the skin. Scientia Pharmaceutica. 2013; **81**(4):1141-1150

[12] Pinnell SR, Fairhurst D, Gillies R, Mitchnick MA, Kollias N. Microfine zinc oxide is a superior sunscreen ingredient to microfine titanium dioxide. Dermatologic Surgery. 2000;**26**(4):309-314

[13] Tyner KM, Wokovich AM, Godar DE, Doub WH, Sadrieh N. The state of nano-sized titanium dioxide (TiO_2) may affect sunscreen performance. International Journal of Cosmetic Science. 2011;**33**(3):234-244

[14] Newman MD, Stotland M, Ellis JI. The safety of nanosized particles in titanium dioxide- and zinc oxide-based sunscreens. Journal of the American Academy of Dermatology. 2009;**61**(4):685-692

[15] Serpone N, Salinaro A, Emeline A. Deleterious effects of sunscreen titanium dioxide nanoparticles on DNA: Efforts to limit DNA damage by particle surface modification. In: BiOS 2001 The International Symposium on Biomedical Optics. Proceedings of SPIE, 2001;**4258**:86-98

[16] Hidaka H, Kobayashi H, Koike T, Serpone N. DNA damage photoinduced by cosmetic pigments and sunscreen agents under solar exposure and artificial UV illumination. Journal of Oleo Science. 2006;**55**(5):249-261

[17] Lademann J, Weigmann HJ, Rickmeyer C, Barthelmes H, Schaefer H, Mueller G, Sterry W. Penetration of titanium dioxide microparticles in a sunscreen formulation into the horny layer and the follicular orifice. Skin Pharmacology and Physiology. 1999;**12**(5):247-256

[18] Pflücker F, Wendel V, Hohenberg H, Gärtner E, Will T, Pfeiffer S, et. al. The human stratum corneum layer: An effective barrier against dermal uptake of different forms of topically applied micronised titanium dioxide. Skin Pharmacology and Physiology. 2001;**14**(Suppl. 1):92-97

[19] Sadrieh N, Wokovich AM, Gopee NV, Zheng J, Haines D, Parmiter D, Siitonen PH, Cozart CR, Patri AK, McNeil SE, Howard PC. Lack of significant dermal penetration of titanium dioxide (TiO_2) from sunscreen formulations containing nano-and sub-micron-size TiO_2 particles. Toxicological Sciences. 2010;**115**(1):156-166

[20] Shi H, Magaye R, Castranova V, Zhao J. Titanium dioxide nanoparticles: A review of current toxicological data. Particle and fibre toxicology. 2013;**10**(1):15

[21] Jansen R, Osterwalder U, Wang SQ, Burnett M, Lim HW. Photoprotection: Part II. Sunscreen: Development, efficacy, and controversies. Journal of the American Academy of Dermatology. 2013;**69**(6):867-e1.

[22] Nedorost S, Wagman A. Positive patch-test reactions to gold: Patients' perception of relevance and the role of titanium dioxide in cosmetics. Dermatitis. 2005;**16**(2):67-70

[23] Danesh M, Murase JE. Titanium dioxide induces eyelid dermatitis in patients allergic to gold. Journal of the American Academy of Dermatology. Pearls. 2015;**73**(1):e21

8

Application of Nanotechnology in Agriculture: Assessment of TiO_2 Nanoparticle Effects on Barley

Alessandro Mattiello and Luca Marchiol

Abstract

Many aspects associated with the application of nanotechnology to agricultural activities are still unknown. In particular, there is not enough information on nanotoxicology in crops and we do not know the fate of nanoparticles in crops. Multiple experiments were carried out to study the effects of titanium oxide nanoparticles ($nTiO_2$) on barley (*Hordeum vulgare*). Germinating seeds were exposed to 0, 500, 1000, and 2000 mg l^{-1} $nTiO_2$. Seed germination percentage, mitotic index, root elongation, and Ti concentration in seedlings were observed. In a greenhouse experiment, plants of barley were grown to physiological maturity in control soil and soil enriched with 500 and 1000 mg $nTiO_2$ mg kg^{-1}, respectively. The duration of the growth cycle and the plant biomass was influenced by $nTiO_2$ compared to control plants. Concentrations of Ti were not very high with the exception of roots. However, the $nTiO_2$ soil amendment had an impact on composition and nutritional quality of barley grains. Concentrations of Ca, Mn, and Zn in kernels were increased by *n*TiO2 treatments. Concentration of amino acids was affected by the treatments as well. $nTiO_2$ treatments have the potential to influence the food chain and processing and economics of barley.

Keywords: nanotechnology, titanium dioxide nanoparticles, agriculture, crop growth

1. Introduction

Nanotechnology is a multidisciplinary field, which includes a wide range of processes, materials, and applications. The main aims of this new discipline are the characterization, fabrication, and manipulation of material at nanoscale level [1]. The reason why these new materials are so widely used is linked to their unique and novel properties principally related with the increase of the surface area to volume ratio. Nowadays, a lot of products are based on nanotechnology; at the beginning, these materials were applied in construction materials, new devices and techniques in electronics, cosmetics, sporting equipment, wastewater treatment, medicine, and more recently in agriculture and the food industry [2]. The agri-food was the last sector in terms of succession to be interested by this technological revolution but at the same time it would be far reaching in the next years [1]. In fact, the nanotechnology has recently emerged as the technological advancement to develop and transform the entire agri-food sector, in terms of increasing global food production and nutritional value, quality and safety of food [3]. The type of nanoparticles (NPs) or nanomaterials (NMs) used in plant science are quite wide, but they could be clustered in two principal groups: the carbon nanomaterial (CBNMs) and the metal-based nanomaterials (MBNMs) [4]. In the group of MBNMs, the most common NPs are TiO_2, CeO_2, Fe_3O_4, ZnO, and $AgNO_3$ [5]. Our experiments were focused on TiO_2 nanoparticles ($nTiO_2$) that represent the most used nanomaterial between the MBNMs. Several papers demonstrate the positive effects of $nTiO_2$ on plants [6–9]. More recently, Dehkourdi and Mosavi [10] used nano-anatase to treat parsley seeds, which resulted in an increase in the percentage of germination, the germination rate index, the root and shoot length, the fresh weight, the vigor index, and the chlorophyll content of the seedlings. Also, Feizi et al. [11] observed that the germination rate of *Salvia officinalis* improved when the seeds were exposed to $nTiO_2$. Previous studies demonstrate a positive effect also during the plant vegetative growth, for example, Hong et al. [7] demonstrate an acceleration in the rate of evolution of oxygen by chloroplasts in spinach plants. Another experiment on spinach demonstrated a gain in the photosynthetic carbon reaction in treated plants [9]. More recently, Qi et al. [12] treated tomato plants with $nTiO_2$ and put them in a mild heat stress, the plants resulted to have an improvement in the photosynthetic rate with respect to the control ones. Currently, the application of nanomaterials in the field of primary production is still under investigation, and therefore, it may take many years before specific nanoproducts for agriculture are commercialized worldwide [1]. Since the studies performed up to now have been conducted in a very simplified experimental condition, we still lack accurate information on what is happening in the soil. Further research is required to ensure complete success for these applications of nanotechnologies [13].

2. Materials and methods

2.1. Nanoparticle characterization

Titanium (IV) oxide anatase nanopowder having a minimal average particle size of 25 nm was purchased from Sigma-Aldrich (product ID 637254). Titanium nanoparticle ($nTiO_2$) characterization was carried out at the Facility for Environmental Nanoscience Analysis and Characterization (FENAC), University of Birmingham (UK). Further details on analytical methods are provided by Marchiol et al. [14].

2.2. Laboratory experiment

2.2.1. Seed germination and root elongation

Ten seeds of spring barley (*Hordeum vulgare* L. var. Tunika) were transferred in sterile conditions into each Petri dishes soaked with distilled water (Ctrl), 500, 1000, and 2000 mg l^{-1} $nTiO_2$

suspensions; each treatment was replicated three times. The Petri dishes were taped and placed in the dark at 21°C for 3 days. The germination percentage was calculated as the ratio of germinated seeds out of the total seeds. The seedlings obtained were used for the measurements of their total root length with ImageJ [15]. Root elongation was calculated as the average length and the sum of all roots emerged from each seed.

2.2.2. *Mitotic index*

Seeds of barley were sterilized and transferred in sterile conditions into Petri dishes soaked with distilled water. After 3 days, the germinated seedlings with actively growing roots (at least 2.5 cm in length) were placed in the $n\mathrm{TiO_2}$ suspensions (0, 500, 1000, 2000 $\mathrm{mgl^{-1}}$) for 24 h. Ten root tips per each treatment were studied to evaluate possible genotoxic effects of $n\mathrm{TiO_2}$. The samples prepared were evaluated for a total of about 10,000 cell observations per treatment. The mitotic index was recorded in Feulgen-stained preparations as the percentage of dividing cells out of the total number of cells scored.

2.2.3. *Titanium uptake*

The treated barley seedlings were rinsed three times with MilliQ water. Subsequently, they were divided into three portions: roots, seeds, and coleoptiles. The seedlings portions were dried at 70°C for 24 h, and they were acid-digested (10 ml HNO_3 65%) in a microwave oven according to USEPA method 3052. Plant extracts were filtered (0.45 mm PTFE), diluted with MilliQ water, and analyzed by ICP-OES.

2.3. Greenhouse experiment

2.3.1. *Soil characterization and $n\mathrm{TiO_2}$ amendment*

The possible effects of $n\mathrm{TiO_2}$ to barley were also evaluated along their entire life cycle; for this purpose, seeds of barley were grown in soil contaminated with $n\mathrm{TiO_2}$ at 500 and 1000 mg $\mathrm{kg^{-1}}$. The soil characterization data were reported by Marchiol et al. [14]. According to Priester et al. [16], the soil was amended with $n\mathrm{TiO_2}$ powder before sowing plants reaching the final concentration of 500 and 1000 mg $\mathrm{kg^{-1}}$ of $n\mathrm{TiO_2}$.

2.3.2. *Plant growth*

In a semi-sealed greenhouse, seeds of spring barley were sown in pots containing the $n\mathrm{TiO_2}$-enriched soils. After 2 weeks, seedlings were thinned to two seedlings per pot. During the plant growth, the pots were watered twice a week to maintain soil at 60% WHC. Phenological stages were monitored by adapting the Decimal Growth Scale [17] throughout the growth cycle and were based on 50% of plants within the treatments at each stage. Plants were harvested at physiological maturity. Plant shoots were severed at the collar and separated into stems, leaves, spikes, and grains. Leaf area was measured using a LI-3100C Area Meter. The plant fractions were oven dried at 105°C for 24 h and weighed.

2.3.3. Spectroscopy analysis

Plant fractions were acid-digested in a microwave oven according to USEPA method 3052. Titanium concentration in plant fractions, such as roots, stems, and leaves, was determined by an ICP-OES, whereas Ti concentration in kernels was determined by an ICP-MS.

2.3.4. TEM observations

Serial ultrathin sections from each species were cut with a diamond knife, mounted on copper grids, stained in uranyl acetate and lead citrate, and then observed under a Philips CM 10 transmission electron microscope (TEM) operating at 80 kV.

2.3.5. Macronutrient and micronutrient concentrations in kernels

Total B, Ca, Cu, Fe, K, Mg, Mn, Na, Ni, P, and Zn contents were determined by an ICP-OES with an internal standard solution of Y. Total Ce and Ti contents were determined by an ICP-MS with an internal standard solution of ^{72}Ge and ^{89}Y. Total N and S content were determined through an Elemental CHNS Analyzer using up to 2.5 mg of finely ground samples.

2.3.6. Amino acids in kernels

Amino acids analysis was performed using a LC 200 Perkin Elmer. More technical details about amino acids analysis were provided by Pošćić et al. [18].

2.3.7. Data analysis

The experiments were carried out in a completely randomized factorial design. Analysis of variance was conducted with a one-way ANOVA. Tukey's Multiple Comparison test at 0.05 p level was used to compare means. Statistical analysis was performed using the SPSS program (SPSS Inc. Chicago, IL, USA, ver. 17).

3. Results

3.1. Characterization of *n*TiO_2

The *n*TiO_2 powder measured with BET has a specific surface area equal to 61.6 $m^2\ g^{-1}$ that average value is slightly higher than the product specifications which declares a specific surface area comprehended between 45 and 55 $m^2\ g^{-1}$. The height distribution measured with AFM of the *n*TiO_2 powder results 41.8 ± 24.3 nm on average (**Figure 1A**). The *n*TiO_2 suspension was characterized at dimensional level also by TEM (**Figure 1B**). In this case, the instrument allows to measure the diameters of nanoparticles and the average size result equal to 24.09 ± 7.22 nm. The DLS instrument gives information about nanoparticles at three different levels: (i) at size level, in fact the instrument displays the zeta average size of nanoparticles, which correspond to the average diameter; (ii) at stability level by measuring the zeta potential of nanoparticles,

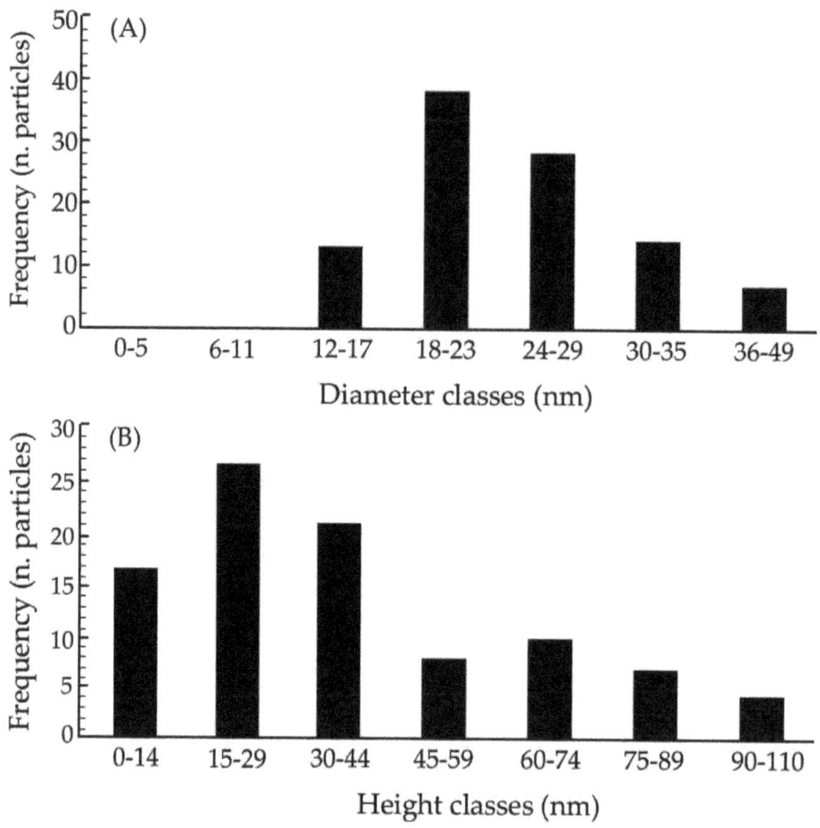

Figure 1. Characterization of: (A) height classes (nm) of $n\text{TiO}_2$ in powder form by AFM. (B) Diameter classes (nm) of $n\text{TiO}_2$ suspension obtained by TEM.

this index gives information about the electric potential in the interfacial double layer, if this value is comprised between −30 and +30 mV, the suspension tends to flocculate; and (iii) at heterogeneity level by the Polydispersity Index (PDI). The $n\text{TiO}_2$ suspensions result to have a zeta average size, zeta potential, and PDI equal to 925 ± 105 nm (**Figure 2A**), 19.9 ± 0.55 mV (**Figure 2B**), and 0.84 ± 0.17 nm, respectively. These values indicate a suspension made by big nanoparticles which tend to aggregate along time, and this brings to a wide-size distribution.

3.2. Seed germination experiments

3.2.1. Germination and root development

After 3 days, the treated seeds (**Figure 3A**) were used for the calculation of germination percentage (**Figure 3B**). The root elongation does not seem affected by $n\text{TiO}_2$ treatments at 1000 and 2000 mg l^{-1}, in fact, the average values are quite comparable with the germinated control seeds, whereas the average total roots value seems to slightly increase with increasing concentrations. The statistical analysis, however, shows there are not significant differences between treatments. The germinated seeds treated for 7 days were used for measuring their root elongation (**Figure 3C**). The root elongation does not seem affected by $n\text{TiO}_2$ treatments at 1000 and 2000 mg l^{-1}; in fact, the average values are quite comparable with the germinated control seeds,

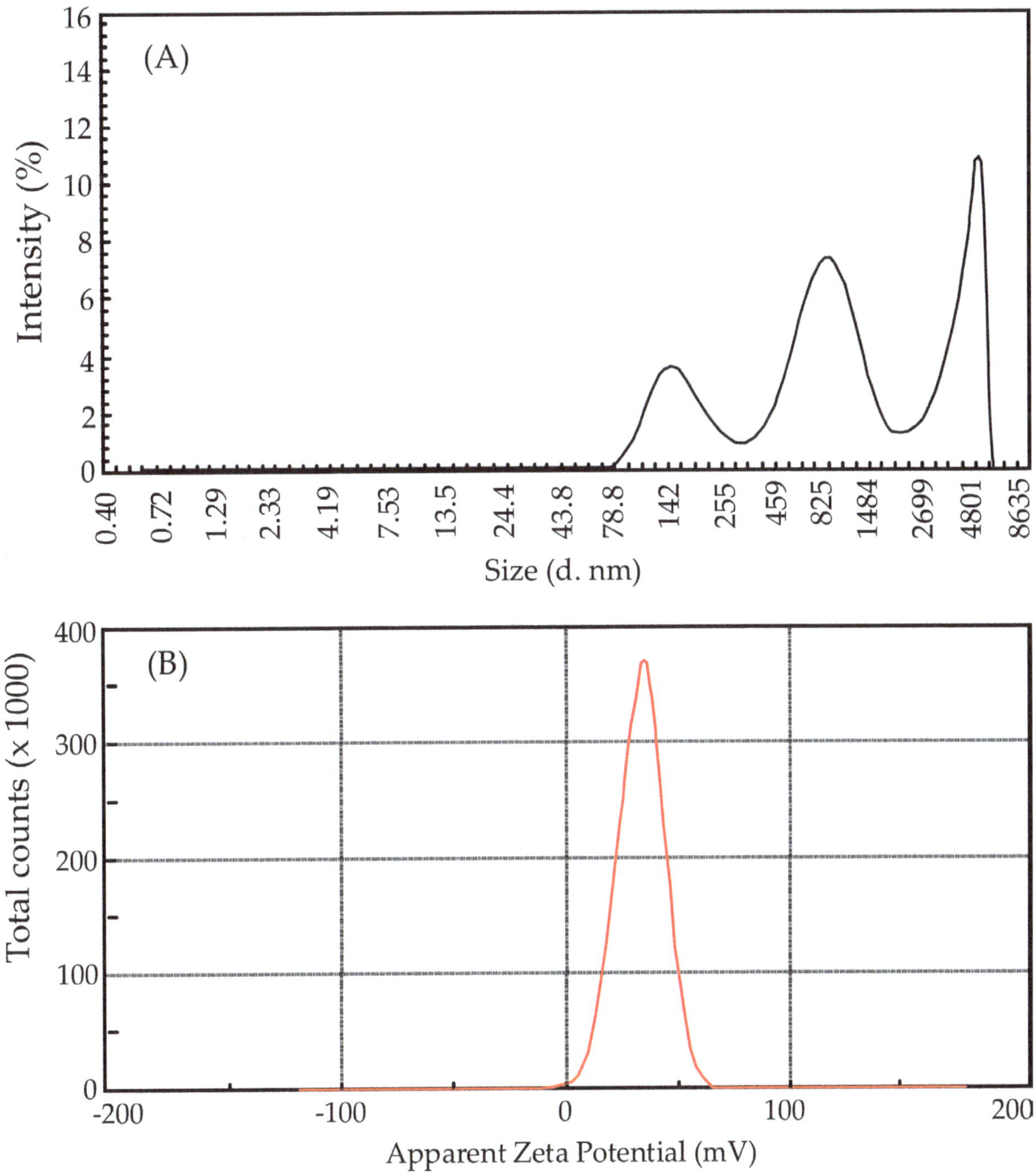

Figure 2. DLS data of $n\mathrm{TiO_2}$ suspension: (A) $n\mathrm{TiO_2}$ zeta average size distribution. (B) $n\mathrm{TiO_2}$ zeta potential.

rather the average total roots value seems to slightly increase with increasing concentrations. Conversely, the seeds treated with $n\mathrm{TiO_2}$ at 500 mg l^{-1} seemed to be affected in a negative way with respect to the other treatments. The statistical analysis has put on evidence a significant negative effect of $n\mathrm{TiO_2}$ at 500 mg l^{-1} for the root elongation, instead there are no significant differences between the other treatments.

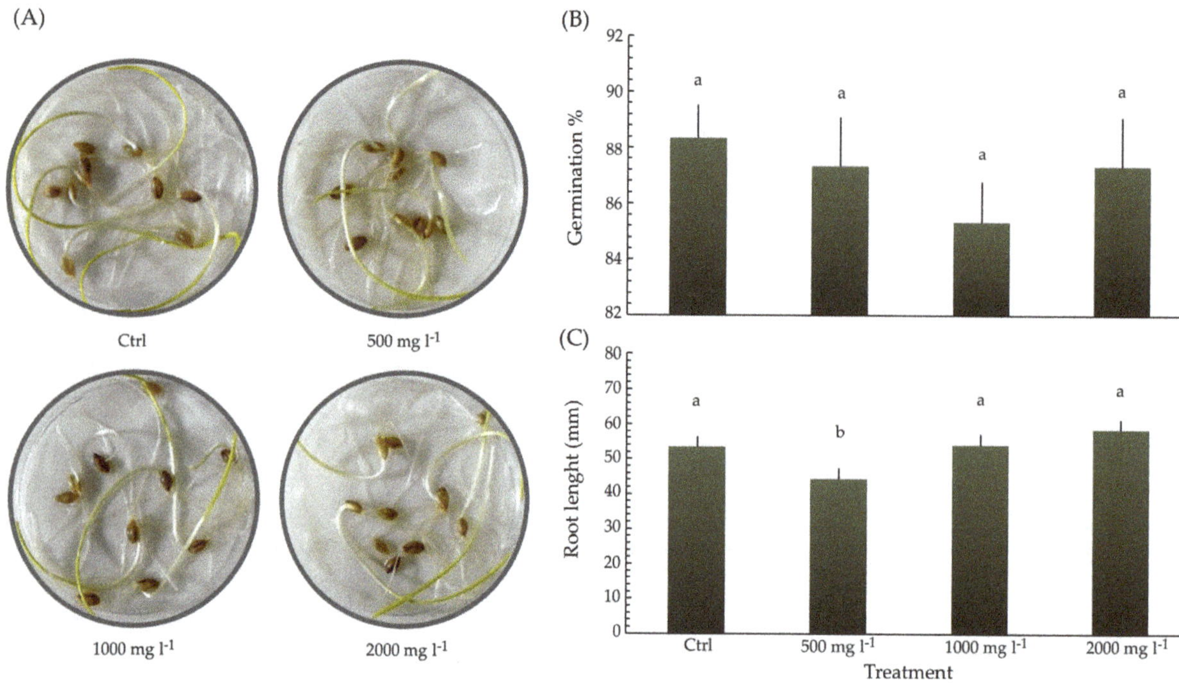

Figure 3. (A) Petri dishes with treated barley seedlings; (B) germination percentage of seeds (mean ± SE; $n = 3$); and (C) total root length in barley seedlings treated with $n\mathrm{TiO_2}$ suspension at 0, 500, 1000, and 2000 mg l^{-1}.

3.2.2. Mitotic index

Figure 4A reports data of mitotic index (MI), which was used as a sensor of genotoxicity. As plant roots grow, the cell division is usually very fast in the apical meristem of root tips. In our case, the control seedlings have a MI lower than the seedlings treated with $n\mathrm{TiO_2}$ suspension at 500 mg l^{-1} but comparable with the other treatments. Like the germination percentage also this parameter results not significantly affected by treatments.

3.2.3. Titanium seedlings uptake

The concentration of total titanium in different portions of barley seedlings is shown in **Figure 4B**. A dose-response was recorded in the accumulation of titanium since the titanium concentration in the seedling fractions increased with the increase of $n\mathrm{TiO_2}$ exposure concentration. In particular, the seedlings grown in the presence of $n\mathrm{TiO_2}$ at 500 mg l^{-1} did not uptake and translocate the titanium in other seedling portions. Instead, the seedlings grown in the presence of $n\mathrm{TiO_2}$ at 1000 mg l^{-1} showed an uptake and a translocation of titanium in each portion. This trend is confirmed by the seedlings grown in the presence of $n\mathrm{TiO_2}$ at 2000 mg l^{-1}; in fact, these seedling portions have the highest concentrations of titanium with respect to the seedling portions of the other treatments. The roots are the most interested area of accumulation; in fact, this portion recorded the highest concentrations of total titanium than the other portions for each treatment. This is particularly evident in the seedlings grown with $n\mathrm{TiO_2}$ at 1000 mg l^{-1} where the concentration results significantly different from the other seedling portions, whereas it is not like that in seedlings grown in the solution with 2000 mg l^{-1}.

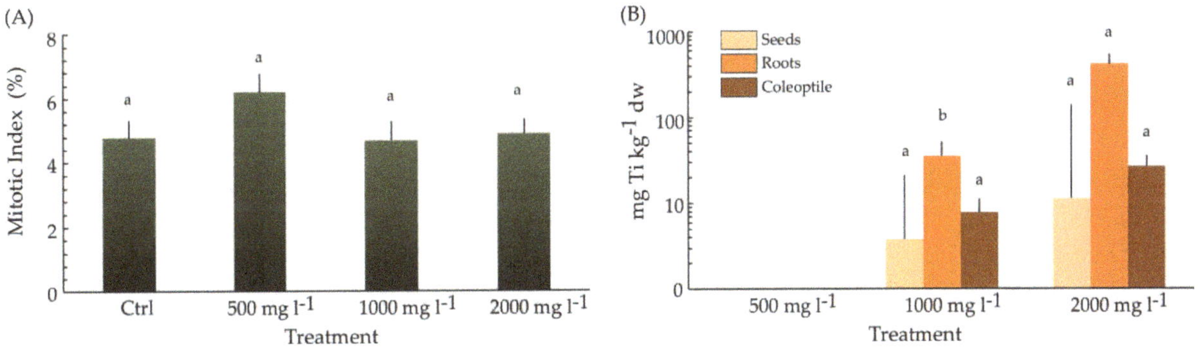

Figure 4. Effects of $n\text{TiO}_2$ on seedlings of *Hordeum vulgare*. (A) Mitotic index (%) (mean ± SE; $n = 3$) observed in root tips of $n\text{TiO}_2$-treated seedlings. (B) Concentration of Ti in seeds, roots, and shoots (mean ± SE; $n = 3$) of $n\text{TiO}_2$ barley-treated seedlings. Different letters indicate statistical difference between treatments at Tukey's test ($p < 0.05$).

3.3. Life cycle study

3.3.1. Plant growth

The data obtained from the phenological observation are shown in **Figure 5**. The barley plants grown in soil spiked with $n\text{TiO}_2$ at 500 and 1000 mg kg^{-1} are in delay with respect to the control barley plants in reaching each physiological maturity; this delay is already appreciable from the second leaf stage. At the end of the physiological maturity, the barley plants were used for measuring the representative parameters of plant growth in particular, the plant height, number of tillers, total leaves area, and grain yield (**Figure 6**). The plant height was not affected in a significant way by the $n\text{TiO}_2$ treatments; however, there is a gradual increment of plant height with the increase of $n\text{TiO}_2$ concentration in the soil. The number of tillers, like the plant height parameter, increases at the increase of $n\text{TiO}_2$ into the soil. Differently from the previous parameter, the average number of tillers of barley plants grown in the soil spiked with $n\text{TiO}_2$ at 2000 mg l^{-1} show almost significant difference from the other treatments. The total leaf area parameter has the same trend of number of tillers parameter, also in this case there is an increase of leaves surface at the increase of $n\text{TiO}_2$ in the soil with a slightly significant difference in the average value obtained for the plants, which were grown in the soil spiked with 1000 mg kg^{-1} of $n\text{TiO}_2$. The last parameter took into account has been the plant yield. This parameter was affected by the treatments in a different way with respect to the other ones; in fact, the control barley plants did not significantly differ from the barley plants treated with 1000 mg kg^{-1} of $n\text{TiO}_2$ except for the barley plants grown in the soil spiked with 500 mg kg^{-1} of $n\text{TiO}_2$ resulted significantly affected.

3.3.2. Spectroscopy analysis

The spiked soil and the barley plant portions were analyzed by ICP-OES and ICP-MS in order to check the total concentration of Ti (**Table 1**). The soils spiked with $n\text{TiO}_2$ at 500 and 1000 mg l^{-1} have a significant difference from the soil without $n\text{TiO}_2$, though the soil spiked with $n\text{TiO}_2$ at 500 mg l^{-1} results slightly different from the control soil. These results confirm that the soil spiking was performed in a correct way. The analyses of the barley plant portions show there

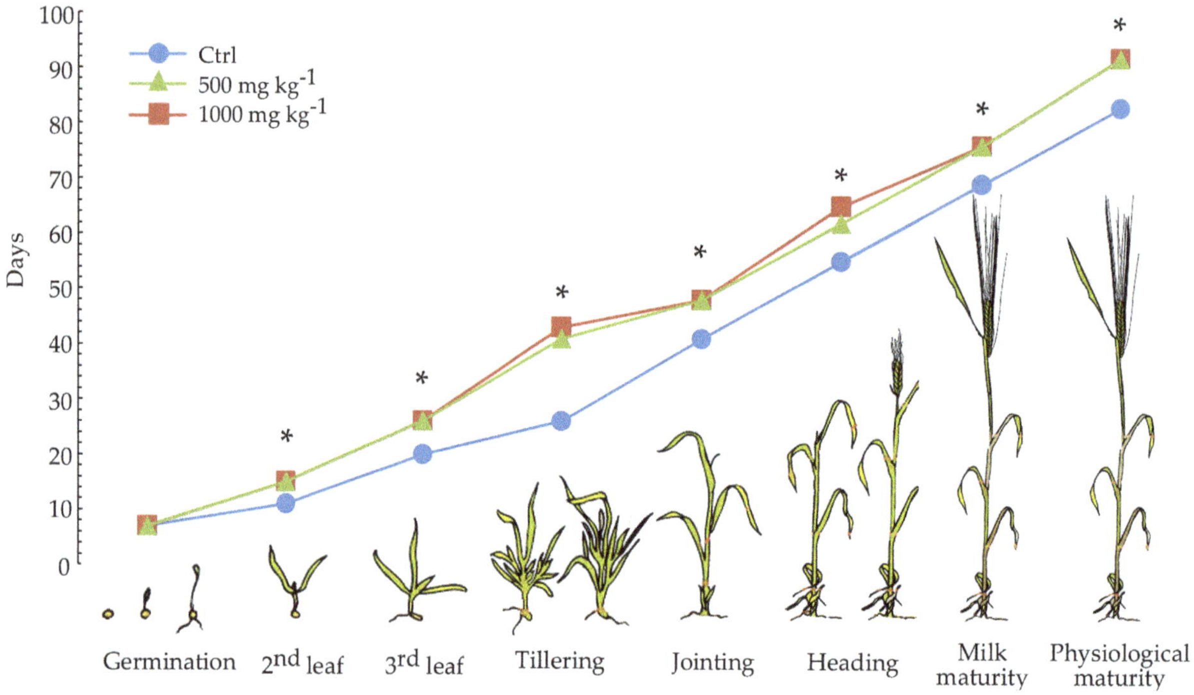

Figure 5. Duration of vegetative and reproductive phenological phases of *Hordeum vulgare* grown in control soil and nTiO$_2$-spiked soil. Asterisk denotes significant differences between control and treated plants ($p \leq 0.05$).

are no significant differences in the Ti concentrations both between the treatments and between treatments and control. The only exception is the total Ti concentration in the stem portion of the plants treated with nTiO$_2$ at 1000 mg kg^{-1} which results slightly different from the same portion of the other plants.

3.3.3. TEM observations

To verify the uptake and subsequent translocations of nTiO$_2$ from roots to aerial plant fractions, ultrastructural analyses on plant leaf tissues were carried out. Rare clusters of nanoparticles were found in leaves sampled from plants grown in soil enriched with the different combinations of nTiO$_2$, at both concentrations (**Figure 7**). nTiO$_2$ were observed in leaf cells and, in particular, in the stroma of the chloroplast and in the vacuoles. Despite the treatment, the chloroplast ultrastructure appeared normal (**Figure 7B**).

3.3.4. Macronutrient and micronutrient concentrations in kernels

The accumulation of macronutrients in barley kernels is shown in **Table 2**. Both N and S concentrations increase at the increment of nTiO$_2$, whereas for Ca there was not the same behavior. Apparently, K, P, and Mg concentrations in kernels did not respond to the treatment. **Table 3** reports the concentrations of micronutrients in kernels. The nTiO$_2$ treatments determined an increase in Fe, Mn, and Zn concentrations in barley kernels, whereas B and Cu concentrations were not influenced by the treatments.

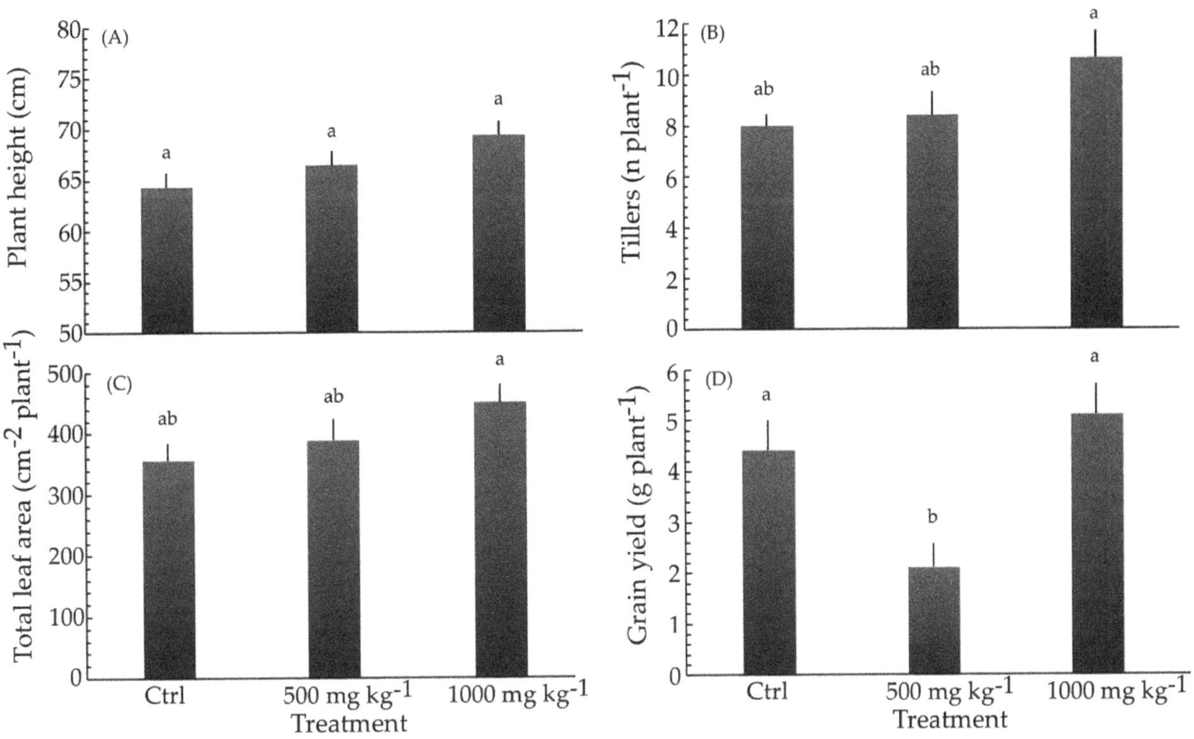

Figure 6. Biometric variables of *Hordeum vulgare* observed in plants grown in control soil and $n\text{TiO}_2$-spiked soils. Variables are respectively: (A) plant height, (B) number of tillers per plant, (C) total leaf area per plant, and (D) grain yield per plant. Bars are mean standard error ($n = 5$). Different letters indicate statistical difference between treatments at Tukey's test ($p < 0.05$).

Treatment	Soil (mg kg^{-1})	Roots (mg kg^{-1})	Stems (mg kg^{-1})	Leaves (mg kg^{-1})	Spike (μg kg^{-1})
Ctrl	1797 ± 119 b	77 ± 3.19 a	0.26 ± 0.04 ab	1.03 ± 0.06 a	2.19 ± 1.2 a
Ti 500	2153 ± 119 ab	66.7 ± 7.49 a	0.28 ± 0.03 ab	1.39 ± 0.35 a	1.71 ± 0.53 a
Ti 1000	2537 ± 56.3 a	81.7 ± 4.96 a	0.39 ± 0.06 a	0.96 ± 0.09 a	1.39 ± 0.27 a

Values are mean ± SE ($n = 5$). Same letters indicated no statistical difference between treatments at Tukey's test ($p \leq 0.05$).

Table 1. Ti concentration observed in soil, roots, stems, leaves, kernels of primary and secondary spikes of barley plants grown in control (Ctrl) and $n\text{TiO}_2$-spiked soil.

3.3.5. Amino acids in kernels

The effects of $n\text{TiO}_2$ treatments on amino acid concentrations in kernels are shown in **Table 4**. Overall, Glu and Pro are the most abundant amino acids in kernels with concentration ranges of 31–43 and 15–21 mg·g^{-1}, respectively. The $n\text{TiO}_2$ treatments did not significantly modify concentrations of Ala, Arg, Asp, His, Ser, and Trp. On the opposite, the concentration of Cys, Glu, Gly, Ile, Leu, Lys, Phen, Pro, Tyr, and Val in kernels significantly increased in response to the $n\text{TiO}_2$ treatments. In the case of Thr, the response to the treatment was less evident. At last, only in the case of Met contradictory results were recorded.

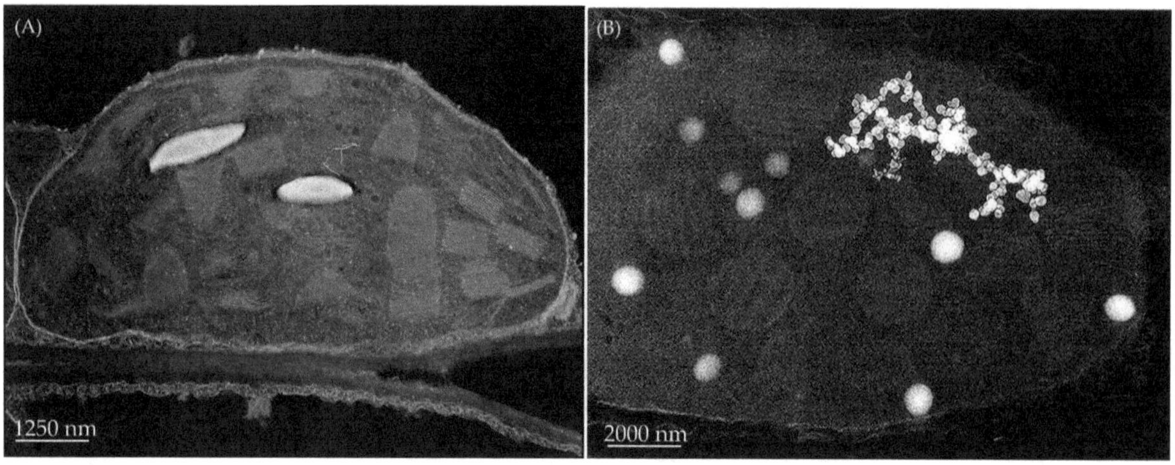

Figure 7. Representative TEM micrograph of leaf tissues of *Hordeum vulgare* plants grown in (A) control soil and (B) nTiO$_2$ 1000 mg kg^{-1}-spiked soils. Clusters of Ti nanoparticles are visible in the stroma of the chloroplasts of nTiO$_2$-treated plants (B).

Element	Ctrl	nTiO$_2$ 500 mg kg^{-1}	nTiO$_2$ 1000 mg kg^{-1}
N (% dw)	2.22 ± 0.08 b	2.62 ± 0.12 a	2.78 ± 0.05 a
K (g kg^{-1})	4.57 ± 0.10 a	4.19 ± 0.16 ab	3.83 ± 0.06 b
P (mg kg^{-1})	4.55 ± 0.33 a	4.87 ± 0.23 a	4.36 ± 0.31 a
Ca (mg kg^{-1})	377 ± 21 b	701 ± 81 a	543 ± 8 ab
Mg (mg kg^{-1})	1881 ± 59 a	1983 ± 124 a	1731 ± 44 a
S (mg kg^{-1})	1814 ± 128 b	2391 ± 47 a	2027 ± 161 b

Values are mean ± SE (n = 5). Same letters indicated no statistical difference between treatments at Tukey's test ($p \leq 0.05$).

Table 2. Nitrogen percentage and concentration of macronutrients in barley kernels at ripening from main shoot grown in control soil (Ctrl) and nTiO$_2$-spiked soil.

Element	Ctrl	nTiO$_2$ 500 mg kg^{-1}	nTiO$_2$ 1000 mg kg^{-1}
B (mg kg^{-1})	8.64 ± 1.02 a	8.01 ± 1.7 a	6.02 ± 1.58 a
Cu (mg kg^{-1})	8.91 ± 1.33 a	7.52 ± 1.16 a	8.24 ± 0.28 a
Fe (mg kg^{-1})	46.4 ± 9.80 b	197 ± 43.7 a	101 ± 27 ab
Mn (mg kg^{-1})	18.8 ± 0.64 b	25.1 ± 1.06 a	21.6 ± 1.23 ab
Zn (mg kg^{-1})	55.7 ± 5.36 b	69.6 ± 2.61 a	59.6 ± 1.34 ab

Values are mean ± SE (n = 5). Same letters indicated no statistical difference between treatments at Tukey's test ($p \leq 0.05$).

Table 3. Concentration of micronutrients in barley kernels at ripening from main shoot grown in control soil (Ctrl) and nTiO$_2$-spiked soil.

Amino acid	Ctrl	nTiO$_2$ 500 mg kg^{-1}	nTiO$_2$ 1000 mg kg^{-1}
Alanine (Ala)	5.65 ± 0.51 a	7.35 ± 1.05 a	6.75 ± 0.16 a
Arginine (Arg)	7.55 ± 1.32 a	9.26 ± 0.56 a	9.12 ± 0.83 a
Aspartic acid (Asp)	7.18 ± 0.67 a	8.58 ± 0.65 a	9.09 ± 0.49 a
Cysteine (Cys)	6.85 ± 0.13 b	8.07 ± 0.01 a	8.42 ± 0.36 a
Glutamic acid (Glu)	31.2 ± 3.56 b	40.7 ± 3.73 a	43 ± 1.83 a
Glycine (Gly)	5.98 ± 0.45 b	7.74 ± 0.29 a	8.01 ± 0.42 a
Histidine (His)	3.14 ± 0.51 a	3.66 ± 0.18 a	3.89 ± 0.18 a
Isoleucine (Ile)	5.29 ± 0.47 b	6.42 ± 0.36 a	6.77 ± 0.25 a
Leucine (Leu)	9.4 ± 0.74 b	11.2 ± 0.81 a	11.7 ± 0.42 a
Lysine (Lys)	3.67 ± 0.31 b	5.85 ± 0.33 a	5.98 ± 0.45 a
Methionine (Met)	2.39 ± 0.13 b	3.08 ± 0.01 a	3 ± 0.20 b
Phenylalanine (Phe)	7.48 ± 0.94 b	9.12 ± 0.65 a	9.37 ± 0.45 a
Proline (Pro)	14.8 ± 1.68 b	20.4 ± 3.04 a	21.4 ± 1.41 a
Serine (Ser)	5.84 ± 0.54 a	6.78 ± 0.36 a	6.84 ± 0.18 a
Threonine (Thr)	4.61 ± 0.31 a	5.11 ± 0.36 ab	5.35 ± 0.18 ab
Tryptophan (Trp)	1.15 ± 0.67 a	0.53 ± 0.01 a	0.75 ± 0.18 a
Tyrosine (Tyr)	3.36 ± 0.42 b	4.34 ± 0.20 a	4.22 ± 0.36 a
Valine (Val)	7.04 ± 0.49 b	8.29 ± 0.65 a	8.68 ± 0.40 a

Table 4. Amino acid (mg·g^{-1}) concentration in barley kernels at ripening from main shoot grown in soil spiked with none (Control), 500 mg nTiO$_2$·kg^{-1}, and 1000 mg nTiO$_2$·kg^{-1}.

4. Discussion

The nTiO$_2$ suspensions did not affect germination of *H. vulgare*. Our results are in agreement with the observations carried out, respectively, on rice [19], lettuce, radish, and cucumber [20], tomato [21], and pea [22]. According to Ref. [11], we demonstrated that nTiO$_2$ treatment did not affect root elongation of seedlings. Other authors published opposite results. In fact, Mushtaq [23] showed an inhibitory effect of nTiO$_2$ on root elongation in cucumber, whereas Fan et al. [22] verified decrease in the number of secondary lateral roots in pea. The nTiO$_2$ treatments did not influence the mitotic index. That is in contrast with Moreno-Olivas et al. [24] which observed a nTiO$_2$-induced genotoxicity in hydroponically cultivated zucchini. Although the size of nTiO$_2$ used in that experiment is comparable to ours, those experiments were carried out in different conditions than ours. This can result in different experimental conditions, with particular regard to the nTiO$_2$ traits (e.g., different grade of agglomeration due to different z-average size and zeta potential). On the other hand, the results obtained by ICP-OES analyses seem to indicate a $n\text{TiO}_2$ uptake by root tissue and a subsequent translocation in the other seedling tissues. This result could be an indication of a real uptake and translocation of $n\text{TiO}_2$. A second hypothesis is that, despite the use of appropriate analytical protocols, the analysis may have been disturbed by sample contamination. In that case, the element concentration in the plant tissues could be significantly overestimated due to a fraction of metal simply adsorbed onto the external sample surface.

With regard to the effects along the entire life cycle, the response of plant phenology was in accordance with previous studies [25, 26]. In fact, the barley plants treated with $n\text{TiO}_2$ result to have a longer vegetative phase. During this phase, the plants keep growing and the leaves continue their photosynthetic activity and consequently the production of photosynthates [27]. Taking into account such evidences, a higher biomass production and grain yield in treated plants respect the control ones it is expected. The analyses of biometric parameters confirm in part the expected results. Except for the plant height and grain yield per plants the other parameters result positively affected. Our results confirm other experimental evidences. In particular, studies carried out on *Spinacia oleracea* have demonstrated that $n\text{TiO}_2$ promotes plant photosynthesis increasing light absorbance and transformation of light energy and enhancing Rubisco activity [28, 29]. The positive effects of $n\text{TiO}_2$ treatments were evidenced also at the grain level. The grains obtained from treated plants result to have a higher content of macro (Na and Ca) and micronutrients (Fe, Mn, and Zn); moreover, a positive effect of $n\text{TiO}_2$ treatment was also observed for several amino acids. The increase of macro/micronutrient and amino acid concentrations in kernels could be an effect related with the longest vegetative phase caused by the $n\text{TiO}_2$ treatments. In order to find the relationship between the effects and treatment, the material obtained at the end of the experiment was analyzed by ICP-OES and observed by TEM. The ICP-OES analyses and the TEM observations were carried out in order to know if the $n\text{TiO}_2$ can enter into plant tissues and subsequently cause the observed effects on plants. The ICP-OES results did not put on evidence an effective uptake of titanium by the plants, but the TEM observations show the presence of $n\text{TiO}_2$ in the stroma of the chloroplast and in the vacuoles of leaf cells. This discrepancy in results could be related with the agglomeration tendency of $n\text{TiO}_2$ in water, previously evidenced by $n\text{TiO}_2$ characterization analyses. The agglomeration makes the $n\text{TiO}_2$ less available because their increased dimension makes difficult the passage of them through the cell wall, this means only the $n\text{TiO}_2$ with the smallest size can pass this plant barrier and consequently small amount of titanium could be uptaken.

5. Conclusions

The amount of products containing nanoparticles will increase in the future years; this will bring an increase of their presence in the environment. In the last years, the majority of literature was focused to investigate the potential negative impact of this new kind of material on human, animals, and plants, but in our study, we put on evidence the potential beneficial effects. At first, we demonstrate the absence of negative impact during the early development stages of barley plants; in fact, each $n\mathrm{TiO_2}$ concentration did not affect the germination percentage and root elongation, except for the lowest concentration ($n\mathrm{TiO_2}$ 500 mg $\mathrm{kg^{-1}}$) which significantly affects in a negative way the last parameter. The analysis focus moves to evaluation of the possible effects at genetic level; for this purpose, the mitotic index was analyzed. The results also show, in this case, the absence of an effect for this parameter. The AFM and DLS analyses give information about the tendency of $n\mathrm{TiO_2}$ to form big agglomerate once dissolved in MilliQ water, this makes the $n\mathrm{TiO_2}$ less available for the seeds/seedlings, and the absence of effects could be related to the incapacity of $n\mathrm{TiO_2}$ to cross the cell wall. However, the ICP-OES and ICP-MS analyses demonstrate the capacity of seeds/seedlings to uptake the titanium, then the absence of effects in the early developmental stages is not due to the absence of titanium in the plant tissues but to this unharmful effect. The experiment set up to evaluate the possible effect of $n\mathrm{TiO_2}$ along the entire barley life cycle demonstrates the positive dose-response effect on vegetative growth, and this has a direct effect on the composition and nutritional value of barley grains.

Acknowledgements

This research was in part supported by Research Grant No. 64 dd. 08-09.2014 (RANDOLPH—Relazioni tra nanoparticelle metalliche e piante superiori) from DI4A, University of Udine, Italy. Anastasios Papadiamantis supervised the characterization of nanomaterials carried out at FENAC (Facility for Environmental Nanoscience Analysis and Characterization, University of Birmingham, UK). We are grateful to Iseult Lynch (School of Geography, Earth and Environmental Sciences, University of Birmingham, UK) as well. We like to thank Giorgio Angelo Lucchini (Department of Agricultural and Environmental Sciences—Production, Landscape, Agro-energy, University of Milan, Italy) for ICP-MS analysis; Barbara Piani and Marta Fontana (DI4A, University of Udine, Italy) for HPLC analysis and CHNS, amylose, and glucans analysis, respectively. We are grateful to Rita Musetti (DI4A, University of Udine) who managed the TEM experiments. Last but not the least, the valuable contributions by Filip Pošćić (Institute for Adriatic Crops and Karst Reclamation, Croatia), Costanza Zavalloni (California State University Stanislaus), Guido Fellet, and Massimo Vischi (DI4A, University of Udine) are acknowledged, as well.

Author details

Alessandro Mattiello and Luca Marchiol*

*Address all correspondence to: marchiol@uniud.it

Department of Agriculture, Food, Environment and Animal Sciences, University of Udine, Udine, Italy

References

[1] Handford CE, Dean M, Henchion M, Spence M, Elliott CT, Campbell K. Implication of nanotechnology for the agri-food industry: Opportunities, benefits and risks. Trends in Food Science & Technology. 2014;**40**:226-241. DOI: http://dx.doi.org/10.1016/j.tifs.2014.09.007

[2] Doyle ME. Nanotechnology: A Brief Literature Review. Food. Madison, WI: University of Wisconsin-Madison; 2006. p. 10

[3] Mousavi SR, Rezaei M. Nanotechnology in agriculture. Journal of Applied Environmental and Biological Sciences. 2011;**1**(10):414-419

[4] Capaldi Arruda SC, Diniz Silva AL, Galazzi RM, Azevedo RA, Zezzi Arruda MA. Nanoparticles applied to plant science: A review. Talanta. 2015;**131**:693-705. DOI: http://dx.doi.org/10.1016/j.talanta.2014.08.050

[5] Rico CM, Majumdar S, Duarte-Gardea M, Peralta-Videa JR, Gardea-Torresdey JL. Interaction of nanoparticles with edible plants and their possible implications in the food chain. Journal of Agricultural and Food Chemistry. 2011;**59**(8):3485-3498. DOI: http://dx.doi.org/10.1021/jf104517j

[6] Zheng L, Hong F, Lu S, Liu C. Effect of nano-TiO_2 on strength of naturally aged seeds and growth of spinach. Biological Trace Element Research. 2005;**104**(1):83-91

[7] Hong F, Zhou J, Liu C, Yang F, Wu C, Zheng L, et al. Effect of nano-TiO_2 on photochemical reaction of chloroplasts of spinach. Biological Trace Element Research. 2005; **105**(1-3):269-279. DOI: http://doi:10.1385/BTER:105:1-3:269

[8] Yang F, Liu C, Gao F, Su M, Wu X, Zheng L, et al. The improvement of spinach growth by nano-anatase TiO_2 treatment is related to nitrogen photoreduction. Biological Trace Element. 2007;**119**(1):77-88. DOI: http://doi:10.1007/s12011-007-0046-4

[9] Gao F, Liu C, Qu C, Zheng L, Yang F, Su M, et al. Was improvement of spinach growth by nano-TiO_2 treatment related to the changes of Rubisco activase? BioMetals. 2008;**21**(2):211-217. DOI: http://doi:10.1007/s10534-007-9110-y

[10] Dehkourdi EH, Mosavi M. Effect of anatase nanoparticles (TiO_2) on parsley seed germination (*Petroselinum crispum*) in vitro. Biological Trace Element Research. 2013;**155** (2):283-286. DOI: http://doi:10.1007/s12011-013-9788-3

[11] Feizi H, Amirmoradi S, Abdollahi F, Pour SJ. Comparative effects of nanosized and bulk titanium dioxide concentrations on medicinal plant *Salvia officinalis* L. Annual Review & Research in Biology. 2013;**3**(4):814-824

[12] Qi M, Liu Y, Li T. Nano-TiO_2 improve the photosynthesis of tomato leaves under mild heat stress. Biological Trace Element Research. 2013;**156**(1-3):323-328. DOI: http://doi:10.1007/s12011-013-9833-2

[13] Khot LR, Sankaran S, Maja JM, Ehsani R, Schuster EW. Applications of nanomaterials in agricultural production and crop protection: A review. Crop Protection. 2012;**35**:64-70. DOI: http://dx.doi.org/10.1016/j.cropro.2012.01.007

[14] Marchiol L, Mattiello A, Pošćić F, Fellet G, Zavalloni C, Carlino E, et al. Changes in physiological and agronomical parameters of barley (*Hordeum vulgare*) exposed to cerium and titanium dioxide nanoparticles. International Journal of Environmental Research and Public Health. 2016;**13**(3):332. DOI: http://doi:10.3390/ijerph13030332

[15] Schneider CA, Rasband WS, Eliceiri K. NIH Image to ImageJ: 25 years of image analysis. Nature Methods. 2012;**9**(7):671-675. DOI: http://doi:10.1038/nmeth.2089

[16] Priester JH, Ge Y, Mielke RE, Horst AM, Moritz SC, Espinosa K, et al. Soybean susceptibility to manufactured nanomaterials with evidence for food quality and soil fertility interruption. Proceedings of the National Academy of Sciences. 2012;**109**(37):14734-14735. DOI: http://doi: 10.1073/pnas.1205431109

[17] Zadoks JC, Chang TT, Konzak CF. A decimal code for the growth stages of cereals. Weed Research. 1974;**14**(6):415-421. DOI: http://doi:10.1111/j.1365-3180.1974.tb01084.x

[18] Pošćić F, Mattiello A, Fellet G, Miceli F, Marchiol L. Effects of cerium and titanium oxide nanoparticles in soil on the nutrient composition of barley (*Hordeum vulgare* L.) kernels. International Journal of Environmental Research and Public Health. 2016;**13**(6):577. DOI: http://doi:10.3390/ijerph13060577

[19] Boonyanitipong B, Kositsup B, Kumar P, Baruah S, Dutta J. Toxicity of ZnO and TiO_2 nanoparticles on germinating rice seed *Oryza sativa* L. International Journal of Bioscience, Biochemistry, and Bioinformatics. 2011;**1**:282-285. DOI: http://doi:10.7763/IJBBB.2011.V1.53

[20] Wu SG, Huang L, Head J, Chen DR, Kong IC, Tang YJ. Phytotoxicity of metal oxide nanoparticles is related to both dissolved metals. Petroleum & Environmental Biotechnology. 2012;**3**(4):126. DOI: http://doi:10.4172/2157-7463.1000126

[21] Song U, Jun H, Waldman B, Roh J, Kim Y, Yi J, et al. Functional analyses of nanoparticle toxicity: A comparative study of the effects of TiO_2 and Ag on tomatoes (*Lycopersicon esculentum*). Ecotoxicology and Environmental Safety. 2013;**93**:60-67. DOI: http://doi:10.1016/j.ecoenv.2013.03.033

[22] Fan R, Huang YC, Grusak MA, Huang CP, Sherrier DJ. Effects of nano-TiO_2 on the agronomically-relevant Rhizobium-legume symbiosis. Science of the Total Environment. 2014;**497-498**:503-512. DOI: http://doi:10.1016/j.scitotenv.2013.07.032

[23] Mushtaq YK. Effect of nanoscale Fe_3O_4, TiO_2 and carbon particles on cucumber seed germination. Journal of Environmental Science and Health. Part A, Toxic/Hazardous Substances and Environmental Engineering. 2011;**46**:1732-1735. DOI: http://doi:10.1080/10934529.2011.633403

[24] Moreno-Olivas F, Gant VU Jr, Johnson KL, Peralta-Videa JR, Gardea-Torresdey JL. Random amplified polymorphic DNA reveals that TiO_2 nanoparticles are genotoxic to *Cucurbita pepo*. Journal of Zhejiang University SCIENCE A. 2014;**15**:618-623. DOI: http://doi:10.1631/jzus.A1400159

[25] Rico CM, Barrios AC, Tan W, Rubenecia R, Lee SC, Varela-Ramirez A, et al. Physiological and biochemical response of soil-grown barley (*Hordeum vulgare* L.) to cerium oxide nanoparticles. Environmental Science and Pollution Research. 2015;**22**(14):10551-10558. DOI: http://doi:10.1007/s11356-015-4243-y

[26] Yoon SJ, Kwaka JI, Lee WM, Holden PA, An YJ. Zinc oxide nanoparticles delay soybean development: A standard soil microcosm study. Ecotoxicology and Environmental Safety. 2013;**100**(1):131-137. DOI: http://dx.doi.org/10.1016/j.ecoenv.2013.10.014

[27] Dofing SM. Phenological development–yield relationships in spring barley in a subarctic environment. Canadian Journal of Plant Science. 1995;**75**(1):93-97. DOI: doi:10.4141/cjps95-015

[28] Yang F, Hong F, You W, Liu C, Gao F, Wu C, et al. Influence of nano-anatase TiO_2 on the nitrogen metabolism of growing spinach. Biological Trace Element Research. 2006;**110**(2):179-190. DOI: http://doi:10.1385/BTER:110:2:179

[29] Gao F, Hong F, Liu C, Zheng L, Su M, Wu X, et al. Mechanism of nano-anatase TiO_2 on promoting photosynthetic carbon reaction of spinach. Biological Trace Element Research. 2006;**111**(1):239-253. DOI: http://doi:10.1385/BTER:111:1:239

9

TiO_2: A Critical Interfacial Material for Incorporating Photosynthetic Protein Complexes and Plasmonic Nanoparticles into Biophotovoltaics

Yiqun Yang and Jun Li

Abstract

TiO_2, a photosensitive semiconducting material, has been widely reported as a good photoanode material in dye-sensitized solar cells and new emerging perovskite cells. Its proper electronic band structure, surface chemistry and hydrophilic nature provide a reactive surface for interfacing with different organic and inorganic photon capturing materials in photovoltaics. Here, we review its enabling role in incorporating two special materials toward biophotovoltaics, including photosynthetic protein complexes extracted from plants and plasmonic nanoparticles (e.g., gold or silver nanoparticles), which interplay to enhance the absorption and utilization of sun light. We will first give a brief introduction to the TiO_2 photoanode, including preparation, optical and electrochemical properties, and then summarize our recent research and other related literature on incorporating photosynthetic light harvest complexes and plasmonic nanoparticles onto anatase TiO_2 photoanodes as a means to tap into the charge separation, electron and energy transfer, and photovoltaic enhancements in the bio-photovoltaics.

Keywords: photoanode, dye-sensitized solar cells, photosynthetic protein complexes, charge separation, plasmonic effect, interface, Schottky barrier, energy transfer, hot electrons

1. Introduction

As a photosensitive semiconductor material with good long-term stability, nontoxicity, low cost and abundance, TiO_2 has been widely used in photocatalysis and photovoltaics [1]. However, due to the wide band gap (i.e., 3.0 eV for rutile and 3.2 eV for anatase TiO_2, respectively), pristine TiO_2 only responds to the irradiation in UV region. It is inefficient for capturing the majority of photons lying in the visible range of the normal solar irradiation spectrum. Decorating visible-light-excitable compounds, so-called photo-sensitizers or dyes, on TiO_2 can effectively overcome this issue, which has been adopted to develop dye-sensitized solar cells (DSSCs) [2].

The operation principle of DSSCs is illustrated in **Figure 1**. Organic dyes anchored on TiO_2 surface are excited by absorbing visible light in the specific wavelength range. The charge separation occurs at the sensitizer/TiO_2 interface by injecting electrons from the excited state of the dye into the TiO_2 conduction band to generate free electrons, which are then diffused through the sintered TiO_2 nanoparticle layer and external circuit to the cathode to generate a photocurrent. Concurrently, the oxidized dye is reduced to its ground state by oxidation of the redox mediator I^- into I_3^-, with I_3^- ions then diffusing through the electrolyte to the cathode and reduced back to I^- by accepting electrons combing back from the external circuit to complete the whole regeneration process. Overall, this system converts solar energy into electricity without any net consumption of chemicals, and thus, the DSSC can continuous supply power under irradiation by sun light. In 2011, a porphyrin sensitized DSSC incorporated with $Co^{II/III}$ tris(bipyridyl) redox electrolyte achieved a record-high power conversion efficiency (PCE) of 12.3% [4]. Recently, a new emerging perovskite solar cell using solid-state mesoscopic TiO_2 photoanode sensitized with lead halide perovskite ($CH_3NH_3PbX_3$) was reported to achieve an exciting PCE of more than 15% [5], and it quickly approached to 20% by post-treatment of mesoporous TiO_2 photoanodes with lithium salts [6].

The prototype of DSSC is analogue to the photosynthesis in plants taking place at thylakoid membrane in the chloroplast composed of various photosynthetic proteins. **Figure 2** depicts the light-dependent reactions in the photosynthesis. Briefly, solar energy is absorbed by exciting the light harvesting complexes (i.e., LHCI and LHCII), a kind of proteins binding a lot of chlorophylls (Chls) as major pigments in the photosystems (PSI and PSII). The excitation

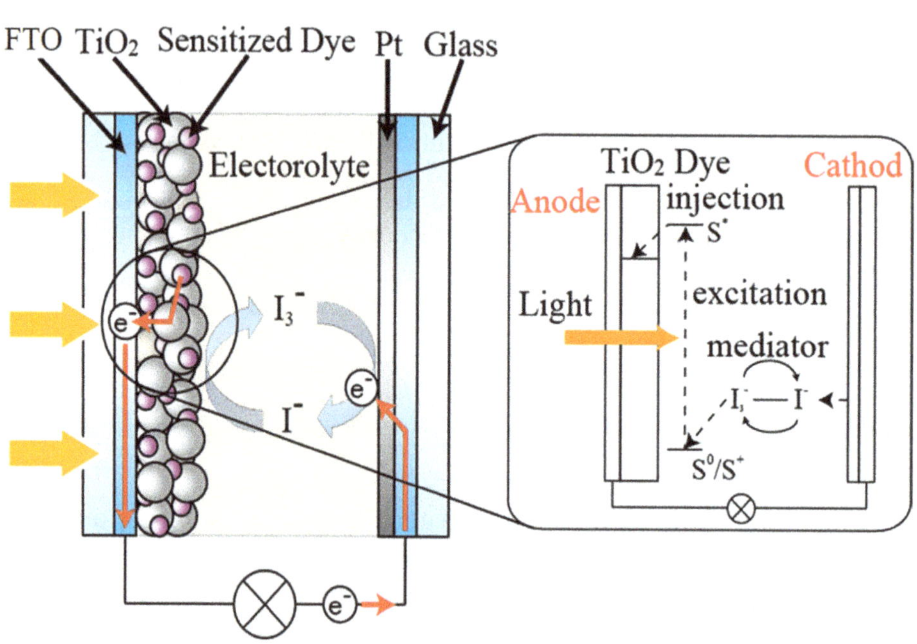

Figure 1. Structure and operating mechanism of a DSSC. Reprinted with permission of Ref. [3].

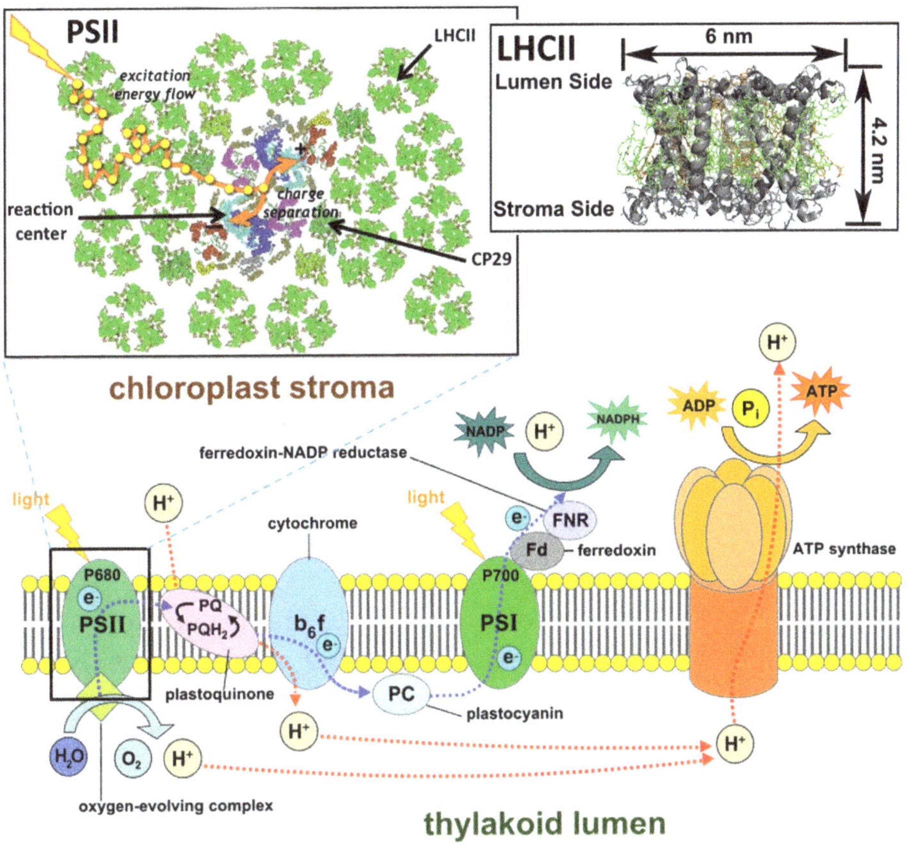

Figure 2. Photoreactions in photosynthesis at the thylakoid membrane of plant cells. Inset: the energy flow in PSII and the structure of LHCII trimer. Adapted with permission of Refs. [7, 8].

energy is resonantly transferred to the associated reaction centers (RCs) where the energy is converted into electrons by exciting a special pair of Chls, triggering a series of chemical reactions, such as water splitting in PSII, reduction of $NADP^+$ to NADPH in PSI and ATP synthesis. Inset of **Figure 2** shows components of PSII and the energy flow therein. The largest PSII supercomplex, $C_2S_2M_2$, consists of a dimeric core complex (C_2) containing RCs, 4 monomeric minor antenna complexes (CP29), 4 strongly attached LHCII trimers (S_2 and M_2), and 3–4 loosely attached LHCII trimers [9]. LHCII trimer is the most abundant Chl-protein complex in nature and the major antenna complex in PSII. The LHCII trimer consists of three monomers each of which comprises a polypeptide of about 232 amino-acid residues, 8 Chl *a* and Chl *b* molecules, 3–4 carotenoids and one phospholipid [10].

It should be noted that both photosynthesis systems and DSSCs utilize separate media for photon capture and energy transfer (executed by the excitation of LHCs and photosensitizers, respectively) and charge separation (occurs in RCs and the dye/TiO_2 interface, respectively). This mechanism has an advantage to reduce the possibility of charge recombination [11]. Owing to such similarity, various DSC architectures have been explored to directly use natural extracted pigments [12, 13] and photosynthetic LHCs [14–17] as photosensitizers to replace synthetic dyes in developing biophotovoltaic cells. Although the biophotovoltaic cells have much lower PCE than normal DSSCs, these hybrid systems serve as a unique platform to study the crucial processes including charge separation and transport at the interface of TiO_2 photoanode with natural photosensitizers and provide insights into the limiting factors. In this chapter, we will summarize our recent research and other related literature on incorporating photosynthetic proteins and plasmonic nanoparticles (PNPs) onto anatase TiO_2 photoanodes as a means to tap into the charge separation, electron and energy transfer processes, and plasmonic enhancement in the biophotovoltaics. The following aspects will be involved: (i) configuration and surface modification of TiO_2 photoanode, (ii) energy state coupling and charge transfer between photosynthetic proteins and TiO_2, (iii) plasmonic effects on biophotovoltaics, and (iv) hot electrons across Schottky barrier at Au/TiO_2 interface.

2. Configuration and surface modification of TiO_2 photoanode

2.1. Fabrication of TiO_2 photoanode

Conventional photoanode in DSSCs is composed of a 10-μm-thick nanostructured TiO_2 film prepared from deposition and sintering of spherical TiO_2 nanoparticles (NPs) on conducting fluorine doped tin oxide (FTO) glass. This mesoporous layer has a large surface area for dye adsorption while maintaining a percolation network for electron transport. Later, a variety of TiO_2 nanorods, nanowires and networks were designed to replace the spherical NPs in photoanode [18–20]. These structures are regarded with more efficient electron transport pathway owing to the inherent well-aligned crystalline domains and the greater electron diffusion length [21, 22]. Typically, the TiO_2 nanomaterials are synthesized by sol-gel method, then dispersed with surfactants into a paste and coated on FTO glass via doctor-blade casting, spin coating or screen printing. This multi-step route is tedious and induces large variances in prepared TiO_2 layers. The electron diffusion in the layer is restrained by large boundary. Instead, direct growth of highly ordered architectures on substrates is among the most exciting developments for novel photoanodes. Vertically aligned TiO_2 nanotube arrays have been successfully produced by potentiostatic anodization of titanium metal in a fluoride containing electrolyte and exhibit larger electron diffusion length [23]. A forest-like photoanode combining efficient light trapping and high surface area for dye absorption was synthesized via fine control of pulse laser deposition, which consists of hierarchical assemblies of nanocrystalline particles of anatase TiO_2 [24]. This hierarchical architecture was demonstrated to suppress electron recombination with tri-iodide along with increase of electron lifetime and perform no hindering in mass transport using ionic liquid electrolyte. Recently, we employed a similar anatase TiO_2 nanotree array as photoanode scaffold for the LHCII sensitized biophotovoltaic cells [25]. This TiO_2 nanotree array can be simply grown on TiO_2 coated FTO glass by one-pot hydrothermal reaction without necessity of high-tech equipment [26]. The transmission electron microscopy (TEM) and scanning electron microscopy (SEM) images in **Figure 3** confirm the morphology of the TiO_2 nanotrees prepared by this method. Each nanotree is composed of 6-μm-long TiO_2 nanowire trunk covered by short and thinner branches extending sideway. The growing model is illustrated in **Figure 4**, showing a hierarchical assembly of TiO_2 nanostructure in which the number and length of branches on the TiO_2 nanowire trunk can be increased with the longer reaction time.

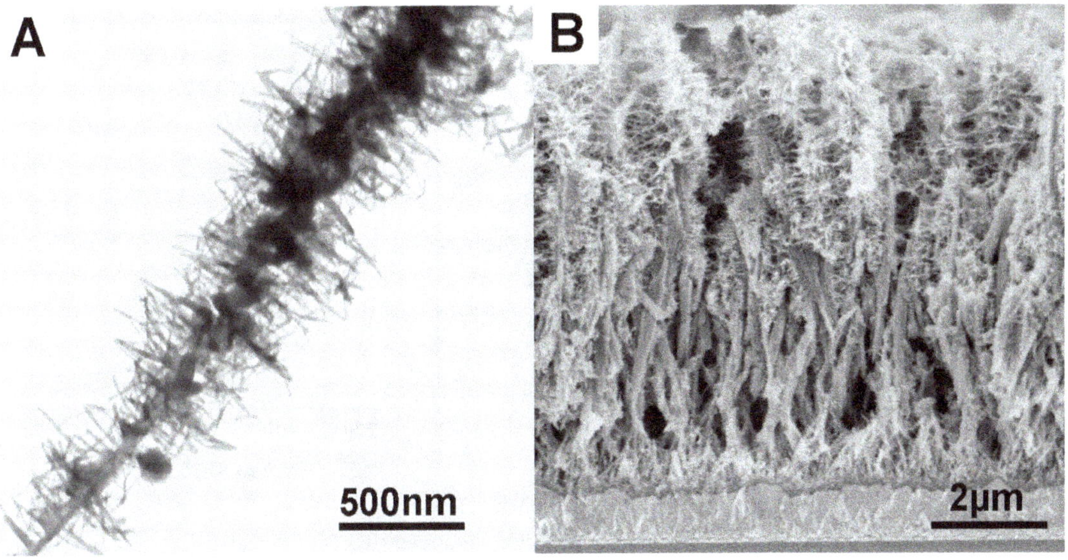

Figure 3. (A) TEM image of the TiO_2 nanotree scraped off from the FTO substrate and (B) the cross-sectional view of TiO_2 nanotree array by SEM. Reprinted with permission of Ref. [27].

In addition, post-synthesis thermal treatment can be applied to attain highly crystalline TiO_2 photoanodes. Due to higher dye loading and faster electron transport rate, the TiO_2 with pure anatase phase is more favorable than rutile phase in photoanode applications [28]. The crystallization strongly depends on annealing process. It normally yields the anatase phase if the annealing temperature is below 550°C but tends to form the thermodynamically stable rutile phase at higher temperatures [29]. The XRD (**Figure 5A**) and Raman (**Figure 5B**) characterizations confirmed that single crystalline anatase phase was attained for the TiO_2 nanotrees subjected to 500 °C calcination for 30 min.

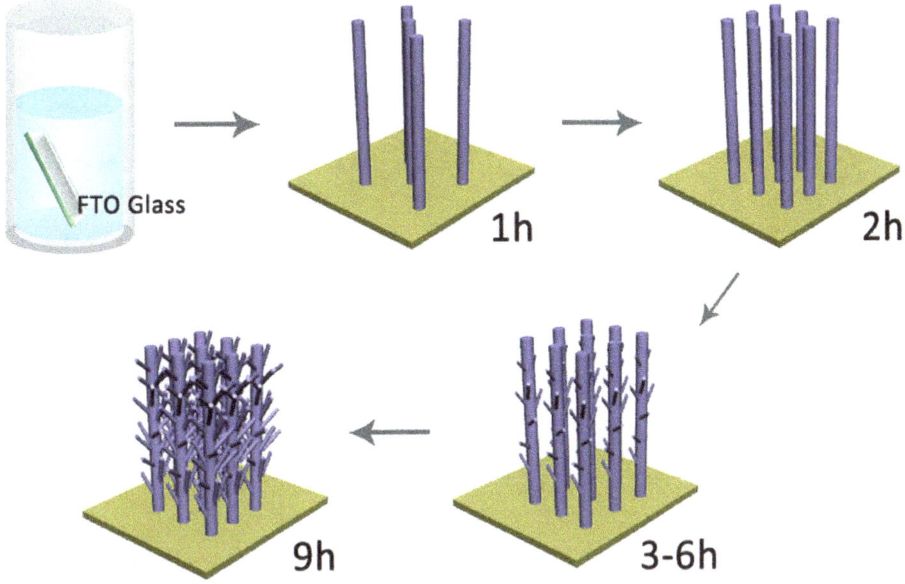

Figure 4. Schematic formation process of the hierarchical anatase TiO_2 nanotree arrays on FTO substrates. Reprinted with permission of Ref. [26].

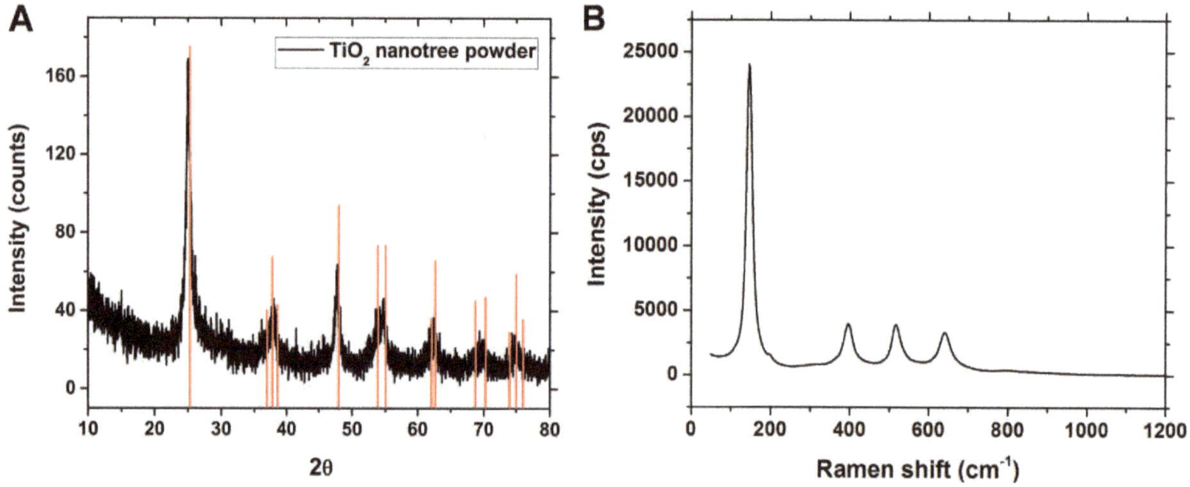

Figure 5. (A) XRD pattern and (B) Raman spectrum of the TiO_2 nanotrees after 500 °C thermal annealing. The standard XRD peak position of anatase TiO_2 (JCPDS card No 71–1166) is indicated as the vertical red lines in (A). Reprinted with permission of Ref. [26].

2.2. TiO_2 barrier layer

In the photoanode of DSSCs, besides using TiO_2 for charge separation and electron transport, a thin compact TiO_2 layer of tens to hundreds of nanometers is usually deposited between mesoporous TiO_2 nanoparticle film and transparent conductive oxide (TCO) coated glass as a barrier layer. This barrier layer was found to be critical in impairing the electron backflow at the TCO/electrolyte interface, increasing the shunt resistance, and therefore increasing the fill factor and overall cell efficiency [30, 31], as depicted in **Figure 6A**.

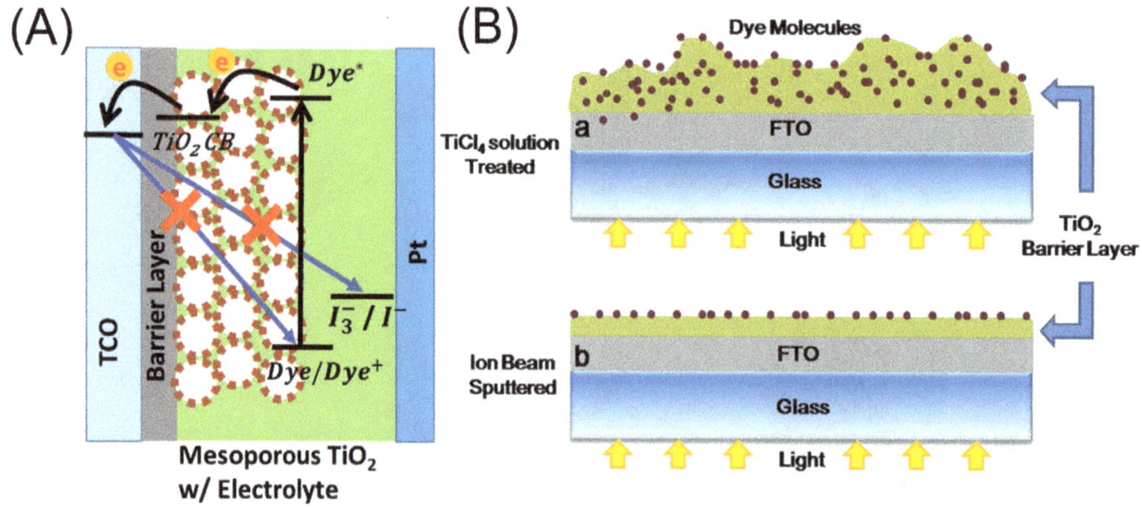

Figure 6. (A) Charge recombinations in DSSC due to the electron back flow from TCO to oxidized dye and redox electrolyte. (B) Illustration of the differences of the TiO_2 barrier layers formed by (a) $TiCl_4$ treatment and (b) Ti sputtering followed by thermal annealing. Adapted with permission of Ref. [32].

To establish a reliable planar photoanode for biophotovoltaics, we systematically studied the TiO_2 barrier layer deposited by two distinct methods and correlated the TiO_2 structure with its barrier properties [32]. **Figure 6B** schematically shows that a porous dye-penetrable TiO_2 film was attained by the $TiCl_4$ solution treatment, while a conformal compact TiO_2 film was obtained by sputtering-annealing Ti. The latter seemed to be an ideal barrier layer since dye molecules can only adsorb on the external TiO_2 surface. However, the performance of DSSCs made with the sputtering-annealing method was worse than those by $TiCl_4$ treatments due to the lower electrical conductivity since anatase structure is mixed with amorphous and rutile phases in the film. In this work, the DSSCs fabricated with photoanodes by 20 min $TiCl_4$ treatment showed the best performance, likely due to the formation of desired anatase crystallites with the optimum thickness. Such thin-film DSSC was used as a model system to test the photovoltaic effects of photosynthetic proteins that cannot easily access the interior pores of traditional mesoporous DSSCs.

2.3. Biosensitizations on surface modified TiO_2 photoanodes

One complication to fabricate biophotovoltaic devices is to integrate biophotosensitizers with artificial semiconductors in photoanode. Unlike synthetic dyes that can easily use various anchoring groups, for example, carboxylate (—COOH), phosphonate (—H_2PO_3) or siloxy moiety (—O—SiR_3), through molecular engineering to increase binding affinity to metal oxide semiconductor [33, 34], chemical modifications on natural extracted photosynthetic protein complexes would cause unfavorable structural changes that impair their intrinsic photoelectric properties. Although the photosynthetic protein complexes contain carboxylate groups in their polypeptide matrix, they are usually extracted and dispersed in aqueous buffer solution in which the strong polar water solvent and the surfactants tend to break the adsorption equilibrium causing desorption from the metal oxide surface. Addition of binding agents is desired to conjugate photosynthetic protein complexes to artificial photoanodes. Mershin et al. bioengineered PSI with a designed peptide surfactant that contains an amino acid sequence with specific high binding affinity to ZnO (**Figure 7**) [35]. In the study, the native electron acceptor subunit PsaE within PSI was substituted with the ZnO-binding peptide tag: RSNTRMTARQHRSANHKSTQRARS to promote attachment and orientation of the PSI on ZnO nanowires. Thus, the modified PSI was preferentially bound to ZnO nanowires by the electron acceptor side, minimizing the electron traveling distance between electron acceptor and electrode and maximizing the electron transfer.

Beyond introducing the specific linkers through delicate bioengineering on photosynthetic protein complexes, another simpler method to improve the protein attachment is to perform surface modifications on photoanode materials with binding molecules. Dihydroxyacetone phosphate was reported as a suitable linker between PSI and metal oxide. The indium-tin oxide (ITO) and titanium suboxide (TiO_x, $x = 1, 2$) substrates covered with a self-assembled monolayer of dihydroxyacetone phosphate can immobilize a densely packed PSI layer by electrostatic and hydrogen bond interactions with the polar stroma and lumen faces of PSI [36]. LHCII of PSII can also be isolated and appointed as photosensitizers in biophotovoltaic cells. However, the physisorption of LHCII on the TiO_2 photoanode was found to be very weak and unstable, as indicated by the long incubation time (96 hours) required to reach saturated adsorption [45]. It was recently reported that strong LHCII attachment can be obtained

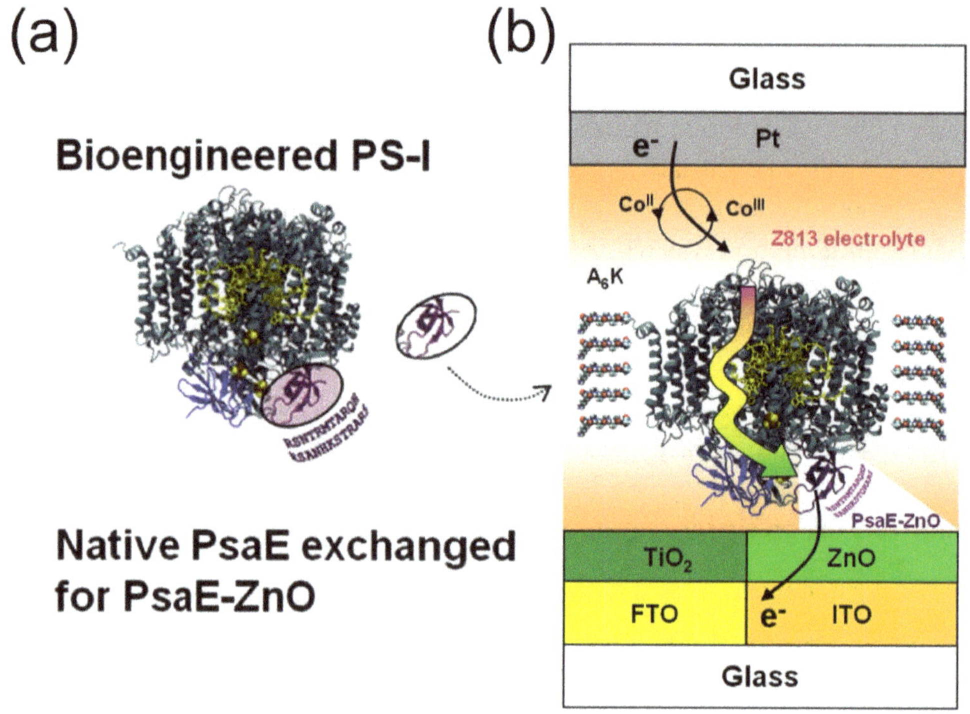

Figure 7. (a) Bioengineered modification of PS I by substitution of native PsaE with PsaE-ZnO. (b) Schematic electron transfer path in PS I-based biophotovoltaic cells. Adapted with permission of Ref. [35].

on a APTES grafted FTO substrate via electrostatic interaction between the anionic residues on the stromal side with cationic $-NH_3^+$ groups [14]. We adopted this approach and used APTES to functionalize the surface of TiO_2 thin film, TiO_2 nanotree array photoanodes and TiO_2-encapsulated plasmonic NPs in LHCII sensitized solar cells [25]. Clear improvement in protein attachment is indicated by the more intense and uniform greenish color on the APTES modified photoanode (**Figure 8a**).

Moreover, adsorption of the photosynthetic proteins onto internal surface of the mesoporous TiO_2 anode is also hindered by its much larger size (4–20 nm) than dye molecules (<1 nm), researchers strive to increase their loading capacity by engineering more open three-dimensional (3D) electrode architecture. The amount of the adsorbed proteins can be extracted into buffer solutions and quantitatively assessed from the absorption spectra of the extracted Chls in **Figure 8b** based on the following equation [37]:

$$\text{Chlsa}(a+b) = 17.6\ \text{A}^{646.6} + 7.34\ \text{A}^{663.6} \quad (1)$$

A is the absorbance at certain wavelength. By this means, the amount of LHCII trimers adsorbed on APTES-treated TiO_2 nanotrees was determined to be 2.5 folds of that on bare TiO_2 nanotrees. It can be derived from basic calculations that LHCII trimers containing 0.2 μg Chls are required to form a hexagonal close-packed monolayer on a flat 1 × 1 cm^2 surface. Since the adsorbed LHCII trimers were equivalent to 6.6 μg Chls, 33 times of that on the flat TiO_2 surface, it is evident that nanoscale LHCII trimers were able to penetrate into the 3D TiO_2 nanotree array and adsorb on a large surface area.

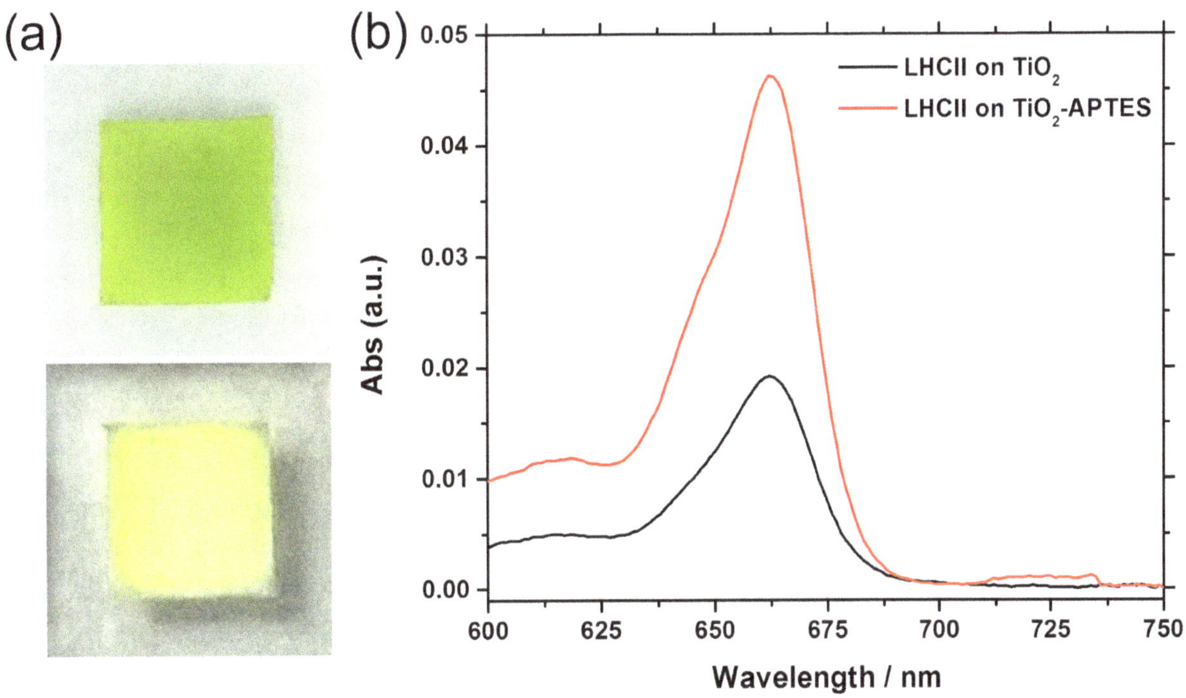

Figure 8. (a) The digital photographs of the LHCII-sensitized region of an APTES-treated TiO_2 nanotree array (top) and a bare TiO_2 nanotree array (bottom) (b) UV-Vis absorption of the chlorophylls pigment extracted from the LHCII trimers adsorbed on TiO_2 nanotree photoanodes with and without APTES functionalization. Reprinted with permission of Ref. [25].

3. Energy state coupling and charge transfer between photosynthetic proteins and TiO_2

3.1. General electron transfer in the biophotovoltaic cells with photosynthetic protein sensitized TiO_2 photoanodes

Similar to DSSCs, photocurrent generation by biophotovoltaic cells based on photosynthetic proteins sensitized TiO_2 photoanode is inextricably linked with the charge separation at the photosynthetic protein/TiO_2 interface. The energy state matching among photosynthetic proteins, semiconductive TiO_2, and redox mediators is crucial to enable the electron injection from photosynthetic proteins to TiO_2 as well as the electron refill from redox mediators to the proteins. The photoinduced electrons originate from Q band excitation of Chls. Since Chl is the major pigment contained in photosynthetic proteins, the energy levels of the photosynthetic proteins can be represented by the ground and excitation states of Chls Q band, and they can be determined by measuring the oxidation potential and the absorption spectrum of the photosynthetic proteins. **Figure 9** shows the electron transfer and energy level scheme of the biophotovoltaic cell based on LHCII aggregates as the sensitizer on thin film TiO_2 photoanode. Potentials are relative to the normal hydrogen electrode (NHE) [38]. Similar to DSSCs, when the chlorophylls in the photosynthetic proteins are photoexcited to the higher energy level Chl*, the electrons are able to inject into the less negative conduction band of TiO_2. In

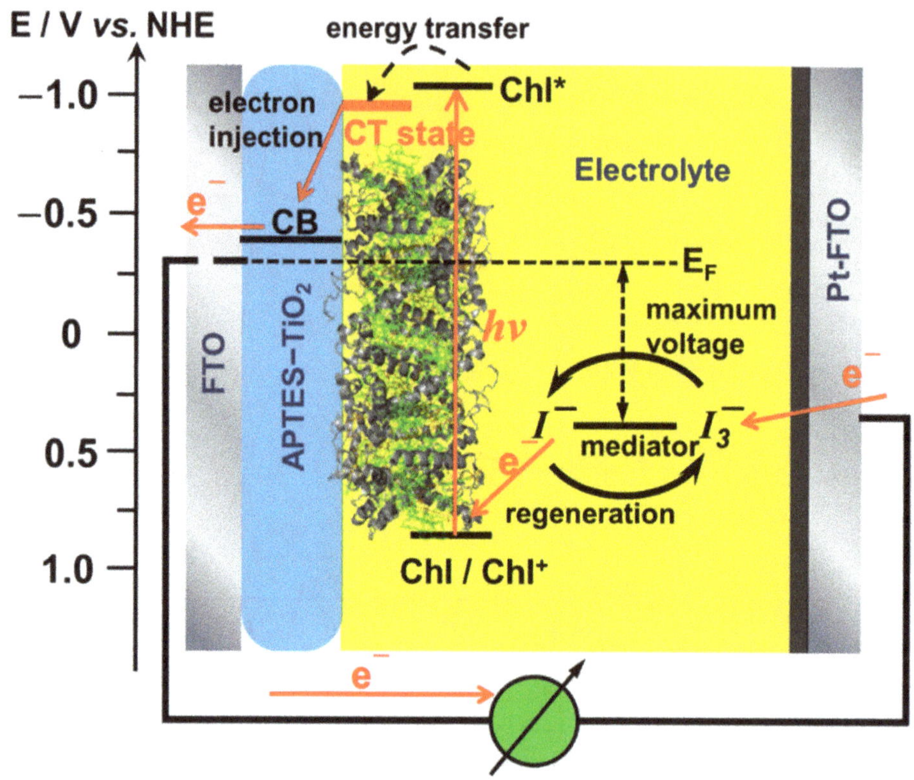

Figure 9. Electron transfer and energy level scheme of a photovoltaic device based on aggregated LHCII complexes. Reprinted with permission from Ref [38].

this step, the harvested solar energy is converted into anodic photocurrent. Meanwhile, the redox potential of the mediator should be more negative (i.e., in lower position in **Figure 9**) than the hole left at the ground state of Chlorophyll Ch^+ so that it can resupply an electron to the oxidized Chl, thus regenerating the sensitizer. The V_{OC} (maximum output voltage of solar cell) corresponds to the difference between the redox potential of the redox mediator and the Fermi level of the FTO current collector.

3.2. Effect of charge transfer state in LHCs

Comparing to artificial DSSCs, the biophotovoltaics involving photosynthesis complexes as sensitizers have two distinct features. First, the captured photon may go through a rapid internal energy transfer process to the charge-separation states. For example, **Figure 10A** shows that the excitation energy of LHCII at 496.5 nm is quickly transferred to the lower-energy Q band around 650–690 nm before giving fluorescence or producing charge separation. Second, the charge transfer process of densely assembled chlorophylls in photosynthetic protein complexes depends on the specific protein environments involving photosynthesis regulation through a photoprotective mechanism called non-photochemical quenching (NPQ) [9, 39–44]. Excess energy in the photo-excited chlorophylls was dissipated through specific LHCII protein aggregation [45]. The Chl excited states in the aggregated LHCII, unlike in isolated LHCII trimers, are severely quenched due to the formation of chlorophyll-chlorophyll coupled charge transfer (CT) states, which has been observed by the high-resolution hole-burning spectroscopy [46].

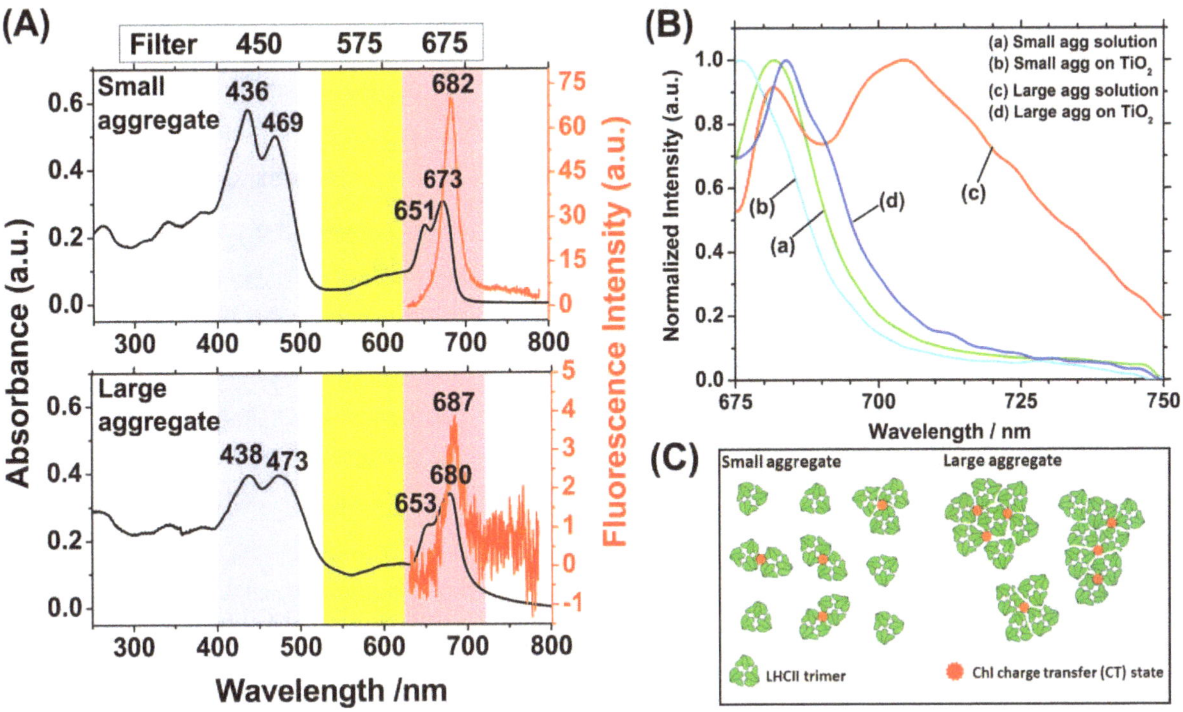

Figure 10. Absorption and emission properties of LHCII aggregates associated with the formation of CT states. (A) Absorption and fluorescence emission spectra (λ_{ex} = 496.5 nm) of small- and large-size LHCII aggregate in tricine buffer. (B) Normalized steady-state fluorescence emission spectra (λ_{ex} = 663 nm) of small- and large-size LHCII aggregate in solutions and deposited on the APTES-TiO_2-FTO photoanode surface, respectively. (C) Schematic illustration of the CT states (dots) formed in small and large aggregates. Reprinted with permission of Ref. [38].

Previous study revealed that the photovoltaic performance of the biophotovoltaic cell was correlated with strong coupling between the extensive CT states formed between the aggregated LHCIIs (schematically depicted in **Figure 10C**) and the TiO_2 conduction band. The CT states have slightly lower oxidation potential (i.e., less negative in energy level in **Figure 9**) than the excited state of chlorophylls due to their more reddish absorption and red tail in UV spectra (**Figure 10A**). The CT states couple with the TiO_2 conduction band more effectively, convinced by severe quench on the fluorescence emission of the CT states when the large LHCII aggregates were anchored on TiO_2 surface (as shown in **Figure 10B**). This strong coupling facilitated more efficient electron injection across LHCII/TiO_2 interface and resulted in larger photocurrent generation in the corresponding biophotovoltaic cells.

3.3. Effect of plasmonic nanoparticles

While the biophotovoltaic cells based on interfacing the artificial DSSC platform with the photosynthetic proteins provide useful insight into the fundamental photon capture and charge separation processes, their PCE is much lower than the conventional DSSCs using organic dye molecules (such as N719 [47]) as light harvesting antennas. Noble metal nanoparticles, that is, gold or silver nanoparticles, have been explored to enhance the solar cell performance utilizing their surface plasmonic resonance (SPR) effects that can enormously alter the optical absorption and emission of photosynthetic proteins near the nanoparticle surface [48]. An example of such solar cells is shown in **Figure 11** [25]. Enhancement of light absorption was observed for

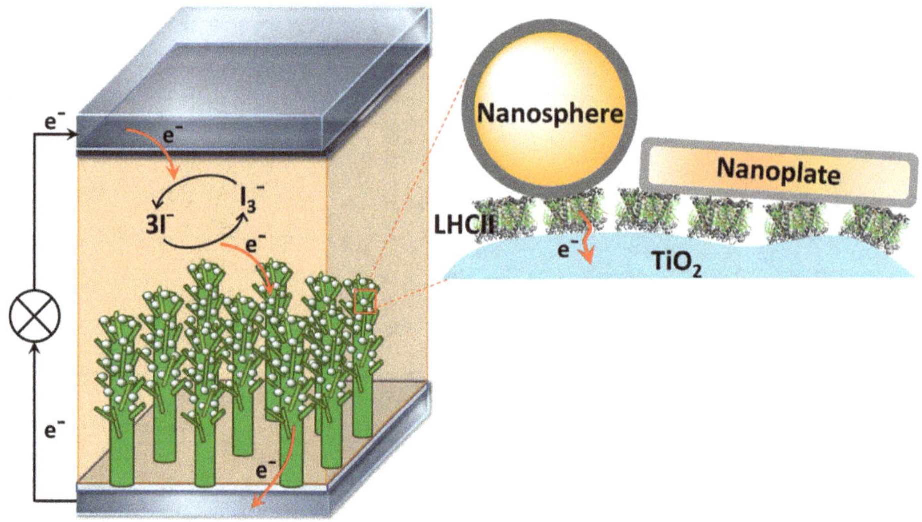

Figure 11. The structure of the plasmonic biophotovoltaic cell. The enlarged portion (in square) shows the binding of different PNPs on LHCII-sensitized TiO_2 nanotrees and the electron injection from LHCII to TiO_2 (curve arrow). Reprinted with permission of Ref. [25].

PSI attached to plasmonic nanoparticles (PNPs) [49]. The LHCs anchored on plasmonic gold or silver island substrates were able to generate 10- to 20-fold of fluorescence emission [50–52]. A theoretical model for SPR-enhanced free electron production and photocurrent generation was proposed based on PSI-RCs bound to Au and Ag nanocrystals [53]. The internal photosynthetic efficiency of PSI-RC was found to be strongly enhanced by the metal nanoparticles, which involved two competing effects, that is, plasmon enhanced light absorption of Chl molecules and energy transfer from Chl to metal nanoparticles [53]. These studies provide useful insights into energy-conversion devices involving the interplay between photosynthetic proteins and PNPs.

PNPs have already been widely employed in photovoltaic devices, including DSSCs [54–56] and emerging Perovskite solar cells [57, 58], to enhance the performance. However, the explanation of the interplays between PNPs and different light harvesting antennas is still ambiguous in each specific cases. Unlike single chromatic synthetic dye, photosynthetic proteins contain multiple pigments whose light harvesting and conversion involve intrinsic energy transfer among pigments and cofactors, inducing additional complexity to understand the plasmonic effects on the whole photovoltaic processes of the bio-hybrid systems. Recently, a bio-solar cell using natural extract graminoids coupled with silver nanoparticles (Ag-NPs) has been reported to achieve larger photocurrent [59]. First, incorporating plasmonic Ag-NPs (~13.8 nm in diameter) enables quenching the emission of the natural graminoid sensitizers and thus enhancing the electron collection efficiency. Meanwhile, the small size of the Ag-NPs barely takes up surface area for attachment of light-harvesting sensitizers. Second, a TiO_2(001) nanosheet structure provides a good surface for collection of solar-driven electrons from graminoids. Third, the ligand tethering enables good attachment between graminoids and Ag-NPs on the (100) face of TiO_2 nanosheet. The enhanced performance of the plasmonic biophotovoltaic cell in this case was attributed to the emission quenching by Ag-NPs for efficient collection of photoinduced electrons from graminoid complexes, as well as the efficient light trapping due to the plasmonic enhanced local electromagnetic field.

We proposed another enhancement mechanism for plasmonic biophotovoltaics with the design shown in **Figure 12** [25]. Core-shell PNPs with a 2–5 nm TiO_2 shell and a plasmonic silver or gold core were hybridized with LHCII and incorporated into the aforementioned 3D TiO_2 nanotree photoanode for the plasmonic enhanced biophotovoltaic cells. Compared with the bare Ag NPs used in the above-discussed work [59], this core-shell structure has multiple functions. First, the hydrophilic nature of TiO_2 shell makes the PNP surface compatible for protein attachment [60]. Second, the semiconductive TiO_2 shell serves as an energy barrier to prevent electron recombination on the metallic core due to unwanted electron flow from the attached proteins to the metallic core [61]. Third, the TiO_2 shell acts as a protective armor to ensure the stability of the metallic core in the corrosive iodide electrolytes in solar cells [62]. These PNPs with different plasmonic resonance bands were able to enhance and manipulate the photon capture of LHCII at specific wavelength ranges. The photocurrent

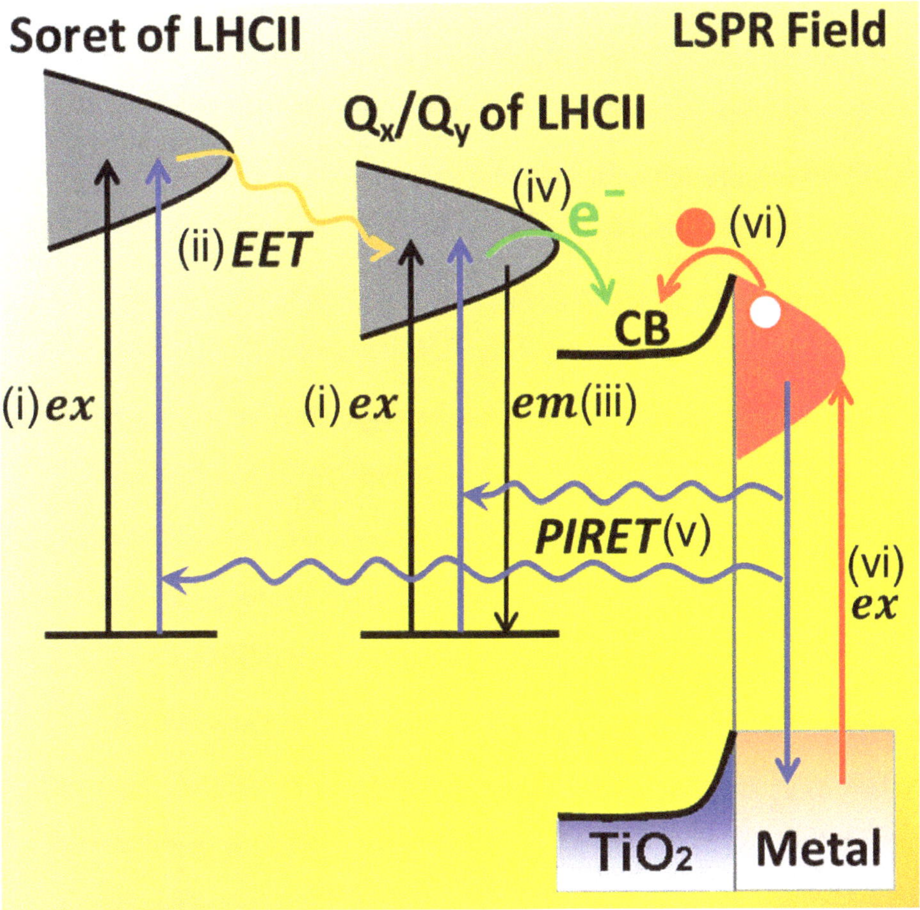

Figure 12. Schematic diagram of the energy and electron pathways in LHCII-PNP hybrid system. The LHCII trimers are excited by strong absorption of Chls' Soret and Q bands (i). The excited Chls go through an ultrafast excitation energy transfer (EET) from Soret band to Q band (ii) and then give fluorescence emission (iii) to return to the ground level. With LHCII attached to TiO_2 surface, a charge transfer process occurs, leading to injection of excited electrons in Q band to the conduction band (CB) of TiO_2 (iv). Thus, the fluorescence intensity is reduced. In the presence of the metallic core, further excitation to LHCII may occur due to plasmon-induced resonance energy transfer (PIRET) from PNPs to LHCII (v), resulting in larger electron injection from LHCII to TiO_2. At the meantime, the injection of hot electrons from the metal core of PNPs across the Schottky barrier (vi) leads to higher charge carrier density in TiO_2. Reprinted with permission from Ref. [25].

and incident photon-to-current efficiency (IPCE) of the plasmonic biophotovoltaic cell were achieved, while the fluorescence emission of excited LHCII was quenched along with shortened lifetime. Obviously, the electrons in the excited LHCII state flow efficiently through the TiO_2 network.

Cushing et al. [63, 64] elaborated that three mechanisms are involved charge generation in a semiconductor incorporated with plasmonic metal NPs, including light trapping based on scattering, hot electron/hole transfer, and plasmon-induced resonance energy transfer (PIRET) based on near-field. These are also applicable to the LHCII-PNP hybrids, with the possible mechanisms illustrated in **Figure 12**. During the excitation process, electrons are pumped from the ground state to the excited states of Soret band or Q band of LHCII. However, the excited Soret band quickly goes through an ultrafast excitation energy transfer (EET) to the Q band, as verified by a theoretical modeling [65]. Thus, all fluorescence emission from LHCIIs is at 683 nm, corresponding to a radiative relaxation for the excited electrons to return to the ground state of Q band. When LHCII is adsorbed on the TiO_2 surface, upon excitation a charge transfer process occurs, that is, the excited electrons in Q band are injected to the conduction band (CB) of TiO_2, resulting in the reduced fluorescence intensity. In the presence of the metallic core, the incident photons by plasmonic absorption generate a strong near-field oscillation with ~10 nm decay length, which can affect all LHCII adsorbed on the surface of the ~2–3-nm-thick TiO_2 shell. A strong PIRET is enabled by the strong dipole-dipole coupling between the plasmon and LHCII, leading to enhanced LHCII excitations at Soret and Q bands. More efficient electron injection from LHCII to TiO_2 is also facilitated by the near-field and thus quenches the fluorescence emission though PIRET induces higher LHCII excitation. These effects collectively enhance the photocurrent of the corresponding plasmonic biophotovoltaic cells. In addition, the plasmonic hot electrons excited at the metal core of PNPs may overcome the Schottky barrier at the metal-TiO_2 interface, raise the charge carrier density in TiO_2 shell, and therefore the charge collection efficiency. Details about such interfacial activity are unravelled in the next section.

4. Hot electron injection from plasmonic metal to TiO_2

Schottky barrier is an energy barrier for electrons formed at the junction of metal and semiconductor where their fermi levels merge together to achieve thermal equilibrium, leading to the band bending and blocking the electron flow across the junction. The energy band diagram of Schottky contact is illustrated in **Figure 13**. The height of Schottky barrier equals to the subtraction of work function of metal with the bandgap of semiconductor, and it is usually much smaller in value (e.g., ~0.9 eV for Au/TiO_2 [66] and ~0.2 eV for Ag/TiO_2 [67]) than the semiconductor bandgap (i.e., 3.2 eV for anatase TiO_2). Recent studies proposed hot electrons from plasmonic-excited metal cores could easily overcome this energy barrier and be injected into the TiO_2 conduction band, resulting in a new mechanism for plasmon enhancement to DSSCs [68]. In general, the injected hot electrons are considered to be either being converted into photocurrent or functioning as charge carriers in the semiconductor matrix [69]. The photocurrent generated by direct hot electron transfer across the Schottky barrier has been

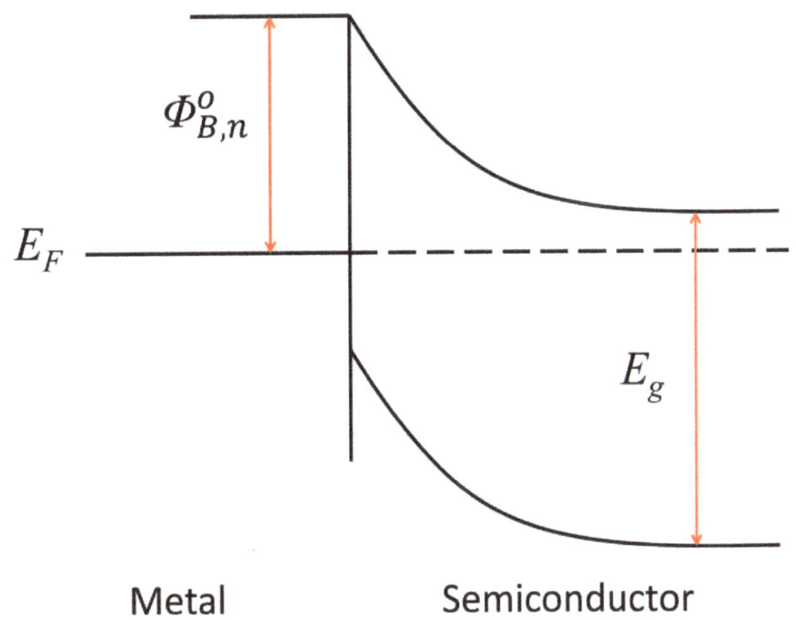

Figure 13. Energy band diagram of Schottky barrier formed at metal/n-type semiconductor interface.

collected and utilized for photodetection and photovoltaics based on well-designed devices with a complete circuit allowing refilling electrons back to the metal [70–73]. However, for the metal@TiO_2 NPs embedded in the mesoporous TiO_2 film in DSSCs, the sustainability of the photocurrent generation from hot electron injection is under debate considering that the metal core is unaccessible to the electron donors or the external circuit, which is needed for charge regeneration. On the other hand, the initially injected hot electrons may be converted into steady-state charge carriers and sufficiently raise the conductivity of the mesoporous TiO_2 frame, as has been indirectly demonstrated by enhanced photoconductivity in metal coupled semiconductors [74, 75]. In addition, a recent study by Cushing et al. reported that metal@TiO_2 and metal@SiO_2@TiO_2 NPs can also enhance DSSCs by exciting surrounding TiO_2 matrix and dye molecules with near-field-based plasmon-induced resonance energy transfer (PIRET) beside hot electron injection [63]. In the previous section, we have also discussed such effects on the plasmonic biophotovoltaic cells based on hybrids of natural LHCII and PNP [25]. Actually, these three effects are mixed in most plasmonic photovoltaic cells.

In order to sort out the contributions of hot electron injection, we propose a strategy by comparing the photoconductivity and the photovoltaic properties of the same material, that is, Au@TiO_2 network in two model devices, that is, a micro-gap electrode and a DSSC [76]. The core-shell structure consisting of isolated Au NPs embedded at the nodes of a nanostructured TiO_2 network was used as the bridging material in the micro-gap between two Au electrodes and as the mesoporous film on a DSSC anode to measure photoconductance and photocurrent, respectively. Enhancements on the photoconductance and the photocurrent were observed on both devices, with distinct dependence on the illumination wavelength (**Figure 14A** and **B**). This difference was explained with the scheme drawn in **Figure 14C** and **D**. The enhanced photoconductance is ascribed to the hot electron injection from Au NPs to TiO_2 that increase the charge carrier density of the TiO_2 network. This interfacial electron injection across the Au/TiO_2

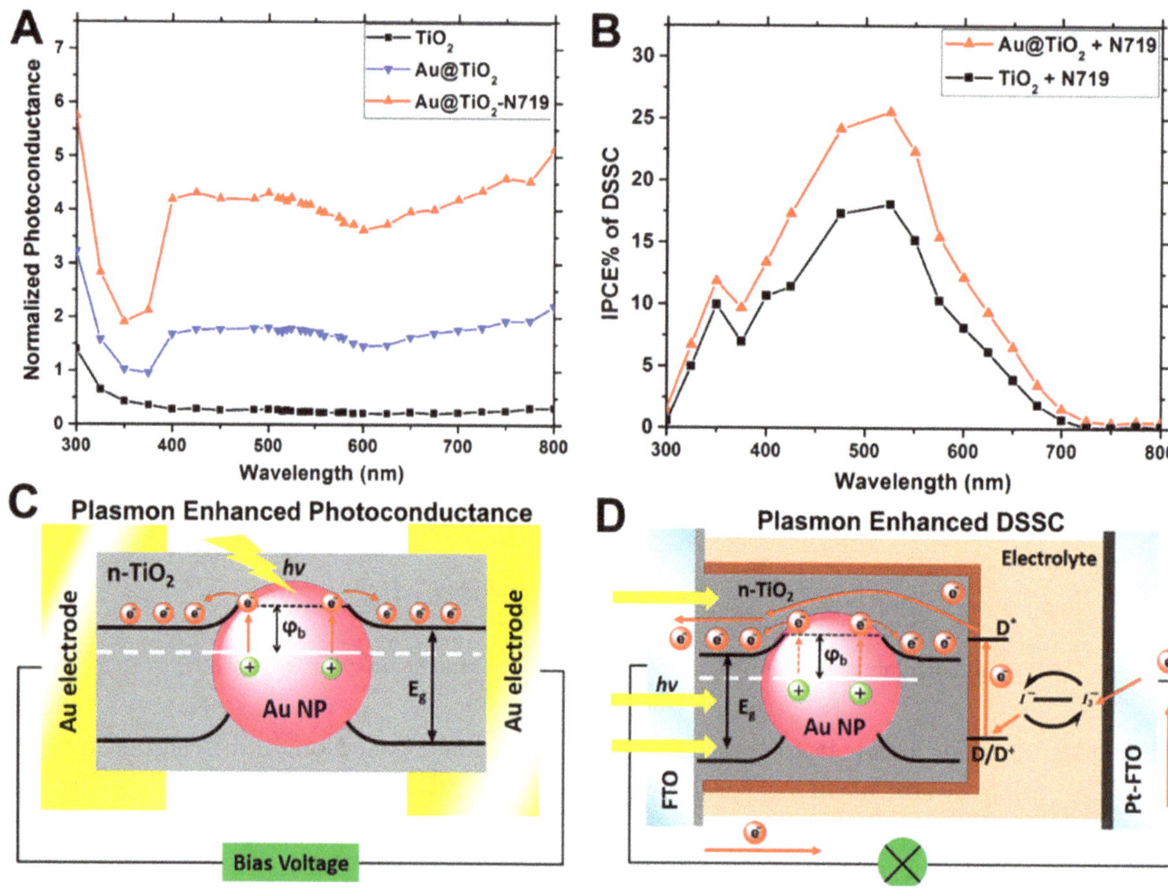

Figure 14. (A and B) Wavelength dependence of photoconductance and incident photon-to-current efficiency (IPCE) studies and (C and D) the schematics of the possible enhancement mechanisms for the Au@TiO_2 network on the micro-gap electrode and in the DSSC, respectively. Reprinted with permission from Ref. [76].

Schottky barrier (~0.9 eV) can be easily realized under illumination over the whole visible range, allowing extending the enhancement effect to the light in the near-infrared region. The photon energies in wavelength larger than 700 nm are smaller than the energy of semiconductor band gap, Au plasmonic band, and dye absorption band. In the DSSC, the plasmonic generated hot electrons cannot be the source of continuous steady-state photocurrent, since the Au NPs embedded within the TiO_2 shell are not accessible by the regenerating agents. The major contribution for the photocurrent enhancement must be the surface plasmonic resonance effect that can only be induced by the illumination in the range where the plasmonic band of Au NPs is resonant with dye absorption band (i.e., band overlap in absorption spectra). However, the injected hot electrons are sufficient to raise the charge carrier density in TiO_2 and reduce the series resistance and charge transfer resistance in the corresponding DSSCs. This facilitates the transport of the photo-induced electrons through the TiO_2 network.

5. Conclusions

This chapter reviews the synergistic interplay among TiO_2 photoanode, biophotosensitizers (e.g., LHCII) and plasmonic nanoparticles in the photovoltaic devices. The effectiveness of

TiO_2 as an interfacial photoanode material compatible with photosynthetic proteins and plasmonic nanoparticles was demonstrated. The electron injection from excited LHCII to TiO_2 conduction band was realized due to the perfect match of energy bands, resulting in the photocurrent generation in the LHCII sensitized TiO_2 solar cells. The charge separation at LHCII/TiO_2 interface can be facilitated by incorporation of PNPs. This effect can be ascribed to the near-field-assisted PIRET from PNPs to LHCII across the TiO_2 interfacial layer. The hot electron injection across Schottky barrier from plasmonic core into TiO_2 network can increase the charge carrier density in TiO_2, leading to the increase of photoconductivity and the improved photovoltaic performance of TiO_2-based photoanode. Understanding of these fundamental energy/charge transfer processes and interface properties will inspire future optoelectronic devices with smart designs for outstanding performance.

Acknowledgements

This work was supported by the National Science Foundation (NSF) EPSCoR Award EPS-0903806, a National Aeronautics and Space Administration (NASA) grant NNX13AD42 A, and the matching funds to these two grants provided by the State of Kansas.

Author details

Yiqun Yang[1] and Jun Li[2*]

*Address all correspondence to: junli@ksu.edu

1 Xavier University of Louisiana, New Orleans, LA, United States

2 Kansas State University, Manhattan, KS, United States

References

[1] Chen X, Mao SS. Titanium dioxide nanomaterials: synthesis, properties, modifications, and applications. Chemical Reviews. 2007;**107**:2891-2959

[2] O' Regan B, Grätzel M. A low-cost, high-efficiency solar-cell based on dye-sensitized colloidal TiO_2 films. Nature. 1991;**353**:737-740. DOI: 10.1038/353737a0

[3] Miki N. Liquid encapsulation technology for microelectromechanical systems. In: Kenichi Takahata editor, Advances in Micro/Nano Electromechanical Systems and Fabrication Technologies, In Tech, Croatia-European Union; 2013. DOI: 10.5772/55514

[4] Yella A, Lee H-W, Tsao HN, Yi C, Chandiran AK, Nazeeruddin MK, Diau E, W-G, Yeh C-Y, Zakeeruddin SM, Grätzel M. Porphyrin-sensitized solar cells with cobalt (II/III)–based Redox electrolyte exceed 12 percent efficiency. Science. 2011;**334**:629-634. DOI: 10.1126/science.1209688

[5] Burschka J, Pellet N, Moon S-J, Humphry-Baker R, Gao P, Nazeeruddin M. K Gratzel M. Sequential deposition as a route to high-performance perovskite-sensitized solar cells. Nature. 2013;**499**:316-319. DOI: 10.1038/nature12340

[6] Giordano F, Abate A, Correa Baena JP, Saliba M, Matsui T, Im SH, Zakeeruddin SM, Nazeeruddin MK, Hagfeldt A, Graetzel M. Enhanced electronic properties in mesoporous TiO_2 via lithium doping for high-efficiency perovskite solar cells. Nature Communications. 2016;**7**:10379. DOI: 10.1038/ncomms10379

[7] Various Authors, Photosystem, Wikipedia [Internet]. 2015. Available from: https://commons.wikimedia.org/wiki/File%3AThylakoid_membrane_3.svg

[8] Fleming GR, Schlau-Cohen GS, Amarnath K, Zaks J. Design principles of photosynthetic light-harvesting. Faraday Discussions. 2012;**155**:27-41. DOI: 10.1039/C1FD00078K

[9] Caffarri S, Kouřil R, Kereïche S, Boekema EJ, Croce R. Functional architecture of higher plant photosystem II supercomplexes. EMBO J. 2009;**28**:3052-3063. DOI: 10.1038/emboj.2009.232

[10] Liu Z, Yan H, Wang K, Kuang T, Zhang J, Gui L, An X, Chang W. Crystal structure of spinach major light-harvesting complex at 2.72 Å resolution. Nature. 2004;**428**:287-292. DOI: 10.1038/nature02373

[11] Ondersma JW, Hamann TW. Recombination and redox couples in dye-sensitized solar cells. Coordination Chemistry Reviews 2013;**257**:1533-1543. DOI: 10.1016/j.ccr.2012.09.010

[12] Ludin NA, Al-Alwani Mahmoud AM, Bakar Mohamad A, Kadhum AAH, Sopian K, Abdul Karim NS. Review on the development of natural dye photosensitizer for dye-sensitized solar cells. Renewable & Sustainable Energy Reviews. 2014;**31**:386-396. DOI: 10.1016/j.rser.2013.12.001

[13] Hug H, Bader M, Mair P, Glatzel T. Biophotovoltaics: Natural pigments in dye-sensitized solar cells. Applied Energy. 2014;**115**:216-225. DOI: 10.1016/j.apenergy.2013.10.055

[14] Nagata M, Amano M, Joke T, Fujii K, Okuda A, Kondo M, Ishigure S, Dewa T, Iida K, Secundo F, Amao Y, Hashimoto H, Nango M. Immobilization and photocurrent activity of a light-harvesting antenna complex II, LHCII, isolated from a plant on electrodes. ACS Macro Letters. 2012;**1**:296-299. DOI: 10.1021/mz200163e

[15] Narayan MR. Review: Dye sensitized solar cells based on natural photosensitizers. Renewable & Sustainable Energy Reviews. 2012;**16**:208-215. DOI: 10.1016/j.rser.2011.07.148

[16] Yu D; Zhu G, Liu S, Ge B, Huang F. Photocurrent activity of light-harvesting complex II isolated from spinach and its pigments in dye-sensitized TiO_2 solar cell. International Journal of Hydrogen Energy. 2013;**38**:16740-16748. DOI: 10.1016/j.ijhydene.2013.02.114

[17] Yu D, Wang M, Zhu G, Ge B, Liu S, Huang F. Enhanced photocurrent production by bio-dyes of photosynthetic macromolecules on designed TiO_2 film. Scientific Reports. 2015;**5**:9375. DOI: 10.1038/srep09375

[18] Wang M, Bai J, Le Formal F, Moon S-J, Cevey-Ha L, Humphry-Baker R, Grätzel C, Zakeeruddin SM, Grätzel M. Solid-State dye-eensitized solar cells using ordered TiO_2 nanorods on transparent conductive oxide as photoanodes. The Journal of Physical Chemistry C. 2012;**116**:3266-3273. DOI: 10.1021/jp209130x

[19] Tétreault N, Horváth E, Moehl T, Brillet J, Smajda R, Bungener S, Cai N, Wang P, Zakeeruddin SM, Forró L, Magrez A, Grätzel M. High-efficiency solid-state dye-sensitized solar cells: fast charge extraction through self-assembled 3D fibrous network of crystalline TiO_2 nanowires. ACS Nano. 2010;**4**:7644-7650. DOI: 10.1021/nn1024434

[20] Yu H, Zhang S, Zhao H, Xue B, Liu P, Will G. High-performance TiO_2 photoanode with an efficient electron transport network for dye-sensitized solar cells. The Journal of Physical Chemistry C. 2009;**113**:16277-16282. DOI: 10.1021/jp9041974

[21] Sheng X, He D, Yang J, Zhu K, Feng X. Oriented assembled TiO_2 hierarchical nanowire arrays with fast electron transport properties. Nano letters. 2014;**14**:1848-1852. DOI: 10.1021/nl4046262

[22] Tetreault N, Gratzel M. Novel nanostructures for next generation dye-sensitized solar cells. Energy and Environmental Science. 2012;**5**:8506-8516. DOI: 10.1039/C2EE03242B

[23] Shankar K, Mor GK, Prakasam HE, Yoriya S, Paulose M, Varghese OK, Grimes CA. Highly-ordered TiO_2 nanotube arrays up to 220 μm in length: Use in water photoelectrolysis and dye-sensitized solar cells. Nanotechnology. 2007;**18**:065707

[24] Sauvage F, Di Fonzo F, Li Bassi A, Casari C. S, Russo V, Divitini G, Ducati C, Bottani CE, Comte P, Graetzel M. Hierarchical TiO_2 photoanode for dye-sensitized solar cells. Nano Letters. 2010;**10**:2562-2567. DOI: 10.1021/nl101198b

[25] Yang Y, Gobeze HB, D'Souza F, Jankowiak R, Li J. Plasmonic enhancement of biosolar cells employing light harvesting complex II incorporated with core–shell metal@ TiO_2 nanoparticles. Advanced Materials Interfaces. 2016;**3**:1600371. DOI: 10.1002/admi.201600371

[26] Wu W-Q, Lei B.-X, Rao H-S, Xu Y-F, Wang Y-F, Su C-Y, Kuang D-B. Hydrothermal fabrication of hierarchically Anatase TiO_2 nanowire arrays on FTO glass for dye-sensitized solar cells. Scientific Reports. 2013;**3**:1352. DOI: 10.1038/srep01352

[27] Yang Y. Integration of photosynthetic pigment-protein complexes in dye sensitized solar cells towards plasmonic-enhanced biophotovoltaics [Thesis]. Proquest: Kansas State University, **2016**

[28] Park NG, van de Lagemaat J, Frank AJ. Comparison of dye-sensitized rutile- and Anatase-based TiO_2 solar cells. The Journal of Physical Chemistry B. 2000;**104**:8989-8994. DOI: 10.1021/jp994365l

[29] Schattauer S, Reinhold B, Albrecht S, Fahrenson C, Schubert M, Janietz S, Neher D. Influence of sintering on the structural and electronic properties of TiO_2 nanoporous

layers prepared via a non–sol–gel approach. Colloid and Polymer Science 2012;**290**:1843-1854. DOI: 10.1007/s00396-012-2708-9

[30] Burke A, Ito S, Snaith H, Bach U, Kwiatkowski J, Grätzel M. The function of a TiO_2 compact layer in dye-sensitized solar cells incorporating "Planar" organic dyes. Nano Letters. 2008;**8**:977-981. DOI: 10.1021/nl071588b

[31] Hore S, Kern R. Implication of device functioning due to back reaction of electrons via the conducting glass substrate in dye sensitized solar cells. Applied Physics Letters. 2005;**87**:263504. DOI: 10.1063/1.2149215

[32] Zheng Y, Klankowski S, Yang Y, Li J. Preparation and characterization of TiO_2 barrier layers for dye-sensitized solar cells. ACS applied materials and interfaces. 2014;**6**:10679-10686. DOI: 10.1021/am502421w

[33] Brennan BJ, Llansola Portoles MJ, Liddell PA, Moore TA, Moore AL, Gust D. Comparison of silatrane, phosphonic acid, and carboxylic acid functional groups for attachment of porphyrin sensitizers to TiO_2 in photoelectrochemical cells. Physical Chemistry Chemical Physics: PCCP. 2013;**15**:16605-16614. DOI: 10.1039/C3CP52156G

[34] Zhang L, Cole JM. Anchoring groups for dye-sensitized solar cells. ACS Applied Materials and Interfaces. 2015;**7**:3427-3455. DOI: 10.1021/am507334m

[35] Mershin A, Matsumoto K, Kaiser L, Yu D, Vaughn M, Nazeeruddin MK, Bruce BD, Graetzel M, Zhang S. Self-assembled photosystem-I biophotovoltaics on nanostructured TiO_2 and ZnO. Scientific Reports. 2012;**2**:234. DOI: 10.1038/srep00234

[36] Gordiichuk PI, Wetzelaer G-J, AH, Rimmerman D, Gruszka A, de Vries JW, Saller M, Gautier DA, Catarci S, Pesce D, Richter S, Blom PWM, Herrmann A. Solid-state biophotovoltaic cells containing photosystem I. Advanced Materials. 2014;**26**:4863-4869. DOI: 10.1002/adma.201401135

[37] Porra RJ, Thompson WA, Kriedemann PE. Determination of accurate extinction coefficients and simultaneous equations for assaying chlorophylls a and b extracted with four different solvents: verification of the concentration of chlorophyll standards by atomic absorption spectroscopy. Biochimica et Biophysica Acta (BBA)—Bioenergetics. 1989;**975**:384-394. DOI: 10.1016/S0005-2728(89)80347-0

[38] Yang Y, Jankowiak R, Lin C, Pawlak K, Reus M, Holzwarth AR, Li J. Effect of the LHCII pigment-protein complex aggregation on photovoltaic properties of sensitized TiO_2 solar cells. Physical Chemistry Chemical Physics. 2014;**16**:20856-20865. DOI: 10.1039/C4CP03112A

[39] Johnson M. P, Goral TK, Duffy CDP, Brain APR, Mullineaux CW, Ruban AV. Photoprotective energy dissipation involves the reorganization of photosystem II light-harvesting complexes in the grana membranes of spinach chloroplasts. Plant Cell. 2011;**23**:1468-1479. DOI: 10.1105/tpc.110.081646

[40] Holzwarth AR, Miloslavina Y, Nilkens M, Jahns P. Identification of two quenching sites active in the regulation of photosynthetic light-harvesting studied by time-resolved fluorescence. Chemical Physics Letters. 2009;**483**:262-267. DOI: 10.1016/j.cplett.2009.10.085

[41] Belgio E, Duffy CDP, Ruban AV. Switching light harvesting complex II into photoprotective state involves the lumen-facing apoprotein loop. Physical Chemistry Chemical Physics: PCCP. 2013;**15**:12253-12261. DOI: 10.1039/c3cp51925b

[42] Ruban AV, Berera R, Ilioaia C, van Stokkum IHM, Kennis JTM, Pascal AA, van Ameron gen H, Robert B, Horton P, van Grondelle R. Identification of a mechanism of photoprotective energy dissipation in higher plants. Nature. 2007;**450**:575-578. DOI: 10.1038/nature06262

[43] Holzwarth AR, Jahns P: NPQ mechanisms in intact organisms as derived from ultrafast fluorescence kinetics studies. In: Govindjee, Sharkey, Thomas D, editors. Advances in Photosynthesis and Respiration, Springer Science, Berlin, Germany; **2014**. pp. 129-156

[44] Miloslavina Y, Wehner A, Lambrev PH, Wientjes E, Reus M, Garab G, Croce R, Holzwarth AR. Far-red fluorescence: A direct spectroscopic marker for LHCII oligomer formation in non-photochemical quenching. FEBS Letters. 2008;**582**:3625-3631. DOI: 10.1016/j.febslet.2008.09.044

[45] Magdaong N, Enriquez M, LaFountain A, Rafka L, Frank H. Effect of protein aggregation on the spectroscopic properties and excited state kinetics of the LHCII pigment-protein complex from green plants. Photosynthesis Research. 2013;**118**:18. DOI: 10.1007/s11120-013-9924-0

[46] Kell A, Feng X, Lin C, Yang Y, Li J Reus M, Holzwarth AR, Jankowiak R. Charge-transfer character of the low-energy Chl a Qy absorption band in aggregated light harvesting complexes II. The Journal of Physical Chemistry B. 2014;**118**:6086-6091. DOI: 10.1021/jp501735p

[47] Gratzel M. Recent advances in sensitized mesoscopic solar cells. Accounts of Chemical Research. 2009;**42**:1788-1798. DOI: 10.1021/ar900141y

[48] Mackowski S. Hybrid nanostructures for efficient light harvesting. Journal of Physics: Condensed Matter. 2010;**22**:193102. DOI: 10.1088/0953-8984/22/19/193102

[49] Carmeli I, Lieberman I, Kraversky L, Fan Z, Govorov AO, Markovich G, Richter S. Broad band enhancement of light absorption in photosystem I by metal nanoparticle antennas. Nano Letters. 2010;**10**:2069-2074. DOI: 10.1021/nl100254j

[50] Czechowski N, Nyga P, Schmidt M, Brotosudarmo TP, Scheer H, Piatkowski D, Mackowski S. Absorption enhancement in peridinin-chlorophyll-protein light-harvesting complexes coupled to semicontinuous silver film. Plasmonics. 2012;**7**:115-121. DOI: 10.1007/s11468-011-9283-7

[51] Mackowski S, Wörmke S, Maier AJ, Brotosudarmo THP, Harutyunyan H, Hartschuh A, Govorov A. O, Scheer H, Bräuchle C. Metal-enhanced fluorescence of chlorophylls in light-harvesting complexes. Nano Letters. 2008;**8**:558-564. DOI: 10.1021/nl072854o

[52] Beyer SR, Ullrich S, Kudera S, Gardiner AT, Cogdell R. J, Köhler J. Hybrid nanostructures for enhanced light-harvesting: Plasmon Induced Increase in Fluorescence from Individual Photosynthetic Pigment-Protein Complexes. Nano Letters. 2011;**11**:4897-4901. DOI: 10.1021/nl202772h

[53] Govorov AO, Carmeli I. Hybrid structures composed of photosynthetic system and metal nanoparticles: Plasmon Enhancement Effect. Nano Letters. 2007;**7**:620-625. DOI: 10.1021/nl062528t

[54] Atwater HA, Polman A. Plasmonics for improved photovoltaic devices. Nature Materials. 2010;**9**:205-213. DOI: 10.1038/nmat2629

[55] Dang X, Qi J, Klug MT, Chen P-Y, Yun DS, Fang NX, Hammond PT, Belcher AM. Tunable localized surface plasmon-enabled broadband light-harvesting enhancement for high-efficiency panchromatic dye-sensitized solar cells. Nano Letters. 2013;**13**:637-642. DOI: 10.1021/nl3043823

[56] Ding B, Lee BJ, Yang MJ, Jung HS, Lee JK. Surface-plasmon assisted energy conversion in dye-sensitized solar cells. Advanced Energy Materials. 2011;**1**:415-421. DOI: 10.1002/aenm.201000080

[57] Zhang W, Saliba M, Stranks SD, Sun Y, Shi X, Wiesner U, Snaith HJ. Enhancement of perovskite-based solar cells employing core–shell metal nanoparticles. Nano Letters. 2013;**13**:4505-4510. DOI: 10.1021/nl4024287

[58] Saliba M, Zhang W, Burlakov VM, Stranks SD, Sun Y, Ball JM, Johnston MB, Goriely A, Wiesner U, Snaith HJ. Plasmonic-induced photon recycling in metal halide perovskite solar cells. Advanced Functional Materials. 2015;**25**:5038-5046. DOI: 10.1002/adfm.201500669

[59] Adhyaksa GWP, Prima EC, Lee DK, Ock I, Yatman S, Yuliarto B, Kang JK. A light harvesting antenna using natural extract graminoids coupled with plasmonic metal nanoparticles for bio-photovoltaic cells. Advanced Energy Materials. 2014;**4**:1400470-1400477. DOI: 10.1002/aenm.201400470

[60] Wang Y, Wen C, Hodgson P, Li Y. Biocompatibility of TiO_2 nanotubes with different topographies. Journal of Biomedical Materials Research Part A. 2014;**102**:743-751. DOI: 10.1002/jbm.a.34738

[61] Du J, Qi J, Wang D, Tang Z. Facile synthesis of Au@TiO_2 core-shell hollow spheres for dye-sensitized solar cells with remarkably improved efficiency. Energy & Environmental Science. 2012;**5**:6914-6918. DOI: 10.1039/C2EE21264A

[62] Liu W-L, Lin F-C, Yang Y-C, Huang C-H, Gwo S, Huang MH, Huang J-S. The influence of shell thickness of Au@TiO_2 core-shell nanoparticles on the plasmonic enhancement effect in dye-sensitized solar cells. Nanoscale. 2013;**5**:7953-7962. DOI: 10.1039/C3NR02800C

[63] Cushing SK, Li J, Bright J, Yost BT, Zheng P, Bristow AD, Wu N. Controlling plasmon-induced resonance energy transfer and hot electron injection processes in metal@TiO_2 core–shell nanoparticles. The Journal of Physical Chemistry C. 2015;**119**:16239-16244. DOI: 10.1021/acs.jpcc.5b03955

[64] Cushing SK, Wu N. Progress and perspectives of plasmon-enhanced solar energy conversion. The Journal of Physical Chemistry Letters. 2016;**7**:666-675. DOI: 10.1021/acs.jpclett.5b02393

[65] Götze JP, Kröner D, Banerjee S, Karasulu B, Thiel W. Carotenoids as a shortcut for chlorophyll soret-to-Q band energy flow. ChemPhysChem. 2014;**15**:3392-3401. DOI: 10.1002/cphc.201402233

[66] Lee YK, Park J, Park JY. The effect of dye molecules and surface plasmons in photon-induced hot electron flows detected on Au/TiO_2 Nanodiodes. Journal of Physical Chemistry C. 2012;**116**:18591-18596. DOI: 10.1021/jp303099w

[67] Jiang Z, Zhu J, Liu D, Wei W, Xie J, Chen M. In situ synthesis of bimetallic Ag/Pt loaded single-crystalline anatase TiO_2 hollow nano-hemispheres and their improved photocatalytic properties. CrystEngComm. 2014;**16**:2384-2394. DOI: 10.1039/C3CE41949E

[68] Jang YH, Jang YJ, Kochuveedu ST, Byun M, Lin Z, Kim DH. Plasmonic dye-sensitized solar cells incorporated with Au-TiO_2 nanostructures with tailored configurations. Nanoscale. 2014;**6**:1823-1832. DOI: 10.1039/C3NR05012B

[69] Clavero C. Plasmon-induced hot-electron generation at nanoparticle/metal-oxide interfaces for photovoltaic and photocatalytic devices. Nature Photonics. 2014;**8**:95-103. DOI: 10.1038/nphoton.2013.238

[70] Knight MW, Sobhani H, Nordlander P, Halas NJ. Photodetection with active optical antennas. Science. 2011;**332**:702-704. DOI: 10.1126/science.1203056

[71] McFarland EW, Tang J. A photovoltaic device structure based on internal electron emission. Nature. 2003;**421**:616-618

[72] Tian Y, Tatsuma T. Mechanisms and applications of plasmon-induced charge separation at TiO_2 films loaded with gold nanoparticles. Journal of the American Chemical Society. 2005;**127**:7632-7637. DOI: 10.1021/ja042192u

[73] Chen ZH, Tang YB, Liu CP, Leung YH, Yuan GD, Chen LM, Wang YQ, Bello I, Zapien JA, Zhang WJ, Lee CS, Lee ST. Vertically aligned ZnO nanorod arrays sentisized with gold nanoparticles for Schottky barrier photovoltaic cells. The Journal of Physical Chemistry C. 2009;**113**:13433-13437. DOI: 10.1021/jp903153w

[74] Son M-S, Im J-E, Wang K-K, Oh S-L, Kim Y-R, Yoo K-H. Surface plasmon enhanced photoconductance and single electron effects in mesoporous titania nanofibers loaded with gold nanoparticles. Applied Physics Letters. 2010;**96**:023115. DOI: 10.1063/1.3291052

[75] Mubeen S, Hernandez-Sosa G, Moses D, Lee J, Moskovits M. Plasmonic photosensitization of a wide band gap semiconductor: Converting plasmons to charge carriers. Nano Letters. 2011;**11**:5548-5552. DOI: 10.1021/nl203457v

[76] Yang Y, Wu J, Li J. Correlation of the plasmon-enhanced photoconductance and photovoltaic properties of core-shell Au@ TiO_2 network. Applied Physics Letters. 2016;**109**:091604

10

Composite Calcium Phosphate/ Titania Scaffolds in Bone Tissue Engineering

Massimiliano Dapporto, Anna Tampieri and Simone Sprio

Abstract

Titanium and its alloys have been extensively used as implantation materials due to their favorable properties such as lower modulus, good tensile strength, excellent biocompatibility, and enhanced corrosion resistance. However, their intrinsic bioinertness generally prevents a direct bond with the bone on the surface especially at an early stage of implantation. In recent years, bioactive scaffolds for bone regeneration are progressively replacing bioinert prostheses in orthopedic, maxillofacial, and neurosurgery fields. Given the need of enhanced mechanical strength, several combinations of bioactive and reinforcing phases have been studied, but still no convincing solutions have been found so far. In this context, titanium oxides are light and high-resistance bioactive materials widely employed in dental and bone application due to their capacity of forming strong bonds with bone tissue via the formation of a tightly bound apatite layer on their surface. The addition of titania particles to hydroxyapatite has attracted considerable attention based on the assumption that resulting materials can enhance osteoblast adhesion and promote cell growth while also providing high strength and fracture toughness in the final composite material, thus being adequate for load-bearing applications.

Keywords: hydroxyapatite, titania, bioactive composites, porous scaffolds, mechanical strength, load bearing

1. Introduction

The regeneration of critical-size bone defects, particularly in load-bearing site, still represents a remarkable challenge in orthopedics. Indeed, these clinical cases require the use of scaffolds with cell-instructive ability and remarkable strength to cope with the early and complex biomechanical stimuli in vivo [1]. As the use of autografts or allografts suffers many restrictions and drawbacks, particularly in the case of large defects, synthetic biomaterials are today considered as elective in this respect; however, there is still a lack of suitable materials associating bioactivity and high strength. Natural bone is a hybrid nanocomposite capable of outstanding mechanical performance and ability to establish an active dialogue with cells. In particular, the bone is composed of an inorganic compound (60%), a nearly amorphous calcium phosphate with the crystal structure of hydroxyapatite (HA), heterogeneously nucleated on an organic component prevalently made of type I collagen. The collagen phase provides the bone with great flexural strength, while the mineral component increases the bone compression strength.

The unique factors that contribute to the toughness of bone are the presence of nano-size apatite crystals and a dense network of collagen fibers.

For that reason, in the last decades, the research on biomaterials and scaffolds able to favor bone tissue regeneration upon implantation, while also mechanically supporting the anatomic site affected by lack of bone, has been increasing. In this respect elective materials are calcium phosphates, due to their high chemical similarity with the bone mineral. However, they suffer of low mechanical strength that makes them not suitable to be implanted in load-bearing site. Therefore, a new approach was focused on the development of ceramic composites associating high bioactivity and strength.

The present chapter will provide an overview to illustrate novel potential approaches to develop reinforced bioactive scaffolds to assist the regeneration of load-bearing bony sites, considering that serious drawbacks can arise in case of mechanical mismatching at the bone/biomaterial interface. In particular, the chapter will highlight the use of titanium dioxide, which is a well-established biomaterial for bone applications, as a promising nanomaterial with the ability to reinforce calcium phosphate matrixes.

2. The need of mechanically competent bioactive implants for bone regeneration

In most of load-bearing applications, the main target is the achievement of high mechanical strength. However, this approach can limit the success of the implant when it comes to obtain substantial bone regeneration. As a matter of fact, to date bioinert metallic prostheses are implanted upon occurrence of bone impairments or fractures. These devices are for sure mechanically competent in restoring the bone shape and eventually the biomechanical function of joints in relatively short timing [2]. However, their well-known great mechanical performance may be also detrimental, particularly in the long term. In fact, current implants utilized in orthopedic and maxillofacial surgeries suffer various clinical drawbacks, such as implant loosening, wear, and limited compatibility with the bone in permanent metal implants [3]. In this condition, the excessive stiffness exhibited by metallic implants, generally much greater if compared with the elastic behavior of the bone, results in improper prosthesis-to-bone load transfer and stress shielding that can impair the stability of the implant and its long-term performance (**Figure 1**) [4].

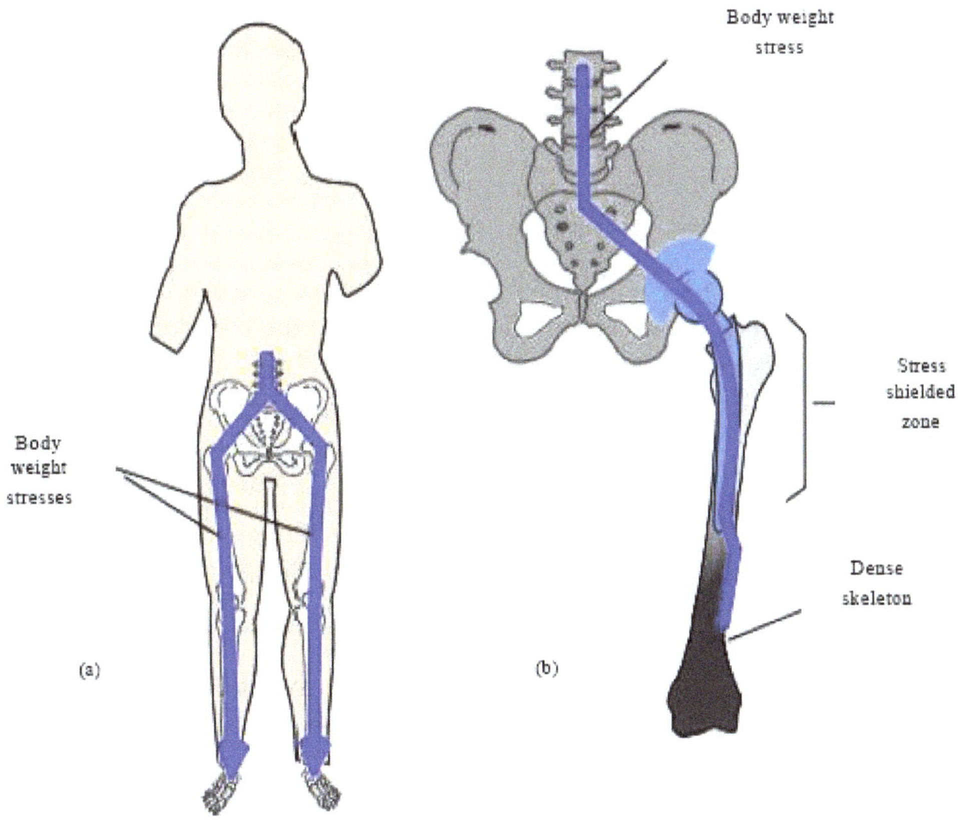

Figure 1. Simple scheme of stress shielding [5].

Briefly, as a simple mechanical rule, considering every composite system composed of two materials where one component is stiffer, the stiffer component will sustain the greater part of the load. In the normal healthy skeleton, the stresses flow symmetrically from there downward through both hip joints, thighbones, knee joints, lower leg bones, and feet onto the floor (**Figure 1a**).

In case of total hip joint replacement, the shaft component generally takes over the majority of the stresses; in this case, the body weight primarily flows down from the joint center and then through the shaft of the device. As a consequence, the upper part of the thighbone is unloaded, thus resulting in weaker areas more susceptible to fracture. Moreover, the skeleton around the tip of the femoral component is overloaded, resulting in a thicker and stronger part. The shaft component of a total hip device is much stiffer than the skeleton and will take the greater part of the body weight load. Consequently, the shaft component is overloaded, whereas the skeleton around the shaft is unloaded (**Figure 1b**) [5].

Unfortunately, the thickening of the skeleton is, in most cases, painful. The patients with cementless shafts of total hip devices often claim about the pain in the thigh, especially during the first years after the surgery [6].

In turn, this can provoke localized osteoporosis and bone resorption, loosening, and detachment of the prosthetic device [7], thus impacting on the course of patient rehabilitation and on the need of repeated revision/correction surgery.

Commonly observed complications after prosthesis removal are infections, impaired wound healing, secondary fractures, tissue and nerve damage, and postoperative bleeding. There is some evidence indicating that the postoperative complication rate depends on the specific localization of the implanted material [8].

Indeed, the above reported drawbacks mostly occur as the used bone/implant systems are often integrated only at the surface [1].

In this respect, bone implants should exhibit substantial cell-instructive ability in order to trigger and sustain the cascade of cell-based phenomena at the basis of new bone formation and organization [9]. Key phenomena in this respect are protein adhesion including the formation of bonds between cell surface receptors (integrins) and the protein functional groups (ligands) (**Figure 2**) [10]. Then, cytoskeletal reorganization with progressive cell spreading on the substrate can take place. Upon implantation in vivo, there are several factors affecting how the proteins will adhere to the material, for example, surface chemistry, surface energy/tension/wettability, roughness, crystallinity, surface charge, and mechanical properties. After this first-stage extensive implant, colonization should take place, driven by a diffuse porosity enabling cell penetration and new bone formation into the inner part of the implant. In this respect, the implant nanotopography can influence the attachment and function of bone cells by modulating key signaling effects essential for their survival [11].

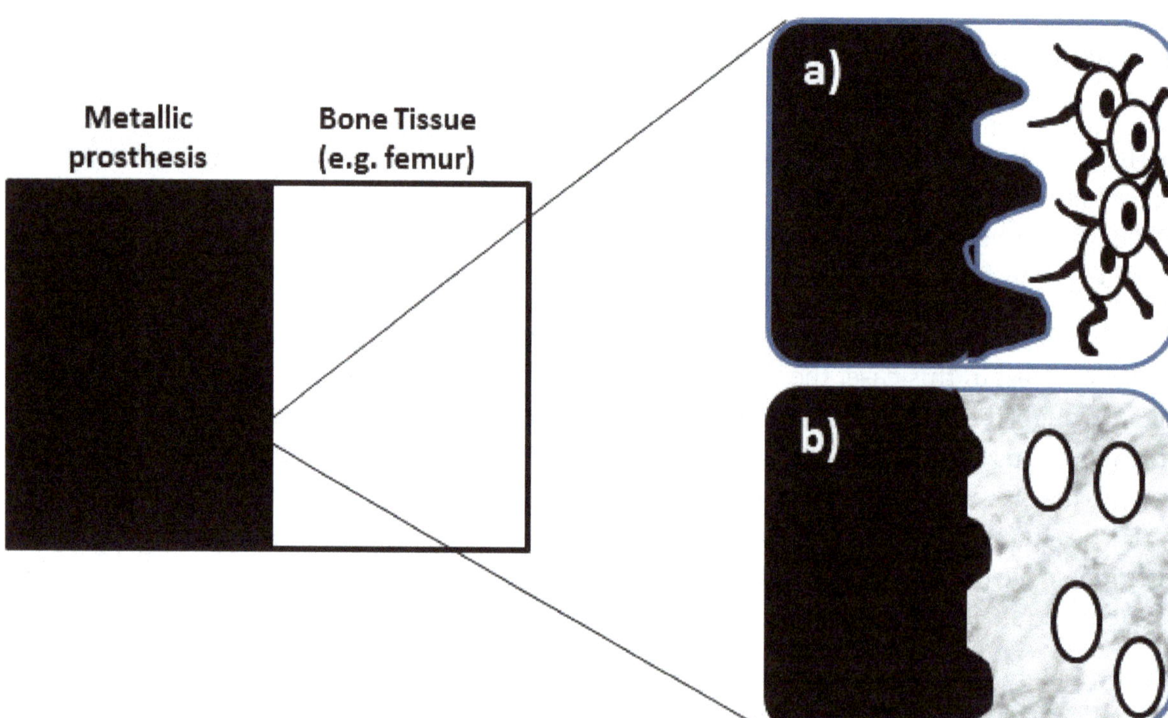

Figure 2. Graphical overview of the effect of surface microstructure on the interaction between a metallic prosthesis and bone tissue: (a) osseointegration, the surface features are able to induce bone formation, leading to long-lasting interaction without side effects; (b) short-loss implant, the surface is now detrimental for bone tissue regeneration, leading to dead bone cells.

Nowadays, it is widely accepted that substantial mimesis of the physicochemical, morphological, and mechanical features of the bone are crucial requisites for regenerative bone scaffolds, particularly in case of repair of long and load-bearing bone segments [12–14]. Indeed, such features can properly drive physiological processes of bone regeneration, to obtain the full recovery of the diseased tissue with all its function. In this respect, recent progresses in materials science research developed a variety of bioactive scaffolds for the healing and repair of damaged or missing bone parts, which are progressively replacing bioinert implants in an increasing number of applications in orthopedic, maxillofacial, and neurosurgery fields [9].

Biomaterials based on hydroxyapatite (HA, $Ca_{10}(PO_4)_6(OH)_2$) or β-tricalcium phosphate (TCP: β-$Ca_3(PO_4)_2$) have attracted considerable interest for orthopedic and dental applications, thanks to their noticeable chemical resemblance to the mineral component of the bone which provides intrinsic biocompatibility and osteointegration ability [15, 16].

Particularly, tricalcium phosphate has been used in clinics to repair bone defects for many years [17, 18]. As well, a wide range of bioactive materials has been investigated so far, in alternative to calcium phosphates, including bioglasses and apatite-wollastonite glass-ceramics [19–21]. The main attractive feature of such bioceramics is their ability to form a direct bond with the host bone resulting in a strong interface compared to bioinert or biotolerant materials that form a fibrous interface [22]. For biomedical applications the incorporation of biomimetic foreign ions in the HA structure (CO_3^{2-}, Mg^{2+}, SiO_4^{4-}) is needed to increase its functionality in terms of stimulation of the natural bone regeneration processes [23].

However, applications of these materials for long bone replacement are hindered by their insufficient strength and toughness [24].

Also, some calcium phosphates can suffer a relatively high dissolution rate in simulated body fluid that affects their long-term stability [25]. In this context, a great deal of research effort has been devoted so far to develop methods of processing hydroxyapatite with good mechanical properties and high resistance to corrosion [26]. As a general rule, ceramic oxides or metallic dispersions have been introduced as reinforcing agents [27, 28]. In respect to the use of reinforcing ceramics, several attempts have been performed by the addition of aluminum or zirconium oxide to calcium phosphate matrices [29, 30]. The main problems arising when developing such materials mainly concern phase decomposition as a consequence of the chemical interactions between HA and the reinforcing phases at high temperatures. In fact, ceramic materials have to be subjected to sintering process for physical consolidation; in the case of ceramic composites, the phenomena of grain coalescence induced by the thermal treatment can thus coexist with solid-state reactions between the ceramic components which often gives rise to formation of undesired phases and phase decomposition. In this context, HA largely decomposes into tricalcium phosphate, and although in many cases the presence of zirconia improves the mechanical resistance of the final composite, secondary phases depressed the bioactivity and bioresorbability of the scaffolds. In various cases, the formation of secondary phases also resulted into volume modifications in the ceramic body, thus possibly inducing microcracks in the final scaffold after the sintering treatments [31, 32].

Among the most interesting ceramics for composite scaffolds, bioactive calcium silicates were also explored as biomaterials for hard tissue repair and replacement since the early 1970s, when Hench and coworkers invented Bioglass®, a silico-phosphate-based glass with composition close to that of bone mineral [20]. However, due to their nature, bioactive glasses were not indicated for scaffold reinforcement; however, the presence of silicon in bone scaffolds has always been addressed as promoter of new bone formation in vivo, due to its ability to be a center for nucleation of apatite phase in physiological environment [33]. On this basis, calcium silicate phases such as dicalcium and tricalcium silicate, as well as wollastonite or pseudo-wollastonite ($CaSiO_3$), were widely investigated as scaffolds or cements [34, 35]. In particular, the development of composites made of HA reinforced with dicalcium silicate was investigated [36], on the basis of its high flexure strength (≈200 MPa) and reduced elastic modulus (≈40 GPa) [37], compared with HA, thus resulting as promising compositions for bone scaffolding. As a main drawback, calcium silicates exist in a variety of polymorphs stable in different conditions of temperature [38], thus making difficult to obtain pure phases and avoid decomposition detrimental for the mechanical properties.

A different system that recently attracted the interest of scientists is given by titanium (Ti) and its alloys, particularly titanium dioxide (TiO_2), which have been already validated and extensively used as implantation materials due to their favorable properties such as lower modulus, good tensile strength, excellent biocompatibility, and enhanced corrosion resistance [39].

3. Strengths of titanium dioxide in bone tissue engineering

Titanium oxide has been extensively investigated as a biomaterial due to its excellent biocompatibility and superior corrosion/erosion resistance, as well as high stability [40]. The addition of titania particulates to HA has attracted considerable attention based on the assumption that titania is capable of enhancing osteoblast adhesion and inducing cell growth [41].

Titanium oxides (anatase and rutile) are light and high-resistance bioactive materials widely employed in dental and bone application due to their capacity of forming strong bonds with bone tissue via the formation of a tightly bound apatite layer on their surface [33, 42–44].

In particular, rutile is a very interesting biomaterial for developing bioactive ceramic composites with improved mechanical performances [45].

In this respect, in spite of the numerous studies and applications of HA/TiO_2 composites as bioactive coatings for load-bearing titanium prostheses [46–49], only a few studies were reported so far about the development of bulk TiO_2-containing composites addressed to the development of bone scaffolds [50, 51].

The use of spark plasma sintering and hot pressing to obtain TCP/TiO_2 composites from hydroxyapatite and titania nanopowders has been previously reported [51, 52], whereas Nath et al. used metallic titanium by traditional sintering at different maximum temperatures [50]. In these works, an accurate physicochemical description of the phenomena occurring after thermal treatment of HA/TiO_2 mixtures pointed out the capability of TiO_2 to favor the decomposition of HA, with subsequent formation of β-TCP and $CaTiO_3$, according to the following reaction:

$$Ca_{10}(PO_4)_6(OH)_2 + TiO_2 \rightarrow 3\,Ca_3(PO_4)_2 + CaTiO_3 + H_2O\uparrow \quad (1)$$

In particular, good cell adhesion and proliferation in contact with bulk TCP/TiO_2 composites were reported [50]. This finding was also confirmed by Hu et al. [47] and Sato et al. [48], which reported good cell behavior in contact with coatings of similar compositions.

4. A focus on calcium phosphate/titania bulk composite materials

Most studies have been devoted so far to the investigation of calcium phosphate sintering and the mechanical properties of pure TCP or the pure TiO_2. However, a little work has been reported on the performances of TCP-TiO_2 composites [28, 52, 53]. These papers focused on the synthesis of TCP-TiO_2 composites where titania nanoparticles could enhance the mechanical properties of calcium phosphate matrices, without penalizing biocompatibility.

In particular, Sprio et al. proposed a pressureless air sintering of mixed hydroxyapatite and titania (TiO_2) powders [28]; the sintering process was optimized to achieve dense ceramic bodies consisting in a bioactive/bioresorbable β-TCP matrix reinforced with defined amounts of submicron-sized titania particles.

A crucial step in the development of ceramic composites is the control of particle size and the driving energy for thermal consolidation processes [54]. Indeed, homogeneous ceramic composites come from adequately prepared powder mixtures, possibly preventing particle agglomeration. In this respect, HA powder was calcined at 900°C to increase the particle size whereas reducing surface activity possibly promoting the formation of particle clusters and to promote the achievement of composites with homogeneous microstructure [55]. On the other hand, an excessive increase of the HA particle size can reduce the driving energy for further HA grain growth during sintering, thus resulting in limited consolidation [54]. This comes very relevant when designing materials for load-bearing applications which need improved mechanical properties.

Therefore, a detailed study of the phase composition of HA/TiO_2 mixtures with temperature was mandatory; several mixtures were prepared (HA/TiO_2 = 90:10, 80:20, 70:30 vol%) and treated at different temperatures. As a general rule, the starting phase composition remained unchanged upon firing at temperatures up to 700°C, where the transformation of anatase into rutile started to take place; at higher temperatures, anatase underwent progressively increasing transformation in rutile and completely disappeared at 850°C. Therefore, even though titanium dioxide is present in different polymorphs, this does not result as a drawback, as above certain temperatures, of interest for ceramic sintering; the thermodynamically stable phase is always rutile.

As induced by the presence of rutile, the decomposition of HA phase into β-TCP occurred at relatively low temperatures (950°C); in the same temperature range, the formation of perovskite ($CaTiO_3$) was also detected. At higher temperatures, the phase composition of the mixture resulted unchanged up to 1250°C, when part of β-TCP was converted into the high-temperature polymorph α-TCP [56]; the raising of the firing temperature up to 1300°C promoted a further increase of the α-TCP content. Therefore, despite the highest volume shrinkage was detected at 1300°C by dilatometric analysis, the final sintering temperature was limited to 1250°C. Indeed, as α-TCP is characterized by very high solubility in physiological environment [56], its presence may result in excessively fast resorption in vivo, hindering an adequate bone regeneration process. Moreover, the transformation of β-TCP in α-TCP is associated with an average 10% volume increase, which potentially penalizes the mechanical performances by micro-damages. A full consolidation of the composites was obtained by applying a dwell time of 1 h.

On this basis, the reinforcing mechanisms of titania particles embedded in the sintered composites were investigated by scanning electron microscopy, thus revealing that the presence of different amounts of titania does not strongly influence grain growth. Moreover, the spatial distribution of the submicron grains of titania (the brighter areas) could be still recognized also showing that, in high concentrations, they tended to coalesce in an interconnected framework (**Figure 3**).

The increase of mechanical properties was shown to depend strongly on the amount of titania particles introduced in the calcium phosphate matrix (**Figure 4**) [28].

Due to the different mechanical and thermal properties of the constituent phases, possible toughening mechanisms operating in these composites are crack deflection [57], crack bowing [58], residual stress [59], and microcracking toughening [60].

The Knoop hardness increased almost linearly with the content of TiO_2, as this phase is much harder than β-TCP. In literature, hardness is reported to be 10 GPa for TiO_2 [61] and 3.43 GPa for β-TCP [62].

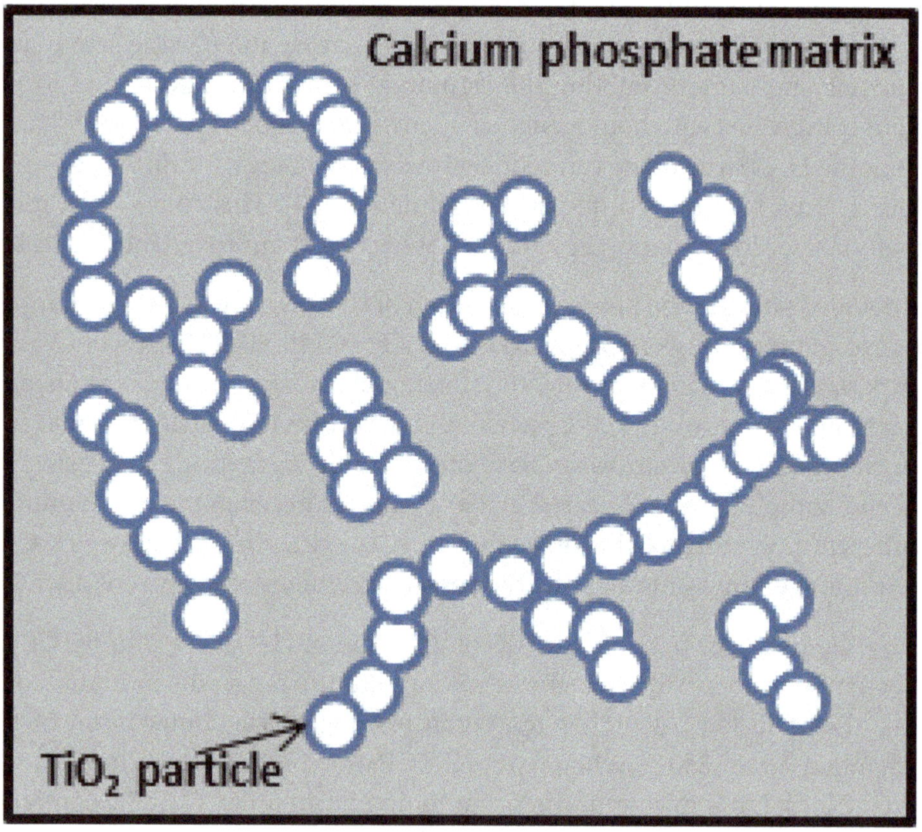

Figure 3. Schematic representation of the TCP/TiO_2 composite microstructure, evidencing the interconnection of TiO_2 grains.

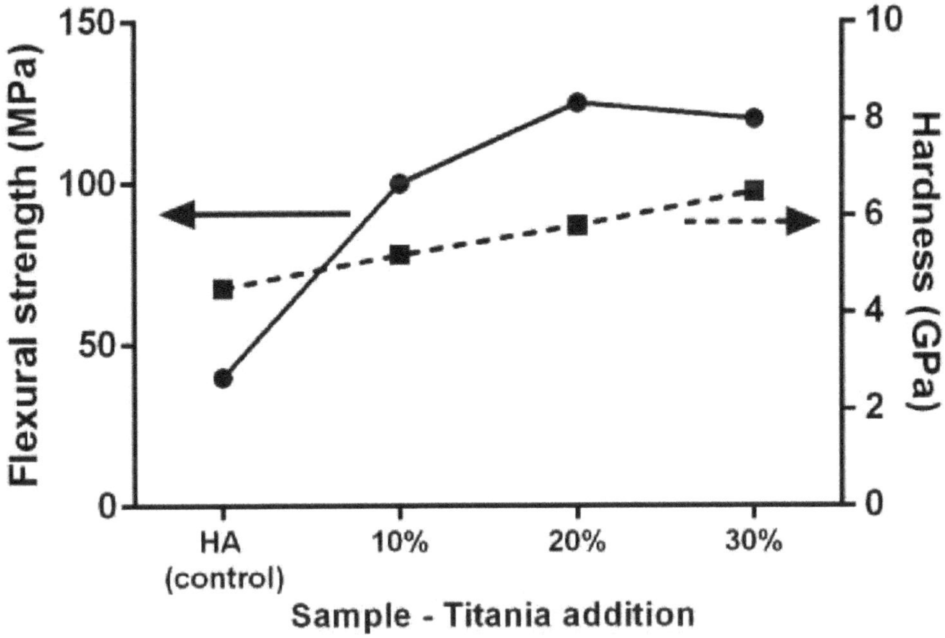

Figure 4. Plot of the flexural strength and Knoop hardness as a function of initial TiO_2 content.

TiO_2-based composites exhibited mechanical properties compliant with those of human cortical bone [33]. In this respect, the sample containing about 20 vol% of TiO_2 was of particular interest, as it represented the maximum level for successful strengthening of the final composite, at least in the range 0–30 vol% (**Figure 4**). This was attributed to a reduced number of microstructural defects, unavoidably generated by adding excessive amounts of titania. With the aim to develop porous bioactive scaffolds, the achievement of good mechanical properties by introducing limited amounts of bioactive, but nonresorbable, reinforcing components is a relevant point that place TCP/TiO_2 composites as very promising materials for the regeneration of load-bearing bone segments.

5. Designing porous calcium phosphate/titania scaffolds exploiting direct foaming method

Regeneration of load-bearing bone segments is still an open challenge due to the lack of biomaterials mimicking natural bone with suitable physicochemical and mechanical performance. Additionally, bone scaffolds should exhibit wide open and interconnected porosity, which however could strongly penalize the mechanical strength. Therefore, the research on adequate methods for porous ceramic development is today a hot topic in materials science [9].

Among the several processes proposed in literature to produce porous ceramics [63, 64], template-free foaming techniques are particularly promising, especially due to the absence of large amounts of organic phases to be eliminated during thermal consolidation. Indeed, forming techniques, making use of sacrificial templates, require long and slow thermal treatments to eliminate the organic component, possibly yielding structural damage and penalization of the final mechanical properties.

In particular, the direct foaming method was stated to be a low-cost and easy process that can provide pore volumes in the range 40–97% by incorporating air into a ceramic suspension that is subsequently dried and sintered. It was also reported that cellular structures prepared by direct foaming usually exhibit considerably higher mechanical strength than those obtained by other template-based techniques, mainly due to the strongly reduced occurrence of flaws in the cell struts [64, 65]. The decisive step in direct foaming methods is related to the development of ceramic slurries with optimal rheological properties, so that they can be dried and physically stabilized upon pouring into preshaped containers, while maintaining the shape, size, and distribution of the air bubbles.

This method was successfully applied to the synthesis of ceramic bone scaffolds made of β-TCP and TiO_2, developed from hydroxyapatite (HA) and TiO_2 powders, on the basis of the approach carried out by [23] exhibiting high and interconnected macroporosity (>70 vol%).

As the foaming process is based on a concentrated ceramic suspension, rheological properties are a critical issue. Indeed, when applied on a simple mixture of HA and TiO_2 powders (see previous paragraph), phase separation occurred in the green ceramic body, thus resulting in scaffolds characterized by reduced structural homogeneity. Therefore, an alternative approach was needed to obtain homogeneous blends, for which the mixture was subjected to a thermal treatment at 1000°C with a dwell time of 1 h before applying the foaming process, to obtain a powder with homogeneous composition.

With this, the application of direct foaming process was feasible and successful in obtaining ceramic bone scaffolds exhibiting high and interconnected macroporosity (>70 vol%).

Direct foaming process resulted very interesting to generate highly porous ceramics, as described in Ref. [66]. As observed by scanning electron microscopy, the microstructure of foamed scaffolds was characterized by large pores in the range 700–900 μm, in turn containing smaller pores, which provide interconnection throughout the whole scaffold (**Figure 5**).

Such microstructure is ideal when it comes to enable extensive penetration of new bone and expressing, at the same time, remarkable strength. In fact the spheroidal pore morphology provides enhanced resistance against fracturing, whereas smaller pores can aid to develop an effective vascular network. In this respect, the lack of vascularization in critical-size bone defects was reported as among the most critical issues limiting the extent of bone regeneration [1].

Besides morphology, compositional aspects can play a relevant role in determining the mechanical properties. Indeed the compressive strength of the TCP/TiO_2 composite scaffolds resulted about 8 MPa, with 75% porosity, i.e., thrice than the reference HA scaffolds. Together with Young's modulus, these composites thus exhibited mechanical properties in the range of cancellous bones (i.e., compression strength, 2–12 MPa; Young's modulus, 0.05–0.5 GPa). The enhanced mechanical competence was also associated to superior biological performance in vitro. Osteoblast-like cells (MG63) cultivated on the scaffold surface for 7 days covered almost completely the external surfaces of the scaffolds, and most of the macropores were completely infiltrated by cells, demonstrating high biocompatibility and osteointegrative potential as well [33].

Furthermore, an increase in cell proliferation was detected during 2 weeks of analysis, whereas the analysis of alkaline phosphatase (ALP) activity revealed a higher osteogenic activity for β-TCP/TiO_2 scaffolds (**Figure 6**). This improvement could be related to both the higher solubility of TCP that yielded enhanced release of calcium ions to cells and also the presence of TiO_2 that in physiological environment can be promptly covered by a layer of HA.

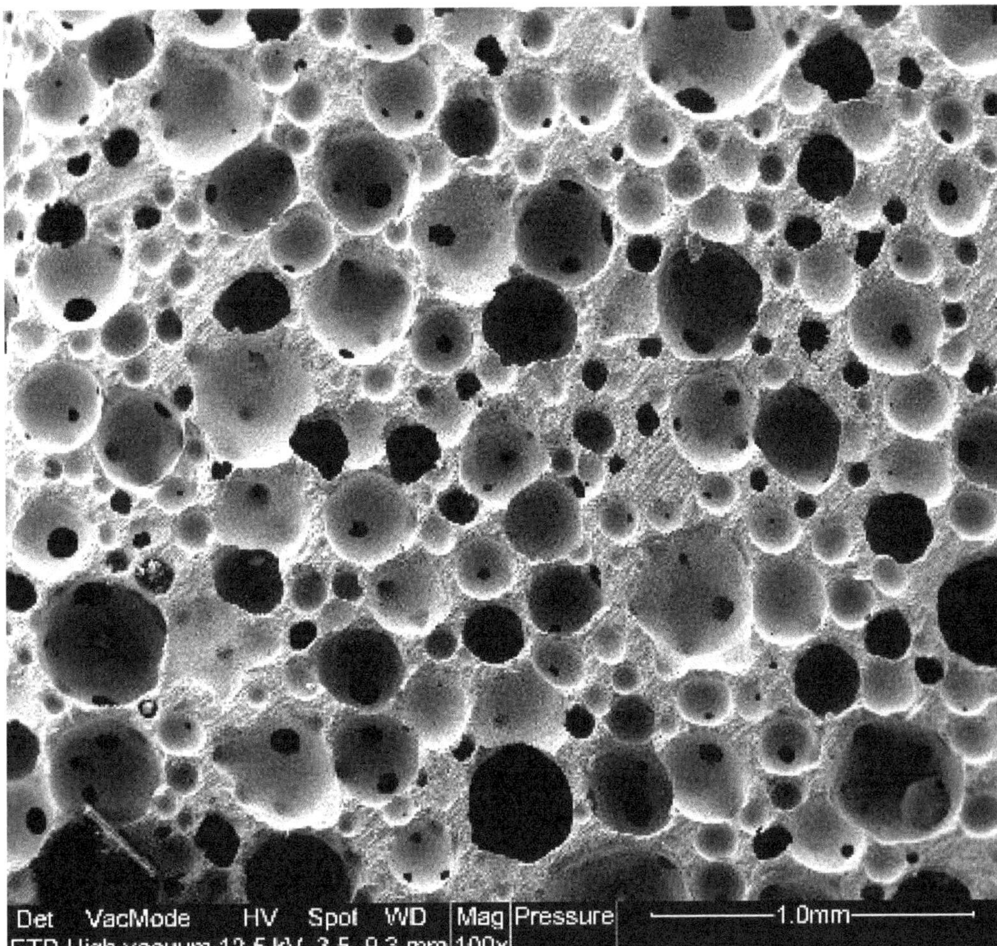

Figure 5. Porous microstructure of sintered porous TCP/TiO_2.

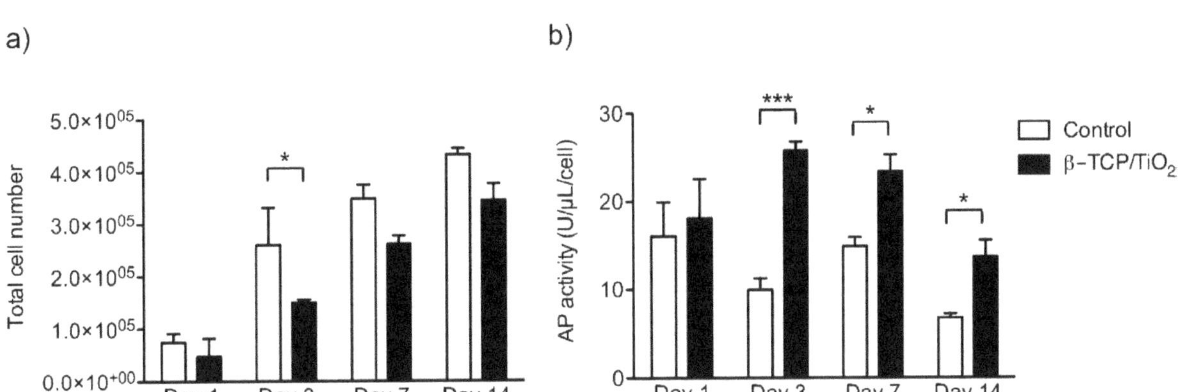

Figure 6. Proliferation of MG-63 osteoblast-like cells (a) and alkaline phosphatase (ALP) activity (b) when seeded on β-TCP/TiO_2 scaffolds and HA control [66].

6. Conclusions and future perspectives

The presented results show that porous scaffolds with bone-like composition and strength can be developed, by following approaches based on ceramic composite development. In particular, TiO_2 is a promising material as bioactive reinforcing phase for calcium phosphate matrices, giving its high biocompatibility. In respect to the design and development of adequate ceramic compositions, the challenge is still open as the settling of ceramic systems requires optimization of a variety of parameters related to initial composition, preliminary powder processing, forming methods, and sintering, all of which are crucial for the final biologic and mechanical properties. In this respect, direct foaming is a very promising method for porous scaffold development that can be flexibly applied to a variety of compositions. Therefore, the application of such process can be decisive for the development of reinforced scaffolds; in this respect, TiO_2-based scaffolds were still investigated in a limited extent, in spite of their potential to chemically and mechanically assist the regeneration of bone tissue defects, particularly load-bearing bone segments, as no regenerative solutions still exist in this field.

Author details

Massimiliano Dapporto, Anna Tampieri and Simone Sprio*

*Address all correspondence to: simone.sprio@istec.cnr.it

Laboratory of Bioceramics and Bio-hybrid Composites, Institute of Science and Technology for Ceramics, National Research Council of Italy, Faenza, Italy

References

[1] Sprio S, Sandri M, Iafisco M, Panseri S, Filardo G, Kon E, et al. Composite biomedical foams for engineering bone tissue A2. In: Netti PA, editor. Biomedical Foams for Tissue Engineering Applications: Woodhead Publishing; Cambridge, UK, 2014. pp. 249-280

[2] Shan L, Shan B, Graham D, Saxena A. Total hip replacement: A systematic review and meta-analysis on mid-term quality of life. Osteoarthritis and Cartilage. 2014;**22**:389-406. DOI: 10.1016/j.joca.2013.12.006

[3] Zilberman M. Active Implants and Scaffolds for Tissue Regeneration. Berlin Heidelberg: Springer; 2011

[4] Ridzwan M, Shuib S, Hassan A, Shokri A, Ibrahim MM. Problem of stress shielding and improvement to the hip implant designs: A review. Journal of Medical Science. 2007;**7**:460-467

[5] Surin HBV. Stress Shielding Effect of the Shaft Component. 2005. http://www.bananarepublican.info/Stress_shielding.htm

[6] Forster-Horvath C, Egloff C, Valderrabano V, Nowakowski AM. The painful primary hip replacement—Review of the literature. Swiss Medical Weekly. 2014;**144**:w13974. DOI: 10.4414/smw.2014.13974

[7] Dai K. Rational utilization of the stress shielding effect of implants. In: Poitout DG, editor. Biomechanics and Biomaterials in Orthopedics. London: Springer London; 2004. pp. 208-215

[8] Reith G, Schmitz-Greven V, Hensel KO, Schneider MM, Tinschmann T, Bouillon B, et al. Metal implant removal: benefits and drawbacks—A patient survey. BMC Surgery. 2015;**15**:96. DOI: 10.1186/s12893-015-0081-6

[9] Sprio S, Sandri M, Iafisco M, Panseri S, Cunha C, Ruffini A, et al. Biomimetic materials in regenerative medicine A2. In: Ruys AJ. Biomimetic Biomaterials: Woodhead Publishing; Cambridge, UK, 2013. pp. 3-45

[10] Schwartz Z, Kieswetter K, Dean DD, Boyan BD. Underlying mechanisms at the bone-surface interface during regeneration. Journal of Periodontal Research. 1997;**32**:166-171. DOI: 10.1111/j.1600-0765.1997.tb01399.x

[11] Su Y, Komasa S, Li P, Nishizaki M, Chen L, Terada C, et al. Synergistic effect of nanotopography and bioactive ions on peri-implant bone response. International Journal of Nanomedicine. 2017;**12**:925-934. DOI: 10.2147/ijn.s126248

[12] Ingber DE. Tensegrity-based mechanosensing from macro to micro. Progress in Biophysics and Molecular Biology. 2008;**97**:163-179. DOI: 10.1016/j.pbiomolbio.2008.02.005

[13] Sikavitsas VI, Temenoff JS, Mikos AG. Biomaterials and bone mechanotransduction. Biomaterials. 2001;**22**:2581-2593

[14] Poitout DG. Biomechanics and Biomaterials in Orthopedics. London: Springer; 2016

[15] Bouslama N, Ben Ayed F, Bouaziz J. Sintering and mechanical properties of tricalcium phosphate–fluorapatite composites. Ceramics International. 2009;**35**:1909-1917. DOI: http://dx.doi.org/10.1016/j.ceramint.2008.10.030

[16] Destainville A, Champion E, Bernache-Assollant D, Laborde E. Synthesis, characterization and thermal behavior of apatitic tricalcium phosphate. Materials Chemistry and Physics. 2003;**80**:269-277. DOI: http://dx.doi.org/10.1016/S0254-0584(02)00466-2

[17] Hench LL, Wilson J. An Introduction to Bioceramics: World Scientific; Publishing, London, UK, 1993

[18] Jensen SS, Broggini N, Hjørting-Hansen E, Schenk R, Buser D. Bone healing and graft resorption of autograft, anorganic bovine bone and β-tricalcium phosphate. A histologic and histomorphometric study in the mandibles of minipigs. Clinical Oral Implants Research. 2006;**17**:237-243. DOI: 10.1111/j.1600-0501.2005.01257.x

[19] Chen QZ, Thompson ID, Boccaccini AR. 45S5 Bioglass®-derived glass–ceramic scaffolds for bone tissue engineering. Biomaterials. 2006;**27**:2414-2425. DOI: http://dx.doi.org/10.1016/j.biomaterials.2005.11.025

[20] Hench LL, Splinter RJ, Allen WC, Greenlee TK. Bonding mechanisms at the interface of ceramic prosthetic materials. Journal of Biomedical Materials Research. 1971;**5**:117-141. DOI: 10.1002/jbm.820050611

[21] Kokubo T, Kim H-M, Kawashita M. Novel bioactive materials with different mechanical properties. Biomaterials. 2003;**24**:2161-2175. DOI: http://dx.doi.org/10.1016/S0142-9612(03)00044-9

[22] Daculsi G, LeGeros R. Biphasic calcium phosphate (BCP) bioceramics: Chemical, physical, and biological properties. Encyclopedia of Biomaterials and Biomedical Engineering, Second Edition (Online Version): CRC Press; Boca Raton, USA, 2008. pp. 359-366

[23] LeGeros RZ. Calcium phosphates in oral biology and medicine. Monographs in Oral Science. 1991;**15**:1-201

[24] Un S, Durucan C. Preparation of hydroxyapatite-titania hybrid coatings on titanium alloy. Journal of Biomedical Materials Research Part B, Applied Biomaterials. 2009;**90**:574-583. DOI: 10.1002/jbm.b.31319

[25] Wang J, Chao Y, Wan Q, Yan K, Meng Y. Fluoridated hydroxyapatite/titanium dioxide nanocomposite coating fabricated by a modified electrochemical deposition. Journal of Materials Science: Materials in Medicine. 2009;**20**:1047-1055. DOI: 10.1007/s10856-008-3673-1

[26] Silva CC, Graça MPF, Valente MA, Sombra ASB. AC and DC conductivity analysis of hydroxyapatite and titanium calcium phosphate formed by dry ball milling. Journal of Non-Crystalline Solids. 2006;**352**:1490-1494. DOI: http://dx.doi.org/10.1016/j.jnoncrysol.2006.01.028

[27] Georgiou G, Knowles JC. Glass reinforced hydroxyapatite for hard tissue surgery—Part 1: mechanical properties. Biomaterials. 2001;**22**:2811-2815. DOI: http://dx.doi.org/10.1016/S0142-9612(01)00025-4

[28] Sprio S, Guicciardi S, Dapporto M, Melandri C, Tampieri A. Synthesis and mechanical behavior of beta-tricalcium phosphate/titania composites addressed to regeneration of long bone segments. Journal of the Mechanical Behavior of Biomedical Materials. 2013;**17**:1-10. DOI: 10.1016/j.jmbbm.2012.07.013

[29] Khalil KA, Kim SW, Kim HY. Consolidation and mechanical properties of nanostructured hydroxyapatite–(ZrO_2 + 3 mol% Y_2O_3) bioceramics by high-frequency induction heat sintering. Materials Science and Engineering: A. 2007;**456**:368-372. DOI: http://dx.doi.org/10.1016/j.msea.2006.12.005

[30] Yun-Mo S, Young-Keun S, Jae-Jun R. Preparation of hydroxyapatite/zirconia bioceramic nanocomposites for orthopaedic and dental prosthesis applications. Nanotechnology. 2007;**18**:065602

[31] Sakka S, Bouaziz J, Ben AF. Sintering and mechanical properties of the alumina–tricalcium phosphate–titania composites. Materials Science and Engineering: C. 2014;**40**:92-101. DOI: http://dx.doi.org/10.1016/j.msec.2014.03.036

[32] Sallemi I, Bouaziz J, Ben AF. The effect of adding magnesium oxide on the mechanical properties of the tricalcium phosphate-zirconia composites. Materials Chemistry and Physics. 2015;**151**:50-59. DOI: http://dx.doi.org/10.1016/j.matchemphys.2014.11.027

[33] Hench LL. Bioceramics. Journal of the American Ceramic Society. 1998;**81**:1705-1728. DOI: 10.1111/j.1151-2916.1998.tb02540.x

[34] Hughes E, Yanni T, Jamshidi P, Grover LM. Inorganic cements for biomedical application: calcium phosphate, calcium sulphate and calcium silicate. Advances in Applied Ceramics. 2015;**114**:65-76. DOI: 10.1179/1743676114Y.0000000219

[35] Wu C, Chang J. A review of bioactive silicate ceramics. Biomedical Materials. 2013;**8**:032001. DOI: 10.1088/1748-6041/8/3/032001

[36] Sprio S, Tampieri A, Celotti G, Landi E. Development of hydroxyapatite/calcium silicate composites addressed to the design of load-bearing bone scaffolds. Journal of the Mechanical Behavior of Biomedical Materials. 2009;**2**:147-155. DOI: 10.1016/j.jmbbm.2008.05.006

[37] Gou Z, Chang J. Synthesis and in vitro bioactivity of dicalcium silicate powders. Journal of the European Ceramic Society. 2004;**24**:93-99. DOI: http://dx.doi.org/10.1016/S0955-2219(03)00320-0

[38] Chan CJ, Kriven WM, Young JF. Physical stabilization of the $\beta \rightarrow \gamma$ transformation in dicalcium silicate. Journal of the American Ceramic Society. 1992;**75**:1621-1627. DOI: 10.1111/j.1151-2916.1992.tb04234.x

[39] Jung S, Kim JH. Sintering characteristics of TiO2 nanoparticles by microwave processing. Korean Journal of Chemical Engineering. 2010;**27**:645-650. DOI: 10.1007/s11814-010-0057-2

[40] He G, Hu J, Wei SC, Li JH, Liang XH, Luo E. Surface modification of titanium by nano-TiO_2/HA bioceramic coating. Applied Surface Science. 2008;**255**:442-445. DOI: http://dx.doi.org/10.1016/j.apsusc.2008.06.088

[41] Li H, Khor KA, Cheang P. Impact formation and microstructure characterization of thermal sprayed hydroxyapatite/titania composite coatings. Biomaterials. 2003;**24**:949-957. DOI: http://dx.doi.org/10.1016/S0142-9612(02)00431-3

[42] Ducheyne P, Qiu Q. Bioactive ceramics: The effect of surface reactivity on bone formation and bone cell function. Biomaterials. 1999;**20**:2287-2303

[43] Li P, Ohtsuki C, Kokubo T, Nakanishi K, Soga N, de Groot K. The role of hydrated silica, titania, and alumina in inducing apatite on implants. Journal of Biomedical Materials Research. 1994;**28**:7-15. DOI: 10.1002/jbm.820280103

[44] Gerhardt LC, Jell GM, Boccaccini AR. Titanium dioxide (TiO(2)) nanoparticles filled poly(D,L lactid acid) (PDLLA) matrix composites for bone tissue engineering. Journal of Materials Science Materials in Medicine. 2007;**18**:1287-1298. DOI: 10.1007/s10856-006-0062-5

[45] Rumian L, Reczynska K, Wrona M, Tiainen H, Haugen HJ, Pamula E. The influence of sintering conditions on microstructure and mechanical properties of titanium dioxide scaffolds for the treatment of bone tissue defects. Acta of Bioengineering and Biomechanics. 2015;**17**:3-9

[46] Kim H-W, Koh Y-H, Li L-H, Lee S, Kim H-E. Hydroxyapatite coating on titanium substrate with titania buffer layer processed by sol–gel method. Biomaterials. 2004;**25**:2533-2538. DOI: http://dx.doi.org/10.1016/j.biomaterials.2003.09.041

[47] Hu H, Liu X, Ding C. Preparation and in vitro evaluation of nanostructured TiO2/TCP composite coating by plasma electrolytic oxidation. Journal of Alloys and Compounds. 2010;**498**:172-178. DOI: http://dx.doi.org/10.1016/j.jallcom.2010.03.147

[48] Sato M, Aslani A, Sambito MA, Kalkhoran NM, Slamovich EB, Webster TJ. Nanocrystalline hydroxyapatite/titania coatings on titanium improves osteoblast adhesion. Journal of Biomedical Materials Research Part A. 2008;**84**:265-272. DOI: 10.1002/jbm.a.31469

[49] Stoch A, Brożek A, Kmita G, Stoch J, Jastrzbski W, Rakowska A. Electrophoretic coating of hydroxyapatite on titanium implants. Journal of Molecular Structure. 2001;**596**:191-200. DOI: http://dx.doi.org/10.1016/S0022-2860(01)00716-5

[50] Nath S, Tripathi R, Basu B. Understanding phase stability, microstructure development and biocompatibility in calcium phosphate–titania composites, synthesized from hydroxyapatite and titanium powder mix. Materials Science and Engineering: C. 2009;**29**:97-107. DOI: http://dx.doi.org/10.1016/j.msec.2008.05.019

[51] Que W, Khor KA, Xu JL, Yu LG. Hydroxyapatite/titania nanocomposites derived by combining high-energy ball milling with spark plasma sintering processes. Journal of the European Ceramic Society. 2008;**28**:3083-3090. DOI: http://dx.doi.org/10.1016/j.jeurceramsoc.2008.05.016

[52] Lee JK, Jung HC, Seo DS, Kim H, Hwang KH. Preparation of β-TCP/TiO2 composite by hot-pressing. Solid State Phenomena. 2007;**121-123**:983-986

[53] Caroff F, Oh K-S, Famery R, Boch P. Sintering of TCP–TiO_2 biocomposites: Influence of secondary phases. Biomaterials. 1998;**19**:1451-1454. DOI: http://dx.doi.org/10.1016/S0142-9612(98)00057-X

[54] Landi E, Tampieri A, Celotti G, Sprio S. Densification behaviour and mechanisms of synthetic hydroxyapatites. Journal of the European Ceramic Society. 2000;**20**:2377-2387. DOI: http://dx.doi.org/10.1016/S0955-2219(00)00154-0

[55] Juang HY, Hon MH. Effect of calcination on sintering of hydroxyapatite. Biomaterials. 1996;**17**:2059-2064

[56] Amjad Z. Calcium Phosphates in Biological and Industrial Systems. US: Springer; 1997

[57] Faber KT, Evans AG. Crack deflection processes—I. Theory. Acta Metallurgica. 1983;**31**:565-576. DOI: http://dx.doi.org/10.1016/0001-6160(83)90046-9

[58] Lange FF. The interaction of a crack front with a second-phase dispersion. Philosophical Magazine. 1970;**22**:0983-0992. DOI: 10.1080/14786437008221068

[59] Taya M, Hayashi S, Kobayashi AS, Yoon HS. Toughening of a particulate-reinforced ceramic-matrix composite by thermal residual stress. Journal of the American Ceramic Society. 1990;**73**:1382-1391. DOI: 10.1111/j.1151-2916.1990.tb05209.x

[60] Evans AG, Faber KT. Toughening of ceramics by circumferential microcracking. Journal of the American Ceramic Society. 1981;**64**:394-398. DOI: 10.1111/j.1151-2916.1981.tb09877.x

[61] Barsoum MW. Fundamentals of Ceramics. Bristol, UK; Philadelphia, PA: Institute of Physics Pub.; 2003

[62] Park J, Lakes RS. Biomaterials: An Introduction. New York: Springer; 2007

[63] Deville S. Freeze-casting of porous ceramics: A review of current achievements and issues. Advanced Engineering Materials. 2008;**10**:155-169. DOI: 10.1002/adem.200700270

[64] Studart AR, Gonzenbach UT, Tervoort E, Gauckler LJ. Processing routes to macroporous ceramics: A review. Journal of the American Ceramic Society. 2006;**89**:1771-1789. DOI: 10.1111/j.1551-2916.2006.01044.x

[65] Dapporto M, Sprio S, Fabbi C, Figallo E, Tampieri A. A novel route for the synthesis of macroporous bioceramics for bone regeneration. Journal of the European Ceramic Society. 2016;**36**:2383-2388. DOI: http://dx.doi.org/10.1016/j.jeurceramsoc.2015.10.020

[66] Cunha C, Sprio S, Panseri S, Dapporto M, Marcacci M, Tampieri A. High biocompatibility and improved osteogenic potential of novel Ca-P/titania composite scaffolds designed for regeneration of load-bearing segmental bone defects. Journal of Biomedical Materials Research Part A. 2013;**101**:1612-1619. DOI: 10.1002/jbm.a.34479

Permissions

All chapters in this book were first published in ATD, by InTech Open; hereby published with permission under the Creative Commons Attribution License or equivalent. Every chapter published in this book has been scrutinized by our experts. Their significance has been extensively debated. The topics covered herein carry significant findings which will fuel the growth of the discipline. They may even be implemented as practical applications or may be referred to as a beginning point for another development.

The contributors of this book come from diverse backgrounds, making this book a truly international effort. This book will bring forth new frontiers with its revolutionizing research information and detailed analysis of the nascent developments around the world.

We would like to thank all the contributing authors for lending their expertise to make the book truly unique. They have played a crucial role in the development of this book. Without their invaluable contributions this book wouldn't have been possible. They have made vital efforts to compile up to date information on the varied aspects of this subject to make this book a valuable addition to the collection of many professionals and students.

This book was conceptualized with the vision of imparting up-to-date information and advanced data in this field. To ensure the same, a matchless editorial board was set up. Every individual on the board went through rigorous rounds of assessment to prove their worth. After which they invested a large part of their time researching and compiling the most relevant data for our readers.

The editorial board has been involved in producing this book since its inception. They have spent rigorous hours researching and exploring the diverse topics which have resulted in the successful publishing of this book. They have passed on their knowledge of decades through this book. To expedite this challenging task, the publisher supported the team at every step. A small team of assistant editors was also appointed to further simplify the editing procedure and attain best results for the readers.

Apart from the editorial board, the designing team has also invested a significant amount of their time in understanding the subject and creating the most relevant covers. They scrutinized every image to scout for the most suitable representation of the subject and create an appropriate cover for the book.

The publishing team has been an ardent support to the editorial, designing and production team. Their endless efforts to recruit the best for this project, has resulted in the accomplishment of this book. They are a veteran in the field of academics and their pool of knowledge is as vast as their experience in printing. Their expertise and guidance has proved useful at every step. Their uncompromising quality standards have made this book an exceptional effort. Their encouragement from time to time has been an inspiration for everyone.

The publisher and the editorial board hope that this book will prove to be a valuable piece of knowledge for researchers, students, practitioners and scholars across the globe.

List of Contributors

Aleksandra Radtke
Faculty of Chemistry, Nicolaus Copernicus University in Toruń, Torun, Poland

Hsi-Kai Tsou
Functional Neurosurgery Division, Neurological Institute, Taichung Veterans General Hospital, Taichung City, Taiwan, ROC
Department of Rehabilitation, Jen-Teh Junior College of Medicine, Nursing and Management, Miaoli County, Taiwan, ROC

Ping-Yen Hsieh
Department of Materials Science and Engineering, National Tsing Hua University, Hsinchu City, Taiwan, ROC

António Armando Lima Sampaio Duarte
Department of Civil Engineering, University of Minho, Portugal

Maria Teresa Pessoa Amorim
Department of Textile Engineering, University of Minho, Portugal

Nuria Vela
Applied Technology Group to Environmental Health, Faculty of Health Science, Catholic University of Murcia, Murcia, Spain

Gabriel Pérez-Lucas and Simón Navarro
Department of Agricultural Chemistry, Geology and Pedology, Faculty of Chemistry, University of Murcia, Murcia, Spain

José Fenoll
Sustainability and Quality Group of Fruit and Vegetable Products, Murcia Institute of Agri- Food Research and Development, Murcia, Spain

Jenny Murase
Department of Dermatology, University of California, San Francisco, CA, USA
Department of Dermatology, Palo Alto Medical Foundation, CA, USA

Alessandro Mattiello and Luca Marchiol
Department of Agriculture, Food, Environment and Animal Sciences, University of Udine, Udine, Italy

Hafiz Muhammad Ali
Mechanical Engineering Department, University of Engineering and Technology, Taxila, Pakistan
Center of Research Excellence in Renewable Energy (CoRE-RE), King Fahd University of Petroleum & Minerals (KFUPM), Dhahran, Saudi Arabia

Muhammad Usman Sajid
Mechanical Engineering Department, University of Engineering and Technology, Taxila, Pakistan

Adeel Arshad
Department of Mechanical Engineering, HITEC University, Taxila, Pakistan
Fluids & Thermal Engineering Research Group, Faulty of Engineering, University of Nottingham, Nottingham, UK

Marie-Hélène Ropers
INRA, Biopolymers Interactions and Assemblies, Nantes, France

Hélène Terrisse and Bernard Humbert
Institute of Materials Jean ROUXEL (IMN), Université de Nantes, CNRS, Nantes, France

Muriel Mercier-Bonin
INRA, Toxalim (Research Centre in Food Toxicology), University of Toulouse, INRA, ENVT, INP-Purpan, UPS, Toulouse, France

Megha Trivedi
University of Michigan Medical School, Ann Arbor, MI, USA
Department of Dermatology, University of California, San Francisco, CA, USA

Yiqun Yang
Xavier University of Louisiana, New Orleans, LA, United States

Jun Li
Kansas State University, Manhattan, KS, United States

Massimiliano Dapporto, Anna Tampieri and Simone Sprio
Laboratory of Bioceramics and Bio-hybrid Composites, Institute of Science and Technology for Ceramics, National Research Council of Italy, Faenza, Italy

Index

A

Anatase Phase, 38, 46, 132, 139, 181
Anisotrophy, 14
Anodic Oxidation, 2, 4, 6, 8, 21, 24
Anodization, 4-5, 7, 10, 14, 24-25, 180
Antibacterial Disinfection, 3
Antibiotic Removal, 51-52, 55-56, 58, 64, 67
Anticorrosive, 31-33, 36, 46
Antimicrobial Activity, 31, 34, 48
Antimicrobial Efficacy, 33, 38, 46

B

Biocompatibility, 1-3, 18-19, 31, 39, 198, 201, 205-207, 210, 212, 216-217
Bioinertness, 32, 201
Biomedical Material, 31
Biomedical Technologies, 1
Biomimetic Immersion, 43
Blood Compatibility, 2-3
Bone Conductivity, 2
Bone Implants, 2, 18, 204
Bone Remodeling, 44-46
Bulk Modification, 38

C

Calcination, 9-10, 181, 216
Cardiovascular Implants, 1
Corrosion Resistance, 2-3, 22, 33, 36, 46, 201, 206
Corrosive Current, 38
Cortical Bone, 45-46, 209
Crystalline Islands, 4, 7

D

Dental Implants, 1, 17, 19, 38, 48
Dielectric Constant, 10
Dye Degradation, 12

E

Elastic Modulus, 38, 206
Electrochemical Etching, 13
Electrospinning, 9-10, 12, 14, 28-29
Electrospun Nanofiber, 11, 28

F

Femtosecond Laser Ablation, 11, 29
Fibroblasts Adhesion, 11
Fibrous Tissue, 45-46

G

Gram positive , 33

H

Human Cancellous Bone, 38
Hydrophobicity, 141, 146
Hydroxyapatite, 3, 11, 15, 20-22, 28, 38, 47, 49-50, 146, 201-202, 205-207, 210, 214-216

I

Implantology, 1-2, 48

K

Kinetic Rates, 12

L

Lamellar Bone, 44-45
Lamellipodia, 11, 15, 42
Langmuir-hinshelwood Kinetic Model, 12

M

Marrow Cavity, 45-46

N

Nanoarchitecture, 1, 9
Nanofibers, 1, 9-13, 16, 28-29, 55, 200
Nanotubes, 1, 3-8, 11, 13, 16, 21, 23-27, 29-30, 55, 84, 198
Nanowires, 1, 3, 13-16, 28-30, 180, 183, 195

O

Organic Pollutant, 8, 12
Orthorhombic Brookite, 3
Osseointegration Capacity, 44, 46
Osseointegration Performance, 44, 50
Osteoblast Compatibility, 40, 43, 45-46, 49
Osteoconductivity, 21, 45
Osteopenia Syndrome, 38

P

Photoactivity, 3, 8, 12, 16, 30, 73, 79, 84

Photocatalytic Activity, 1, 8, 12-13, 16, 21-22, 25, 71, 79-82, 84, 89-90, 98, 102, 133

Photocatalytic Filtration, 51, 53, 56, 59-61, 64-65, 67

Physicochemical Properties, 1, 135, 142

Polymer Concentration, 10

Pore Structure, 12, 79

Potentiodynamic Polarization, 36

Proliferation, 6-7, 11-12, 15, 39-40, 45, 49, 207, 210-211

R

Raman Mapping, 4, 23

Raman Modes, 4

Raman Spectroscopy, 4, 23

Rutile Form, 4, 6-7, 151

S

Shear Strength, 43

Simulated Body Fluid, 11, 20, 25, 205

Sintering, 6, 9, 22, 180, 195, 205-207, 212-216

Spinal Implant, 32, 38-39, 50

Stable Bonding Structure, 31-32

Staphylococcus Aureus, 7, 26, 31-33, 48

Staphylococcus Epidermis, 7, 26

Substrate Dissolution, 31, 38

Surface Functionalization, 13, 129

Surface Sterilization, 8

Surface Topography, 2, 26

T

Thermal Oxidation, 14, 16, 30

Titania Coatings, 3, 11, 15, 22, 216

Titania Nanowires, 3, 13-16

W

Water Treatment Systems, 51-52, 55

Wear Resistance, 2-3

X

X-ray Diffraction, 4, 23, 59-60, 92

www.ingramcontent.com/pod-product-compliance
Lightning Source LLC
LaVergne TN
LVHW080655210926
848485LV00003B/380